# Seventh Workshop on Computational Linguistics and Clinical Psychology: Improving Access (CLPsych 2021)



Online
11 June 2021

ISBN: 978-1-7138-3019-1

**Printed from e-media with permission by:**

Curran Associates, Inc.
57 Morehouse Lane
Red Hook, NY 12571

**Some format issues inherent in the e-media version may also appear in this print version.**



Association for Computational Linguistics
209 N. Eighth Street
Stroudsburg, Pennsylvania 18360

Phone: 1-570-476-8006
Fax: 1-570-476-0860

acl@aclweb.org

**Additional copies of this publication are available from:**

Curran Associates, Inc.
57 Morehouse Lane
Red Hook, NY 12571 USA
Phone: 845-758-0400
Fax: 845-758-2633
Email: curran@proceedings.com
Web: www.proceedings.com

NAACL HLT 2021

# Computational Linguistics and Clinical Psychology: Improving Access

## Proceedings of the Seventh Workshop

June 11, 2021

UMIACS

University of Maryland Institute for Advanced Computer Studies

Silver Sponsor

# Introduction

Mental health is a formidable worldwide challenge. In economic terms, combining direct and indirect costs, the global cost of mental health conditions for 2010 was estimated at \$2.5 trillion dollars and is expected to grow to \$6 trillion by 2030 [Bloom et al. 2021]. In the United States, suicide is the second leading cause of death among those aged 10-34 and the fourth among those aged 35-54, and worldwide 800,000 people are lost to suicide each year [WHO 2014]. Access to professional help is inadequate – in the United States, more than 120 million people live in federally designated Mental Health Care Professional Health Professional Shortage Areas [HRSA 2021]; hence the selection "Improving Access" as this year's workshop theme, with the aim of encouraging submissions and discussion on that subject.

The Seventh Workshop on Computational Linguistics and Clinical Psychology took place, in conjunction with NAACL'21, online on June 11, 2021, and as we wrote this introduction the tragedy of COVID-19 was still ongoing. Many people are experiencing unprecedented pressure – struggling with their health, finances, social isolation, and online work or education. In June 2020, a full 40% of adults in the U.S. reported struggles with mental health or substance abuse, with 31% reporting anxiety or depressive symptoms and 11% reporting having seriously considered suicide [Czeisler et al. 2020]. As of February 2021, those anxiety and depressive symptoms had increased to 41.5% [Vahratian et al. 2021]. Mental health experts have predicted a "tsunami" of need arising from the COVID pandemic [Inkster et al. 2021; Torjesen 2020].

CLPsych has an important role to play in bringing people together to discuss and exchange their recent work and results, with the aim of using human language as a tool to better understand emotional and mental states and reduce emotional suffering and the potential for self-harm. Since 2014, CLPsych has brought together researchers in computational linguistics and NLP, who use computational methods to better understand human language, infer meaning and intention, and predict individuals' characteristics and potential behavior, with psychology researchers and practitioners, including participants who are focused on psychopathology and neurological health and engage directly with the needs of providers and their patients. The workshop's distinctly interdisciplinary nature has improved the exchange of knowledge between computational linguistics and clinical psychology, fostered collaboration, and increased the visibility of mental health and psychological research as a problem domain in NLP.

The potential role of language technology, and AI more generally, in mental health is gaining increasing attention [Lee et al. 2021], leading to corresponding increases in discussion of real-world issues such as the ethics of research and deployment [Benton et al. 2017; Chancellor et al. 2019; Resnik et al. 2021]. At the same time, continued progress on NLP for mental health – indeed, for healthcare in general – is hampered by obstacles to shared, community-level access to relevant data. The 2021 CLPsych Shared Task introduced what is, to our knowledge, the first attempt to address this problem for mental health by conducting a shared task using sensitive data in a secure data enclave, bringing researchers to the data rather than sending the data out to researchers. Participating teams received access to Twitter posts donated for research using Qntfy's OurDataHelps.org platform, including data from users with and without suicide attempts, and did all work with the dataset entirely within a secure computational environment provided by NORC at the University of Chicago. The shared task was organized by Sean MacAvaney, Anjali Mittu, and Philip Resnik, and the overview by MacAvaney, Mittu, Coppersmith, Leintz, and Resnik (2021) discusses the task, team results, and lessons learned to set the stage for future tasks on sensitive or confidential data.

In keeping with CLPsych's traditional interdisciplinary approach, psychology researchers and practicing clinicians were included as part of our program committee along with technological experts, and our call for papers emphasized that communicating ideas and results to a mixed audience would be a very high priority. Submissions to the workshop included 28 papers, of which 6 were accepted for presentation and

12 were accepted to be presented in the poster session. The five shared task papers were also included in the program.

In addition to the submitted papers, CLPsych continued its tradition of superb invited talks and discussions. Keynote talks were delivered by Munmun De Choudhury (Georgia Tech) and Matthew Nock (Harvard), and invited talks by Glen Coppersmith (Qntfy), Carol Espy-Wilson (University of Maryland), and Lyle Ungar (University of Pennsylvania) were followed by a panel discussion that included the three invited speakers, moderated by Dr. Lorenzo Norris, Chief Wellness Officer for George Washington University Hospital and host and editor-in-chief of the MDedge Psychcast, a weekly podcast from MDedge Psychiatry.

The CLPsych organizing committee acknowledges with gratitude the efforts of the many people who helped make the workshop a success. This includes the authors and shared task participants for their insightful contributions, program committee members for their high quality, thoughtful reviews, and our keynote and invited speakers and panel moderator for their valuable insights. The committee is also grateful to Sean MacAvaney and Anjali Mittu for their tremendous efforts organizing and running the shared task, and particularly also to Glen Coppersmith of Qntfy, for his efforts collecting invaluable research data and making it available, as well as Jeff Leintz and his team at NORC at the University of Chicago, for hosting and supporting the shared task in their secure data enclave. Finally, the organizers thank the North American chapter of the Association for Computational Linguistics for making this workshop possible (with special thanks as always to Priscilla Rassmussen), Anjali Mittu for additional help making sure the workshop could run smoothly, Andrea Alessandri of 21am for his late-breaking redesign of the workshop web site, and the University of Maryland Institute for Advanced Computer Studies for its generous sponsorship.

Nazli Goharian, Philip Resnik, Andrew Yates, Molly Ireland, Kate Niederhoffer, & Rebecca Resnik

## References

Benton, Adrian, Glen Coppersmith, and Mark Dredze. Ethical research protocols for social media health research. In Proceedings of the First ACL Workshop on Ethics in Natural Language Processing, pp. 94-102. 2017.

Bloom, D.E., Cafiero, E.T., Jané-Llopis, E., Abrahams-Gessel, S., Bloom, L.R., Fathima, S., Feigl, A.B., Gaziano, T., Mowafi, M., Pandya, A., Prettner, K., Rosenberg, L., Seligman, B., Stein, A.Z., & Weinstein, C. (2011). The Global Economic Burden of Noncommunicable Diseases. Geneva: World Economic Forum.

Chancellor, Stevie, Michael L. Birnbaum, Eric D. Caine, Vincent MB Silenzio, and Munmun De Choudhury. A taxonomy of ethical tensions in inferring mental health states from social media. In Proceedings of the conference on fairness, accountability, and transparency, pp. 79-88. 2019.

Czeisler MÉ , Lane RI, Petrosky E, et al. Mental Health, Substance Use, and Suicidal Ideation During the COVID-19 Pandemic – United States, June 24–30, 2020. MMWR Morb Mortal Wkly Rep 2020;69:1049–1057. DOI: http://dx.doi.org/10.15585/mmwr.mm6932a1

HRSA. Health professional shortage areas dashboard, 2021. URL data.HRSA.gov. Data as of February 6, 2021.

Inkster, Becky and Digital Mental Health Data Insights Group. Early warning signs of a mental health tsunami: A coordinated response to gather initial data insights from multiple digital services providers. Frontiers in Digital Health 2 (2021): 64.

Lee, E.E., Torous, J., De Choudhury, M., Depp, C.A., Graham, S.A., Kim, H.C., Paulus, M.P., Krystal, J.H. and Jeste, D.V. (2021). Artificial Intelligence for Mental Healthcare: Clinical Applications, Barriers, Facilitators, and Artificial Wisdom. In Biological Psychiatry: Cognitive Neuroscience and Neuroimaging.

MacAvaney, Sean, Anjali Mittu, Glen Coppersmith, Jeff Leintz, and Philip Resnik. 2021. Community-level research on suicidality prediction in a secure environment: Overview of the CLPsych 2021 shared task. In Proceedings of the Seventh Workshop on Computational Linguistics and Clinical Psychology (CLPsych 2021). Association for Computational Linguistics.

Resnik, Philip, April Foreman, Michelle Kuchuk, Katherine Musacchio Schafer, and Beau Pinkham. Naturally occurring language as a source of evidence in suicide prevention. Suicide and Life-Threatening Behavior 51, no. 1 (2021): 88-96.

Torjensen, I. (2020). Covid-19: Mental health services must be boosted to deal with 'tsunami' of cases after lockdown. British Medical Journal; 369 doi: https://doi.org/10.1136/bmj.m1994

WHO (World Health Organization), Suicide Fact Sheet, September 2, 2019, URL https://www.who.int/news-room/fact-sheets/detail/suicide, downloaded May 6, 2021.

Vahratian A, Blumberg SJ, Terlizzi EP, Schiller JS. Symptoms of Anxiety or Depressive Disorder and Use of Mental Health Care Among Adults During the COVID-19 Pandemic – United States, August 2020-February 2021. MMWR Morb Mortal Wkly Rep 2021;70:490–494. DOI: http://dx.doi.org/10.15585/mmwr.mm7013e2

**Organizing Committee:**

Nazli Goharian, Georgetown University
Philip Resnik, University of Maryland
Andrew Yates, Max Planck Institute for Informatics
Molly Ireland, Texas Tech University
Kate Niederhoffer, Knowable Research
Rebecca Resnik, Rebecca Resnik and Associates, LLC

**Shared Task Organizers:**

Sean MacAvaney, Georgetown University & University of Glasgow
Anjali Mittu, University of Maryland
Philip Resnik, University of Maryland

**Keynote Speakers:**

Munmun De Choudhury, Georgia Tech
Matthew Nock, Harvard

**Invited Speakers and Panelists:**

Glen Coppersmith, Qntfy
Carol Espy-Wilson, University of Maryland
Lyle Ungar, University of Pennsylvania

**Panel Moderator:**

Lorenzo Norris, George Washington University

**Program Committee:**

Nick Allen, University of Oregon
Steven Bedrick, Oregon Health & Science University
Arman Cohan, Allen Institute for AI
Glen Coppersmith, Qntfy
Tyler Davis, Texas Tech University
Bart Desmet, National Institutes of Health
Samuel Dooley, UMD
Ophir Frieder, Georgetown University
Pranav Goel, University of Maryland
David Hancock, Weill Cornell Medicine
Kristy Hollingshead, IHMC
Loring Ingraham, George Washington University
Andrew Littlefield, Texas Tech University
Sean MacAvaney, IR Lab, Georgetown University
Adam Miner, Stanford University School of Medicine
Taleen Nalabandian, Texas Tech University
Yaakov Ophir, Technion - Israel Institute of Technology
Ted Pedersen, University of Minnesota, Duluth
Daniel Preotiuc-Pietro, Bloomberg LP
Emily Prud'hommeaux, Boston College
Masoud Rouhizadeh, Johns Hopkins University
Jonathan Schler, HIT
H. Andrew Schwartz, Stony Brook University
Han-Chin Shing, University of Maryland at College Park
Luca Soldaini, Amazon
Amelia Talley, Texas Tech University
Jason Van Allen, Texas Tech University
Sarah Wayland, Guiding Exceptional Parents, LLC
Eugene Yang, Georgetown University
Jessica Yu, Livongo
Ayah Zirikly, Johns Hopkins University

# Table of Contents

# Conference Program

**Friday, June 11, 2021 – all times are Central Time**

**8:45am–10:45am Keynotes**

*8:45am–9:00am Conference introduction*
Organizers

*9:00am–9:05am Keynote speaker introductions*
Organizers

*9:05am–9:50am Keynote 1 and Q&A*
Matthew Nock, Harvard

*9:50am–10:35am Keynote 2 and Q&A*
Munmun De Choudhury, Georgia Tech

*10:35am–10:45am Discussion*
Attendees

**10:45am–11:00am** ***Break***

**11:00am–12:00pm Paper session 1**

*Understanding who uses Reddit: Profiling individuals with a self-reported bipolar disorder diagnosis*
Glorianna Jagfeld, Fiona Lobban, Paul Rayson and Steven Jones

*On the State of Social Media Data for Mental Health Research*
Keith Harrigian, Carlos Aguirre and Mark Dredze

*Individual Differences in the Movement-Mood Relationship in Digital Life Data*
Glen Coppersmith, Alex Fine, Patrick Crutchley and Joshua Carroll

*11:30am–12:00pm Discussion*
Attendees

**12:00pm–1:00pm** ***Lunch / Zoom poster session***

**Friday, June 11, 2021 – all times are Central Time (continued)**

**1:00pm–1:55pm Paper session 2**

*Dissociating Semantic and Phonemic Search Strategies in the Phonemic Verbal Fluency Task in early Dementia*
Hali Lindsay, Philipp Müller, Nicklas Linz, Radia Zeghari, Mario Magued Mina, Alexandra Konig and Johannes Tröger

*Demonstrating the Reliability of Self-Annotated Emotion Data*
Anton Malko, Cecile Paris, Andreas Duenser, Maria Kangas, Diego Molla, Ross Sparks and Stephen Wan

*Hebrew Psychological Lexicons*
Natalie Shapira, Dana Atzil-Slonim, Daniel Juravski, Moran Baruch, Dana Stolowicz-Melman, Adar Paz, Tal Alfi-Yogev, Roy Azoulay, Adi Singer, Maayan Revivo, Chen Dahbash, Limor Dayan, Tamar Naim, Lidar Gez, Boaz Yanai, Adva Maman, Adam Nadaf, Elinor Sarfati, Amna Baloum, Tal Naor, Ephraim Mosenkis, Badreya Sarsour, Jany Gelfand Morgenshteyn, Yarden Elias, Liat Braun, Moria Rubin, Matan Kenigsbuch, Noa Bergwerk, Noam Yosef, Sivan Peled, Coral Avigdor, Rahav Obercyger, Rachel Mann, Tomer Alper, Inbal Beka, Ori Shapira and Yoav Goldberg

*1:30pm–1:55pm Discussion*
Attendees

**1:55pm–2:00pm** ***Break***

**2:00pm–4:00pm Invited talks and panel discussion**

*2:00pm–2:20pm Invited talk 1*
Glen Coppersmith, Qntfy

*2:25pm–2:45pm Invited talk 2*
Carol Espy-Wilson, University of Maryland

*2:50pm–3:10pm Invited talk 3*
Lyle Ungar, University of Pennsylvania

*3:10pm–4:10pm Panel*
Speakers & Lorzeno Norris, George Washington University Hospital (moderator)

**Friday, June 11, 2021 – all times are Central Time (continued)**

**4:00pm–5:10pm Shared task**

*Community-level Research on Suicidality Prediction in a Secure Environment: Overview of the CLPsych 2021 Shared Task*
Sean MacAvaney, Anjali Mittu, Glen Coppersmith, Jeff Leintz and Philip Resnik

*Determining a Person's Suicide Risk by Voting on the Short-Term History of Tweets for the CLPsych 2021 Shared Task*
Ulya Bayram and Lamia Benhiba

*Learning Models for Suicide Prediction from Social Media Posts*
Ning Wang, Luo Fan, Yuvraj Shivtare, Varsha Badal, Koduvayur Subbalakshmi, Rajarathnam Chandramouli and Ellen Lee

*Suicide Risk Prediction by Tracking Self-Harm Aspects in Tweets: NUS-IDS at the CLPsych 2021 Shared Task*
Sujatha Das Gollapalli, Guilherme Augusto Zagatti and See-Kiong Ng

*Team 9: A Comparison of Simple vs. Complex Models for Suicide Risk Assessment*
Michelle Morales, Prajjalita Dey and Kriti Kohli

*Using Psychologically-Informed Priors for Suicide Prediction in the CLPsych 2021 Shared Task*
Avi Gamoran, Yonatan Kaplan, Almog Simchon and Michael Gilead

*5:00pm–5:10pm Discussion*
Attendees

**5:10pm–5:20pm** ***Break to make cocktails/obtain a drink***

**5:20pm–6:20pm** ***Zoom cocktail poster session***

*Analysis of Behavior Classification in Motivational Interviewing*
Leili Tavabi, Trang Tran, Kalin Stefanov, Brian Borsari, Joshua Woolley, Stefan Scherer and Mohammad Soleymani

*Automatic Detection and Prediction of Psychiatric Hospitalizations From Social Media Posts*
Zhengping Jiang, Jonathan Zomick, Sarah Ita Levitan, Mark Serper and Julia Hirschberg

**Friday, June 11, 2021 – all times are Central Time (continued)**

*Automatic Identification of Ruptures in Transcribed Psychotherapy Sessions*
Adam Tsakalidis, Dana Atzil-Slonim, Asaf Polakovski, Natalie Shapira, Rivka Tuval-Mashiach and Maria Liakata

*Automated coherence measures fail to index thought disorder in individuals at risk for psychosis*
Kasia Hitczenko, Henry Cowan, Vijay Mittal and Matthew Goldrick

*Detecting Cognitive Distortions from Patient-Therapist Interactions*
Sagarika Shreevastava and Peter Foltz

*Evaluating Automatic Speech Recognition Quality and Its Impact on Counselor Utterance Coding*
Do June Min, Verónica Pérez-Rosas and Rada Mihalcea

*Qualitative Analysis of Depression Models by Demographics*
Carlos Aguirre and Mark Dredze

*Safeguarding against spurious AI-based predictions: The case of automated verbal memory assessment*
Chelsea Chandler, Peter Foltz, Alex Cohen, Terje Holmlund and Brita Elvevåg

*Towards the Development of Speech-Based Measures of Stress Response in Individuals*
Archna Bhatia, Toshiya Miyatsu and Peter Pirolli

*Towards Low-Resource Real-Time Assessment of Empathy in Counselling*
Zixiu Wu, Rim Helaoui, Diego Reforgiato Recupero and Daniele Riboni

*Towards Understanding the Role of Gender in Deploying Social Media-Based Mental Health Surveillance Models*
Eli Sherman, Keith Harrigian, Carlos Aguirre and Mark Dredze

*Understanding Patterns of Anorexia Manifestations in Social Media Data with Deep Learning*
Ana Sabina Uban, Berta Chulvi and Paolo Rosso

**6:20pm–6:30pm** ***Closing remarks***

# Understanding who uses Reddit: Profiling individuals with a self-reported bipolar disorder diagnosis

**Glorianna Jagfeld**[⋆], **Fiona Lobban**[⋆], **Paul Rayson**[▽], **Steven H. Jones**[⋆]
[⋆]Spectrum Centre for Mental Health Research
[▽]School of Computing and Communications
Lancaster University, United Kingdom
{g.jagfeld, f.lobban, p.rayson, s.jones7}@lancaster.ac.uk

## Abstract

Recently, research on mental health conditions using public online data, including Reddit, has surged in NLP and health research but has not reported user characteristics, which are important to judge generalisability of findings. This paper shows how existing NLP methods can yield information on clinical, demographic, and identity characteristics of almost 20K Reddit users who self-report a bipolar disorder diagnosis. This population consists of slightly more feminine- than masculine-gendered mainly young or middle-aged US-based adults who often report additional mental health diagnoses, which is compared with general Reddit statistics and epidemiological studies. Additionally, this paper carefully evaluates all methods and discusses ethical issues.

## 1 Introduction and related work

People who experience extreme mood states that interfere with their functioning, meet the criteria for bipolar disorder (BD) according to the diagnostic manuals Diagnostic and Statistical Manual of Mental Disorders (DSM) (American Psychiatric Association, 2013) and International Classification of Diseases (ICD) (World Health Organisation, 2018). DSM and ICD operationalise extreme mood states in terms of major depressive episodes, 'almost daily depressed mood or diminished interest in activities with additional symptoms for at least 14 days' (World Health Organisation, 2018) and (hypo-)manic episodes, 'a distinct period of abnormally and persistently elevated, expansive, or irritable mood and abnormally and persistently increased goal-directed activity or energy' that lasts at least seven (four) days (American Psychiatric Association, 2013, p. 124).

DSM and ICD distinguish several BD subtypes based on the lifetime frequency and intensity of (hypo-)manic and depressed episodes. The only requirement for a diagnosis of bipolar I disorder (BD-I) is at least one lifetime manic episode, whereas bipolar II disorder (BD-II) requires at least one hypomanic and one major depressive episode (American Psychiatric Association, 2013, pp. 126, 132). Cyclothymic disorder applies to numerous periods of hypomanic and depressive symptoms during at least two years that do not meet criteria for hypomanic or major depressive episodes (American Psychiatric Association, 2013, p. 139).

Bipolar mood episodes are often recurring (Treuer and Tohen, 2010; Gignac et al., 2015), so many individuals living with BD require life-long treatment (Goodwin et al., 2016) and have a heightened suicide risk (Novick et al., 2010). However, characteristics and outcomes of people meeting BD criteria are diverse, with some living well, (e.g., Warwick et al., 2019) and even functioning on a high level (Akers et al., 2019).

### 1.1 Online forums as research data source

Online forums have become an increasingly attractive source for research data, enabling non-reactive data collection, where researchers do not influence data creation, at large scale (Fielding et al., 2016). Natural language processing (NLP) research in this area has focused on predicting people at risk of BD (Coppersmith et al., 2014; Cohan et al., 2018; Sekulić et al., 2018). Health researchers have explored the lived experience of BD with qualitative analyses of online posts (Mandla et al., 2017; Sahota and Sankar, 2019). Unlike in clinical studies, usually little or no demographic information is available for online forum users, so it is unclear to what populations these results generalise (Ruths and Pfeffer, 2014). For example, language differences between Twitter users with self-reported Major depressive disorder (MDD) or Post-traumatic stress disorder (PTSD) correlated highly with their personality and demographic characteristics (Preoţiuc-Pietro et al., 2015). So it is unclear whether these findings really indicate mental health (MH) diagnoses or other user characteristics.



*Proceedings of the Seventh Workshop on Computational Linguistics and Clinical Psychology*, pages 1–14
June 11, 2021. 

### 1.2 The online discussion forum Reddit

Besides MH-specific platforms (Kramer et al., 2004; Vayreda and Antaki, 2009; Bauer et al., 2013; Latalova et al., 2014; Poole et al., 2015; McDonald and Woodward-Kron, 2016; Campbell and Campbell, 2019), blogs (Mandla et al., 2017), and Twitter (Coppersmith et al., 2014; Ji et al., 2015; Saravia et al., 2016; Budenz et al., 2019; Huang et al., 2019), much recent research of user-generated online content in BD has focused on the international online discussion forum Reddit[1] (Gkotsis et al., 2016, 2017; Cohan et al., 2018; Sekulić et al., 2018; Sahota and Sankar, 2019; Yoo et al., 2019).

The platform Reddit is among the most visited internet sites worldwide (Alexa Internet, 2020), hosting a number of subforums ('subreddits') for general topics as well as interest groups. There is a vast and growing amount of BD-related content on Reddit, with more than 50K new posts per month in the four largest BD-related subreddits[2]. Anyone can view posts without registration and the Reddit API offers free access to all historic posts. Reddit profiles do not provide any user characteristics besides the username and sign-up date in a structured format or comparable to a Twitter bio. While some surveys provide general information on Reddit users, none of the BD-specific studies looked at particular user characteristics of their sample, which is important (Amaya et al., 2019).

### 1.3 Research questions and contributions

The above considerations motivate our research questions: What characteristics of Reddit users who disclose a BD diagnosis can be automatically inferred from their public Reddit information and how do they compare to general Reddit users and clinical populations? What are the ethical considerations around determining users' characteristics and ways to minimise potential negative impacts?

This work has two main contributions, both of which may be relevant to different parts of the CLPsych community. Crucially, the authors are an interdisciplinary team of NLP and clinical psychology researchers, as well as practising clinical psychologists, who regularly consult with people with lived experience of BD in an advisory panel.

First, this paper estimates and discusses clinical, demographic and identity characteristics of Reddit users who self-report a BD diagnosis (see Figure 3 for a visual results summary). This has implications for future BD-focused research on Reddit and helps to contextualise previous work. Moreover, this information is relevant for clinicians who may want to recommend certain online forums to clients and to clinical researchers interested in recruiting via Reddit. Second, this work shows how simple rule-based and off-the-shelf state-of-the-art NLP methods can estimate Reddit user characteristics, and carefully discusses ethical considerations and harm-mitigating ways of doing so. These findings and discussions apply to other, also non-clinical, subgroups of Reddit users. The evaluation with manual annotations evaluates published NLP methods in an applied setting.

## 2 Methods

### 2.1 User identification

In this work, the identification of Reddit users with lived experience of BD adapts previous approaches based on self-reported diagnosis statements, e.g., 'I was diagnosed with BD today' (Coppersmith et al., 2015; Cohan et al., 2018; Sekulić et al., 2018). Importantly, this captures self-*reported* diagnoses by a professional and not *self-diagnoses*, which were excluded. Contrary to existing datasets of Reddit posts by people with a self-reported BD diagnosis, all posts of identified people were retained and not only those unrelated to MH concerns. This enables subsequent research on the lived experience of people with BD. All available Reddit posts (January 05 - March 19) that mentioned 'diagnosis' and a BD term (see below) were downloaded from Google BigQuery. User account meta-data (id, username, UTC timestamp of sign-up) for all matching posts was retrieved via the Reddit python API praw[3] to remove posts by users who had deleted their profile after creation of the BigQuery tables. Each of the 170K posts was classified as self-reported diagnosis post after automatically removing quoted content if it met the following criteria adapted from Cohan et al. (2018) (see Table 1 for examples):

- Contains at least one condition term for BD.
- Matches at least one inclusion pattern, i.e., BD diagnosis of any type by a professional.
- Does not match any exclusion pattern, e.g., self-diagnosis.

[1] `https://www.reddit.com/`

[2] r/bipolar, r/BipolarReddit, r/bipolar2, r/bipolarSOs

[3] `https://github.com/praw-dev/praw`

| Component | Number | Examples |
|---|---|---|
| Inclusion patterns | 145 | As someone with a diagnos*, my recent CONDITION diagnos*, I went to a DOCTOR and got diagnos* |
| CONDITION terms | 92 | Bipolar, manic depression, BD-I, BD-II, cyclothymia |
| DOCTOR terms | 18 | Doctor, pdoc, shrink |
| Exclusion patterns | 74 | Not formally diagnos*, self diagnos*, she's diagnos* |

Table 1: Components of patterns to identify English self-reported diagnosis statements; *: wildcard

- The distance between at least one condition term and the beginning or end of an inclusion phrase is less than the experimentally determined threshold of 55 characters.

Subsequently, all posts (id, submissions title, text, subreddit, user id, UTC timestamp of time posted) of the 21K user accounts with at least one self-reported diagnosis post were downloaded via praw. The first author checked the self-reported diagnosis statements of all accounts with more than 1.5K submissions or 200K comments or whose name included 'bot' or 'auto', removing 30 automated user accounts (bots). Finally, 960 user accounts with a self-reported psychotic disorder diagnosis were removed because this constitutes an exclusion criterion for BD (American Psychiatric Association, 2013, pp. 126, 134).

### 2.2 User characteristics extraction/inference

Several NLP methods were applied and compared to extract or infer clinical (MH comorbidities = diagnoses additional to BD), demographic (age, country of residence), and identity (gender) characteristics of Reddit users with a self-reported BD diagnosis. See Appendix A for more details on the age, country, and gender methods and their previously published performance. The first and third author manually annotated self-reported BD diagnoses, age, country, and gender for random included users for evaluation.

#### 2.2.1 Mental health comorbidities

Frequencies for other self-reported MH diagnoses were obtained by matching all dataset posts against inclusion patterns for other diagnoses, in the same way as for identifying self-reported BD diagnoses. Condition terms for nine major DSM-5 and ICD-11 diagnoses were extended from Cohan et al. (2018): Anxiety disorder (Generalised/Social anxiety disorder, Panic disorder), Attention deficit hyperactivity disorder (ADHD), Borderline personality disorder (BPD), MDD, PTSD, Psychotic disorder (Schizophrenia/Schizoaffective disorder), Obsessive compulsive disorder (OCD), Autism spectrum disorder (ASD), and Eating disorder (ED).

#### 2.2.2 Age

Two methods to recognise a user's age relative to one of their posts were compared. An approximate date of birth was calculated from the post timestamp to then calculate the user's age when posting for the first time and their mean age over all posts.

- **Self-reported**: Reddit users sometimes self-report their age and gender in a bracketed format, e.g. 'I [17f] just broke up with bf [18m]'. Regular expressions extracted age and gender from such self-reports in submission titles.

- **Language use**: Tigunova's (2019) neural network model predicts the age group of users with at least ten posts from their contents and language style. Training data for this model came from Tigunova et al. (2020) who automatically labelled Reddit users with their self-reported age (see Appendix A.1).

- **Hybrid**: The Hybrid method assigns the extracted age from the Self-reported method if available, and otherwise the predicted age from the Language use method because evaluation revealed that the Self-reported method had higher accuracy but lower coverage than the Language use method (see Section 4.2).

#### 2.2.3 Country of residence

The only published method for Reddit user localisation to date (Harrigian, 2018) infers a user's country of residence via a dirichlet process mixture model[4]. It uses the distribution of words, posts per subreddit, and posts per hour of the day (timezone proxy) of a user's up to 250 most recent comments.

[4]https://github.com/kharrigian/smgeo

#### 2.2.4 Binary gender

Three methods to recognise binary gender (feminine (f)/masculine (m)) leveraging different types of information were compared. All three methods pertain to a performative gender view, which posits that people understand their and others' gender identity by certain behaviours (including language) and appearances that society stipulates for bodies of a particular sex (Larson, 2017). Non-binary gender identities were not included due to a lack of NLP methods to detect them.

- **Username**: The character-based neural network model of Wang and Jurgens (2018) predicts whether a username strongly performs f or m gender, otherwise it assigns no label.
- **Self-reported**: See Section 2.2.2.
- **Language use**: The neural network model by Tigunova et al. (2019) predicts gender for Reddit users with at least ten posts from the post texts. It was trained on data automatically labelled with self-reported gender provided by Tigunova et al. (2020) (see Appendix A.1).
- **Hybrid**: Evaluation revealed an accuracy ranking of Username > Self-reported > Language use and the inverse for coverage (Section 4.2). The Hybrid method assigns a binary gender identity in a sequential approach, disregarding possible disagreements between methods: If the Username method found the username to perform f or m gender, it takes this prediction, otherwise assumes the self-reported gender if available, and else resorts to the predictions of the Language use method.

## 3 Ethical considerations

At least four main ethical considerations arise for the work presented here: Concerns around (1) consent and (2) anonymity of Reddit users, around the (3) selection, category labels, and assignment of user characteristics (MH diagnoses, age, country, gender), and (4) potentially harmful uses of the presented dataset and methods. The Lancaster University Faculty of Health and Medicine research ethics committee reviewed and approved this study in May 2019 (reference number FHMREC18066).

### 3.1 Consent

If and how research on social media data needs to obtain informed consent is debated (Eysenbach and Till, 2001; Beninger et al., 2014; Paul and Dredze, 2017), mainly because it is not straightforward to determine if posts pertain to a public or private context. Legally, the Reddit privacy policy[5] explicitly allows copying of user contents by third parties via the Reddit API, but it is unclear to what extends users are aware of this (Ahmed et al., 2017). In practice it is often infeasible to seek retrospective consent from hundreds or thousands of social media users. Current ethical guidelines for social media research (Benton et al., 2017; Williams et al., 2017) and practice in comparable research projects (O'Dea et al., 2015; Ahmed et al., 2017), regard it as acceptable to waive explicit consent if users' anonymity is protected. Therefore, Reddit users in this work were not asked for consent.

### 3.2 Anonymity

In line with guidelines for ethical social media health research (Benton et al., 2017), this research only shares anonymised and paraphrased excerpts from posts in publications. Otherwise, it is often possible to recover usernames via a web search with the verbatim post text (see also Section 3.5).

### 3.3 Rationales for user characteristics

As stated in the introduction, user characteristics are important to determine about which populations research on this dataset may generalise. The NLP community increasingly expects data statements for datasets (Bender and Friedman, 2018), which include speaker age and gender specifications. As Section 4.3 shows, characteristics of Reddit users with a self-reported BD diagnosis deviate from both general Reddit user statistics and epidemiological studies, which therefore do not constitute useful proxies. Relying entirely on self-reported information introduces selection biases because not all user groups may be equally inclined to explicitly share certain characteristics. This motivates using statistical methods to infer Reddit users' age, country, and gender here.

The user characteristics comorbid MH issues, age, country, and gender were chosen because they impact peoples' lived experience in BD as discussed in the following. This work identifies users with a self-reported BD diagnosis because collecting posts from BD-specific subreddits does not suffice as carers and people who are unsure if

[5] `https://www.redditinc.com/policies/privacy-policy`

they meet diagnostic criteria also post there. Other self-reported MH diagnoses were extracted because people with BD diagnoses frequently experience additional MH issues (Merikangas et al., 2011). Self-reported diagnoses capture only users who explicitly and publicly share their diagnosis. This research does not infer any users' MH state.

Depp and Jeste (2004), among others, provide evidence for age-related differences in BD symptoms and experiences, also through increasing importance of physical health comorbidities with ageing. Age estimates were grouped in the same way as in a US survey of Reddit users for comparison.

Healthcare systems, including provision of MH care, vastly differ between countries, even within Western countries such as the US, UK, and Germany. The MH services people can access may influence their experience of BD, motivating estimation of their country of residence. While Harrigian (2018) predicts longitude/latitude coordinates in 0.5 steps, these are mapped to countries because more fine-grained user localisations are not needed.

Using a gender variable in NLP deserves special consideration because it concerns people's identity (Larson, 2017). Biological sex can impact on the experience of BD, primarily through issues around childbirth and menopause, also related to mood-impacting hormonal changes (Diflorio and Jones, 2010); Sajatovic et al. (2011) found effects of gender identity on treatment adherence in BD. This work only uses binary m/f gender labels since no NLP method with more diverse categories was available. The gender recognition methods could cause harm to individual users if they were misgendered and then incorrectly addressed or referred to. This project minimises such harm because the labels only serve to estimate the gender distribution and not to target individual users.

### 3.4 Dual use

This research aims to learn more about Reddit users who share their experiences with BD to yield findings that will ultimately lead to new or improved interventions that support living well with BD. However, most research, even when conducted with the best intentions, suffers from the dual-use problem (Jonas, 1984), in that it can be misused or have consequences that affect people's life negatively. Adverse consequences of this study could arise for the Reddit users included in the dataset if they are sought out based on their self-reported BD diagnosis to be targeted with, e.g. medication advertisements. The large number of Reddit posts in this dataset can serve as training data for machine learning systems that assign a likelihood to other Reddit/social media users for meeting BD criteria (e.g., Cohan et al., 2018; Sekulić et al., 2018). For example, health insurance companies could misuse this, using applicants' social media profiles in risk assessments.

### 3.5 Transparency: Dataset and code release

Based on all above considerations, the dataset will only be shared with other researchers upon request and under a data usage agreement that specifies ethical usage of the dataset as detailed in this section. The dataset release necessarily contains the original post texts but with replaced post and user ids. This requires verbatim web searches with the post texts to seek out individual Reddit users and thus complicates automatisation and scaling. User characteristics, including the manually annotated subsets, will only be shared separately with researchers who justify a specific need for them. To aid transparency, the code and patterns to identify self-reported MH diagnoses, age, and gender are released[6].

| Variable | Users | Agreement (%) | Labels (%) |
|---|---|---|---|
| Self-rep. BD diag. | 100 | 97.0 | Yes: 97.0<br>No: 3.0 |
| Date of birth | 116 | 99.1 | Date: 90.5<br>?: 19.5 |
| Country | 100 | 90.0 | US: 46.0<br>CA: 9.0<br>GB: 8.0<br>Other: 25.0<br>?: 12.0 |
| Gender | 116 | 95.7 | F: 51.7<br>M: 34.5<br>Trans: 0.9<br>?: 13.8 |

Table 2: Number of users in manual annotation, raw annotator agreement, and label distributions after resolving disagreements in discussion (?: no label assigned due to lack of user-provided information on Reddit)

[6]https://github.com/glorisonne/reddit_bd_user_characteristics

| Variable | Users$^{test}$ | Method | Accuracy$^{test}$ | Coverage$^{test}$ | Coverage$^{all}$ |
|---|---|---|---|---|---|
| Age group | 105 | Self-reported | 100.0% | 98.1% | 11.5% |
| | | Language use | 60.6% | 94.3% | 66.0% |
| | | Hybrid | 99.0% | 100% | 68.3% |
| Country | 88 | Words, subreddits, timing | 78.4% | 100% | 100% |
| Gender | 100 | Username | 100% | 12.0% | 10.9% |
| | | Self-reported | 97.9% | 94.0% | 11.9% |
| | | Language use | 84.2% | 95.0% | 66.0% |
| | | Hybrid | 97.0% | 100% | 71.5% |

Table 3: Accuracy ($\frac{\text{correct}}{\text{total}}$) for user metadata extraction and inference methods (see Section 2.2) for manually annotated users (test), coverage ($\frac{\text{predicted}}{\text{total}}$) for manually annotated (test) and all (all, n=19,685) users

## 4 Results and discussion

The self-reported BD diagnosis matching method identified 19,685 Reddit users who together had 21,407,595 public Reddit posts between March 2006 and March 2019. Compared to 9K unique user accounts who posted in the four largest BD-related subreddits in May 2020, this likely only constitutes a small fraction of Reddit users with a BD diagnosis that could be reliably automatically identified (see following subsection).

### 4.1 Manual annotation

Two authors manually annotated random subsets of users to evaluate all automatically extracted or inferred information according to the annotation guidelines[7]. As shown in Table 2 agreement for all annotations was above 90%, demonstrating feasibility and high reliability.

The annotators checked all extracted self-reported bipolar disorder diagnosis statements of 100 random included users, disagreeing only for three users (see first line of Table 2)[8]. The pattern matching approach for self-reported diagnosis statements mistakenly identified only three users (subsequently removed from the dataset) based on reports of other MH diagnoses where the word bipolar occurred close to the diagnosis term as well[9].

[7]https://github.com/glorisonne/reddit_bd_user_characteristics/blob/master/ManualAnnotationGuidelines.pdf

[8]No attempt was made to evaluate recall of user identification. Given an international prevalence of meeting BD criteria of about 2% (Merikangas et al., 2011) and expecting numbers of posts per account close to the average of 1,224 in the collected dataset, it was deemed infeasible to manually check all posts of randomly selected user accounts for self-reported bipolar disorder diagnosis statements.

[9]Paraphrased excerpts of incorrectly identified self-reported BD diagnoses: 'clinical depression with bipolar tendencies', 'diagnosed with BPD today, thought it was BD for years', 'diagnosed with depression, but sure I've got bipolar'.

To facilitate manual age and gender annotation, 116 users where randomly selected from the 2854 (14%) of users where the Self-reported age or gender extraction method matched. This explains the discrepancy between the coverage of the Self-reported method in Table 3 for the test set and full dataset. The annotators only checked whether date of birthor gender could be unambiguously extracted from all of a users' posts that matched a self-reported age and gender pattern. The test set for the gender evaluation results in Table 3 comprises only users labelled as m/f and excludes one manually identified transgender person.

### 4.2 Evaluation of NLP methods

Table 3 shows accuracy and coverage for the user characteristics extraction and inference methods described in Section 2.2 against the manually labelled users for which the annotators could determine a label. For age, the Self-reported method outperforms the Language use method for accuracy but not coverage[10]. The Hybrid method, subsequently used in Section 4.3.2, achieves 99% test set accuracy and 68% coverage on the full dataset. Harrigian's (2018) method assigns a country estimate to every user with 78% test set accuracy. For gender, accuracy decreases from the Username, Self-reported, and Language use method, while coverage increases [11]. The Hybrid gender identification method, used in Section 4.3.2, achieves 97% test set accuracy, gender-labelling 72% of users.

[10]The Language use method for age/gender does not have full coverage because it requires at least ten posts per user. The methods agree for 62.6% of the 1,788 users where both assign an age group.

[11]For 195 users where all three methods assign a gender identity, they agree on 73.8% (90.8% agreement between the Username and Self-reported method, 80% between the Language use and Username or Self-reported method).

| Diagnosis | Dataset n=19,685 (%) | SMHD n=6,434 (%) | Epidemiological studies (%) |
|---|---|---|---|
| MDD | 30.2 | 27.4 | N/A |
| Anxiety disorder | 15.8 | 12.8 | 13.3-16.8*, n=921-1,537 |
| ADHD | 12.9 | 9.6 | 17.6†, n=399 |
| BPD | 8.4 | N/A | 16$, n=1,255 |
| PTSD | 6.5 | 5.1 | 10.8*, n=1,185 |
| OCD | 3.9 | 3.4 | 10.7*, n=808 |
| ASD | 2.2 | 2.0 | Unknown |
| ED | 1.0 | 0.8 | 5.3-31⊙, n=51-1,710 |

Table 4: Self-reported comorbid diagnoses with BD in this work, the SMHD dataset, and epidemiological studies: *Nabavi et al. (2015), †(McIntyre et al., 2010), $(Zimmerman and Morgan, 2013), ⊙(Álvarez Ruiz and Gutiérrez-Rojas, 2015)

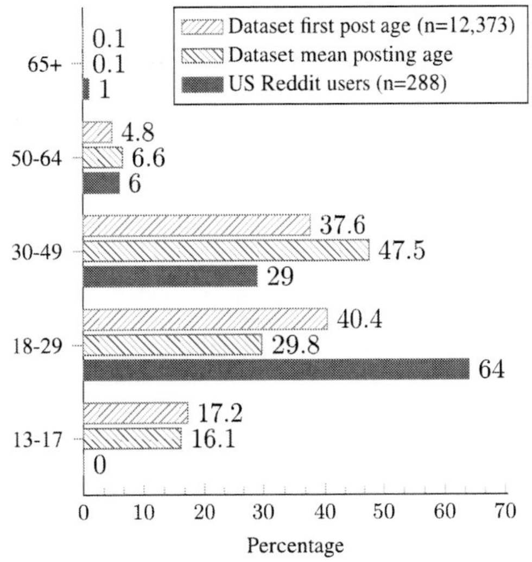


Figure 1: Age of Reddit users

### 4.3 Reddit users' characteristics

The following subsections compare characteristics of Reddit users with a self-reported BD diagnosis to general Reddit users and epidemiological statistics.

#### 4.3.1 Mental health comorbidities

Table 4 shows how many users disclosed other concurrent or lifetime MH diagnoses besides BD. Rates for self-reported MH diagnoses in addition to BD are sightly higher in our dataset compared to the Self-reported MH diagnoses (SMHD) dataset (Cohan et al., 2018), potentially because our dataset covers 27 more months of posts.

Like psychotic disorder (5.2% of users prior to exclusion), a MDD diagnosis is mutually exclusive with BD according to the DSM (American Psychiatric Association, 2013, pp. 126, 134)[12]. A large part of identified self-reported MDD diagnoses were false positives where 'depression' occurred near to a BD diagnosis statement. More conservatively only considering self-reported MDD diagnosis posts that do not also match BD patterns, results in 8.7% users reporting both diagnoses. MDD and Psychotic disorder diagnoses jointly with BD might indicate subsequently changed (mis-)diagnoses or disagreement of professionals. Surveys in Germany (Pfennig et al., 2011) and the US (Hirschfeld et al., 2003) have shown that often more than ten years pass between onset of BD symptoms and receiving the diagnosis, with two thirds of people being misdiagnosed, most frequently with MDD. Moreover, field trials for BD diagnoses with DSM-V criteria only showed moderate clinician agreement (Freedman et al., 2013).

Comorbidity rates for anxiety disorders, BPD and PTSD align with results from epidemiological studies. Rates for comorbid ADHD, OCD, and ED are lower in the Reddit dataset population, which might in part be due to incomplete coverage of the patterns to capture diagnosis self-reports. Additionally, epidemiological studies can be expected to yield higher comorbidity rates because they determine if participants meet criteria for various diagnoses with clinical interviews, whereas Reddit users may not have (or report) diagnoses for every condition they meet the criteria of. Overall, 50.7% of users reported at least one additional MH diagnosis, slightly less than three quarters of surveyed people in the World Mental Health Survey Initiative who met criteria for at least one other DSM-IV disorder besides BD (Merikangas et al., 2011).

More than 2% of users reported an ASD diagnosis in addition to BD, with no epidemiological studies on ASD prevalence with BD yet. Dell'Osso et al. (2019) found significant levels of autistic traits among 43% of people with a BD diagnosis.

#### 4.3.2 Age

As shown in Figure 1, less Reddit users with a self-reported BD diagnosis are 18-29 but more

[12]The dataset includes users with self-reported MDD but not psychotic disorder because depression but not psychosis is a core aspect of extreme mood, our focus of future research.

| Country | Dataset (%) | Reddit.com traffic (%) | 12-months prev. (%) |
|---|---|---|---|
| US | 81.9 | 49.69 | 0.68 |
| UK | 5.6 | 7.93 | 1.11 |
| Canada | 4.9 | 7.85 | 0.75 |
| Australia | 1.7 | 4.32 | 1.15 |
| Germany | 1.4 | 3.17 | 0.83 |

Table 5: Top 5 estimated countries of residence of Reddit users with a self-reported BD diagnosis, location of reddit.com site visitors (Statista.com, 2020) and 12-months prevalence of BD (Global Burden of Disease Collaborative Network, 2018)

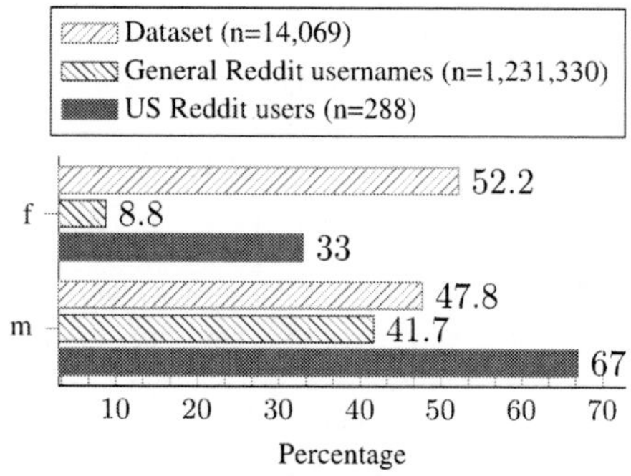


Figure 2: Binary gender of Reddit users

30-49 years old compared to average US Reddit users (Barthel et al., 2016, p. 7)[13]. The age of onset of BD symptoms is most frequently in late adolescence and early adulthood (Pini et al., 2005; Merikangas et al., 2011, p. 6). In line with this, the majority of Reddit users who disclose a BD diagnosis are between 13-29 years old at their first post. In the Global Burden of Disease study 2013, BD 12-months prevalence rates were significantly elated for 20-54 year olds Ferrari et al. (2016, p. 447). In our dataset, almost 80% of the Reddit users are 18-49 years old at their first post.

#### 4.3.3 Country of residence

As shown in Table 5, more than 80% of the Reddit users with a self-reported BD diagnosis are estimated to live in the US, and 95% in one of the English-speaking countries US, UK, Canada, Australia. This ranking aligns with site visitors of the Reddit desktop version (Statista.com, 2020), although US users are even more prevalent in the BD dataset. All of the top-5 countries in the dataset have a 12-months prevalence of BD diagnoses above the global average of 0.62% according to the 2017 Global Burden of Disease Study (Global Burden of Disease Collaborative Network, 2018).

#### 4.3.4 Binary gender

Figure 2 shows that the Hybrid method assigned feminine gender to slightly more than half of the Reddit users for which it ascribed a gender identity. This sharply contrasts with only 9% feminine vs. 41% masculine gender-performing usernames among Reddit users who posted in the top 10K subreddits with most posts (Wang and Jurgens, 2018). A survey of adult US Reddit users (Barthel et al., 2016) found that two thirds were men.

In epidemiological studies, biological men and women are equally likely to meet criteria for BD overall (Pini et al., 2005, American Psychiatric Association, 2013, p. 124) although there is evidence that BD-II is more frequently diagnosed among women (Diflorio and Jones, 2010). Sajatovic et al. (2011) found that biological men with a BD diagnosis scored significantly lower on masculine gender identity than the general male population, while there were no gender identity differences for biological women. Considering a majority of male Reddit users and sex-equal prevalence of the diagnosis, feminine-gender-identifying people with a BD diagnosis seem to be more likely to use Reddit and/or to disclose their diagnosis. The increased rates of female-gender identifying Reddit users with a self-reported BD diagnosis might also point towards a higher relative frequency of BD-II diagnoses (compared to BD-I) in this population.

## 5 Limitations and implications

### 5.1 Limitations

First, unlike in clinical studies with face-to-face interactions, we cannot assume that every Reddit user in the dataset corresponds to one person. Additionally, self-reported diagnoses cannot be confirmed with diagnostic interviews as in clinical research.

Furthermore, there are several limitations to the NLP methods to infer user characteristics. The method to extract self-reported MH diagnoses does not distinguish between actual comorbidities and misdiagnoses or previous diagnoses, for which symptoms may have resolved. Manual evaluation of ten users with BPD comorbidity showed that seven reported concurrent diagnoses, one a BD to BPD change, one a BPD misdiagnosis, and one re-

[13]The Barthel et al. (2016) survey only targeted adults, therefore there are no 13-17-year-old users.

ferred to BD by 'BPD'. Harrigian's (2018) method indicates the predominantly reflected country in a user's most recent posts, disregarding relocations.

The Self-reported age and gender extraction method is fallible to users providing incorrect information, for example disguising themselves as younger than they really are on dating subreddits. Finally, none of the gender inference methods allow us to estimate how many users identify as transgender or non-binary. Such indications were also too diverse to be captured in the regular expressions for self-reported age and gender. Still, four of the subreddits with more than 10K posts by users with a self-reported BD diagnosis target transgender people, indicating that a proportion of the users in this research may not identify with their born sex.

### 5.2 Health research implications

Most importantly this work provides the first large-scale characterisation of Reddit users with a self-reported BD diagnosis, who are on average 27.7 years old at their first post, seem to overwhelmingly live in the US, and are more likely to identify with the feminine gender. Insofar they deviate from general Reddit as well as epidemiological statistics and also from participants in clinical studies.

A large meta-analysis of psychological interventions for BD (Oud et al., 2016) showed that in 55 trials conducted across twelve countries (35% in the US) comprising 6,060 adults with BD, 89% had recruited participants with a mean age higher than the 30 year-average of adult Reddit users with a self-reported BD diagnosis. 67% of the trials recruited a higher percentage of females than the 52% figure in the Reddit dataset (Oud et al., 2016, Table DS2). This cautions against generalising findings from Reddit data to all people with a BD diagnosis, but stresses its complementary role to clinical studies with different selection biases.

Another important implication is that NLP analysis of Reddit social media users largely confirmed high prevalence rates for comorbid MH conditions with BD from epidemiological studies. Besides clinically established comorbidities with, e.g., Anxiety disorder and ADHD, the present analysis also revealed substantial prevalence of ASD, for which there is little clinical research to date. Reddit may constitute a useful platform to learn about the experiences of people with BD with such currently under-researched comorbidities and may be a way to target them for recruitment to clinical studies.

### 5.3 NLP research implications

This work evaluated state-of-the-art methods to infer Reddit user characteristics (Harrigian, 2018; Wang and Jurgens, 2018; Tigunova et al., 2019) and demonstraed their utility in applied research. A hybrid method achieved the best accuracy and coverage for age and gender identity by using high-accuracy information from self-reports (or a gender-performing username) when available, filling in information for more users with less accurate predictions from a neural network language use-based method (Tigunova et al., 2019).

Importantly, gender-inference methods so far are limited to detecting binary gender, although, e.g., 0.4% of the US population identify as transgender (Meerwijk and Sevelius, 2017). Off-the-shelf NLP tools supporting a wider range of gender identities may be more inclusive and give more visibility to these groups of people in research. However, important ethical considerations arise around identifying people with transgender and non-binary gender identities, which are often stigmatised.

## 6 Conclusion

This paper set out to automatically profile Reddit users under consideration of ethical aspects. A combination of pattern-based and previously published NLP methods served to estimate clinical, demographic, and identity characteristics of nearly 20K Reddit users with a self-reported BD diagnosis. Half of the Reddit users disclosed MH diagnoses besides BD and 80% were located in the US. From the users for which age or gender could be estimated, 80% were between 18-49 years old and 52% performed or identified with feminine gender.

These findings indicate about which populations BD-focused research on Reddit may generalise. Additionally, this work may serve as a model for how to provide more information on other specific Reddit populations as requested by recent transparency and accountability movements in NLP.

## Acknowledgements

We would like to thank Anna Tigunova and Keith Harrigian for their assistance in applying their Reddit user profiling NLP tools. We would also like to express our heartfelt thanks to Daisy Harvey, Stephen Mander, and the anonymous reviewers for helpful comments on a draft version of this article, to Andrew Moore for testing the code release, and to Alistair Baron for the initial idea for this work.

## References


Wasim Ahmed, Peter A. Bath, and Gianluca Demartini. 2017. Using Twitter as a data source: an overview of ethical, legal and methodological challenges. In Kandy Woodfield, editor, *The Ethics of Online Research*, pages 79–107. Emerald Books.

Nadia Akers, Fiona Lobban, Claire Hilton, Katerina Panagaki, and Steven H. Jones. 2019. Measuring social and occupational functioning of people with bipolar disorder: A systematic review. *Clinical Psychology Review*, 74.

Alexa Internet. 2020. reddit.com.

Eva M. Álvarez Ruiz and Luis Gutiérrez-Rojas. 2015. Comorbidity of bipolar disorder and eating disorders. *Revista de Psiquiatria y Salud Mental*, 8(4):232–241.

Ashley Amaya, Ruben Bach, Florian Keusch, and Frauke Kreuter. 2019. New Data Sources in Social Science Research: Things to Know Before Working With Reddit Data. *Social Science Computer Review*, pages 1–18.

American Psychiatric Association. 2013. *DSM-5*. Washington, DC.

Michael Barthel, Galen Stocking, Jesse Holcomb, and Amy Mitchell. 2016. Nearly Eight-in-Ten Reddit Users Get News on the Site. Technical report.

Rita Bauer, Michael Bauer, Hermann Spiessl, and Tanja Kagerbauer. 2013. Cyber-support: An analysis of online self-help forums (online self-help forums in bipolar disorder). *Nordic Journal of Psychiatry*, 67(3):185–190.

Emily M. Bender and Batya Friedman. 2018. Data Statements for Natural Language Processing: Toward Mitigating System Bias and Enabling Better Science. *Transactions of the Association for Computational Linguistics*, 6:587–604.

Kelsey Beninger, Alexandra Fry, Natalie Jago, Hayley Lepps, Laura Nass, and Hannah Silvester. 2014. Research using Social Media; Users' Views.

Adrian Benton, Glen Coppersmith, and Mark Dredze. 2017. Ethical Research Protocols for Social Media Health Research. In *Proceedings of the First Workshop on Ethics in Natural Language Processing*, pages 94–102.

Alexandra Budenz, Ann Klassen, Jonathan Purtle, Elad Yom Tov, Michael Yudell, and Philip Massey. 2019. Mental illness and bipolar disorder on Twitter: implications for stigma and social support. *Journal of Mental Health*, 29(2):191–199.

Iain H. Campbell and Harry Campbell. 2019. Ketosis and bipolar disorder: controlled analytic study of online reports. *BJPsych Open*, 5(4):1–6.

Arman Cohan, Bart Desmet, Sean Macavaney, Andrew Yates, Luca Soldaini, Sean Macavaney, and Nazli Goharian. 2018. SMHD: A Large-Scale Resource for Exploring Online Language Usage for Multiple Mental Health Conditions. In *Proceedings of the 27th International Conference on Computational Linguistics (COLING)*, pages 1485–1497.

Glen Coppersmith, Mark Dredze, and Craig Harman. 2014. Quantifying Mental Health Signals in Twitter. In *Proceedings of the Workshop on Computational Linguistics and Clinical Psychology: From Linguistic Signal to Clinical Reality*, pages 51–60.

Glen Coppersmith, Mark Dredze, Craig Harman, and Kristy Hollingshead. 2015. From ADHD to SAD: Analyzing the Language of Mental Health on Twitter through Self-Reported Diagnoses. In *Proceedings of the 2nd Workshop on Computational Linguistics and Clinical Psychology: From Linguistic Signal to Clinical Reality*.

Liliana Dell'Osso, Barbara Carpita, Carlo Antonio Bertelloni, Elisa Diadema, Filippo Maria Barberi, Camilla Gesi, and Claudia Carmassi. 2019. Subthreshold autism spectrum in bipolar disorder: Prevalence and clinical correlates. *Psychiatry Research*, 281(October):112605.

Colin A. Depp and Dilip V. Jeste. 2004. Bipolar disorder in older adults: A critical review. *Bipolar Disorders*, 6(5):343–367.

Arianna Diflorio and Ian Jones. 2010. Is sex important Gender differences in bipolar disorder. *International Review of Psychiatry*, 22(5):437–452.

Gunther Eysenbach and James E. Till. 2001. Ethical issues in qualitative research on internet communities. *BMJ*, 323(7055):1103–1105.

Alize J. Ferrari, Emily Stockings, Jon Paul Khoo, Holly E. Erskine, Louisa Degenhardt, Theo Vos, and Harvey A. Whiteford. 2016. The prevalence and burden of bipolar disorder: findings from the Global Burden of Disease Study 2013. *Bipolar Disorders*, 18(5):440–450.

Nigel G. Fielding, Raymond M. Lee, Grant Blank, and Dietmar Janetzko. 2016. Nonreactive Data Collection Online. *The SAGE Handbook of Online Research Methods*, pages 76–91.

Robert Freedman, David A. Lewis, Robert Michels, Daniel S. Pine, Susan K. Schultz, Carol A. Tamminga, Glen O. Gabbard, Susan Shur-Fen Gau, Daniel C. Javitt, Maria A. Oquendo, Patrick E. Shrout, Eduard Vieta, and Joel Yager. 2013. The Initial Field Trials of DSM-5: New Blooms and Old Thorns. *American Journal of Psychiatry*, 170(1):1–5.

Andréanne Gignac, Alexander McGirr, Raymond W Lam, and Lakshmi N Yatham. 2015. Recovery and recurrence following a first episode of mania: a systematic review and meta-analysis of prospectively characterized cohorts. *The Journal of clinical psychiatry*, 76(9):1241–1248.

George Gkotsis, Anika Oellrich, Tim Hubbard, Richard Dobson, Maria Liakata, Sumithra Velupillai, and Rina Dutta. 2016. The language of mental health problems in social media. In *Proceedings of the Third Workshop on Computational Lingusitics and Clinical Psychology*, pages 63–73.

George Gkotsis, Anika Oellrich, Sumithra Velupillai, Maria Liakata, Tim J.P. Hubbard, Richard J.B. Dobson, and Rina Dutta. 2017. Characterisation of mental health conditions in social media using Informed Deep Learning. *Scientific Reports*, 7:1–11.

Global Burden of Disease Collaborative Network. 2018. Global Burden of Disease Study 2017 (GBD 2017) Results. Technical report, Institute for Health Metrics and Evaluation (IHME), Seattle, United States.

G. M. Goodwin, P. M. Haddad, I. N. Ferrier, J. K. Aronson, T. R.H. Barnes, A. Cipriani, D. R. Coghill, S. Fazel, J. R. Geddes, H. Grunze, E. A. Holmes, O. Howes, S. Hudson, N. Hunt, I. Jones, I. C. MacMillan, H. McAllister-Williams, D. R. Miklowitz, R. Morriss, M. Munafò, C. Paton, B. J. Saharkian, K. E.A. Saunders, J. M.A. Sinclair, D. Taylor, E. Vieta, and A. H. Young. 2016. Evidence-based guidelines for treating bipolar disorder: Revised third edition recommendations from the British Association for Psychopharmacology. *Journal of Psychopharmacology*, 30(6):495–553.

Keith Harrigian. 2018. Geocoding without geotags: a text-based approach for reddit. In *Proceedings of the 2018 EMNLP Workshop W-NUT: The 4th Workshop on Noisy User-generated Text*, pages 17–27.

Robert M. A. Hirschfeld, Lydia Lewis, and Lana A. Vornik. 2003. Perceptions and impact of bipolar disorder: How far have we really come? Results of the National Depressive and Manic-Depressive Association 2000 survey of individuals with bipolar disorder. *The Journal Of Clinical Psychiatry*, 64(2):161–174.

Yen-Hao Huang, Yi-Hsin Chen, Fernando H. Calderon, Ssu-Rui Lee, Shu-I Wu, Yuwen Lai, and Yi-Shin Chen. 2019. Leveraging Linguistic Characteristics for Bipolar Disorder Recognition with Gender Differences. In *Proceedings of the 2019 KDD Workshop on Applied Data Science for Healthcare (DSHealth '19)*.

Xiang Ji, Soon Ae Chun, Zhi Wei, and James Geller. 2015. Twitter sentiment classification for measuring public health concerns. *Social Network Analysis and Mining*, 5(1):1–25.

Hans Jonas. 1984. *The Imperative of Responsibility: Foundations of an Ethics for the Technological Age*. University of Chicago Press, Chicago.

Adam D.I. Kramer, Susan R. Fussell, and Leslie D. Setlock. 2004. Text analysis as a tool for analyzing conversation in online support groups. *Conference on Human Factors in Computing Systems - Proceedings*, pages 1485–1488.

Brian Larson. 2017. Gender as a Variable in Natural-Language Processing: Ethical Considerations. In *Proceedings of the First Workshop on Ethics in Natural Language Processing*, pages 1–11.

Klara Latalova, Jan Prasko, Dana Kamaradova, Katerina Ivanova, Lubica Jurickova, Latalova K., Prasko J., Kamaradova D., and Ivanova K. 2014. Bad on the net, or bipolars' lives on the web: Analyzing discussion web pages for individuals with bipolar affective disorder. *Neuro Endocrinology Letters*, 35(3):206–212.

Anika Mandla, Jo Billings, and Joanna Moncrieff. 2017. "Being Bipolar": A Qualitative Analysis of the Experience of Bipolar Disorder as Described in Internet Blogs. *Issues in Mental Health Nursing*, 38(10):858–864.

D McDonald and R Woodward-Kron. 2016. Member roles and identities in online support groups: Perspectives from corpus and systemic functional linguistics. *Discourse and Communication*, 10(2):157–175.

Roger S McIntyre, Sidney H Kennedy, Joanna K Soczynska, Ha T T Nguyen, Timothy S Bilkey, Hanna O Woldeyohannes, Jay A Nathanson, Shikha Joshi, Jenny S H Cheng, Kathleen M Benson, and David J Muzina. 2010. Attention-deficit/hyperactivity disorder in adults with bipolar disorder or major depressive disorder: results from the international mood disorders collaborative project. *Primary Care Companion To The Journal Of Clinical Psychiatry*, 12(3).

Esther L. Meerwijk and Jae M. Sevelius. 2017. Transgender population size in the United States: A meta-regression of population-based probability samples. *American Journal of Public Health*, 107(2):e1–e8.

Kathleen R. Merikangas, Robert Jin, Jian-ping He, Ronald C. Kessler, Sing Lee, Nancy A. Sampson, Maria Carmen Viana, Laura Helena Andrade, Chiyi Hu, Elie G. Karam, Maria Ladea, Maria Elena Medina Mora, Mark Oakley Browne, Yutaka Ono, Jose Posada-Villa, Rajesh Sagar, and Zahari Zarkov. 2011. Prevalence and correlates of bipolar spectrum disorder in the world mental health survey initiative. *Archives of general psychiatry*, 68(3):241–251.

Behrouz Nabavi, Alex J. Mitchell, and David Nutt. 2015. A Lifetime Prevalence of Comorbidity Between Bipolar Affective Disorder and Anxiety Disorders: A Meta-analysis of 52 Interview-based Studies of Psychiatric Population. *EBioMedicine*, 2(10):1405–1419.

Danielle M. Novick, Holly A. Swartz, and Ellen Frank. 2010. Suicide attempts in bipolar I and bipolar II disorder: A review and meta-analysis of the evidence. *Bipolar Disorders*, 12(1):1–9.

Bridianne O'Dea, Stephen Wan, Philip J. Batterham, Alison L. Calear, Cecile Paris, and Helen Christensen. 2015. Detecting suicidality on Twitter. *Internet Interventions*, 2(2):183–188.

Matthijs Oud, Evan Mayo-Wilson, Ruth Braidwood, Peter Schulte, Steven H. Jones, Richard Morriss, Ralph Kupka, Pim Cuijpers, and Tim Kendall. 2016. Psychological interventions for adults with bipolar disorder: Systematic review and meta-analysis. *British Journal of Psychiatry*, 208(3):213–222.

Michael J. Paul and Mark Dredze. 2017. Social Monitoring for Public Health. *Synthesis Lectures on Information Concepts, Retrieval, and Services*, 9(5):1–183.

Andrea Pfennig, B. Jabs, S. Pfeiffer, B. Weikert, K. Leopold, and M. Bauer. 2011. Health care service experiences of bipolar patients in Germany - Survey prior to the introduction of the S3 Guideline for diagnostics and treatment of bipolar disorders. *Nervenheilkunde*, 30(5):333–340.

Stefano Pini, Valéria De Queiroz, Daniel Pagnin, Lukas Pezawas, Jules Angst, Giovanni B. Cassano, and Hans Ulrich Wittchen. 2005. Prevalence and burden of bipolar disorders in European countries. *European Neuropsychopharmacology*, 15(4):425–434.

Ria Poole, Daniel Smith, and Sharon Simpson. 2015. How Patients Contribute to an Online Psychoeducation Forum for Bipolar Disorder: A Virtual Participant Observation Study. *JMIR Mental Health*, 2(3:e21).

Daniel Preoţiuc-Pietro, Johannes Eichstaedt, Gregory Park, Maarten Sap, Laura Smith, Victoria Tobolsky, H. Andrew Schwartz, and Lyle Ungar. 2015. The role of personality, age, and gender in tweeting about mental illness. In *Proceedings ofthe 2nd Workshop on Computational Linguistics and Clinical Psychology: From Linguistic Signal to Clinical Reality*, pages 21–30.

Derek Ruths and Jürgen Pfeffer. 2014. Social media for large studies of behavior. *Science*, 346(6213):1063–1064.

Puneet K.C. Sahota and Pamela L. Sankar. 2019. Bipolar Disorder, Genetic Risk, and Reproductive Decision-Making: A Qualitative Study of Social Media Discussion Boards. *Qualitative Health Research*.

Martha Sajatovic, Weronika Micula-Gondek, Curtis Tatsuoka, and Christopher Bialko. 2011. The relationship of gender and gender identity to treatment adherence among individuals with bipolar disorder. *Gender Medicine*, 8(4):261–268.

Elvis Saravia, Chun Hao Chang, Renaud Jollet De Lorenzo, and Yi Shin Chen. 2016. MIDAS: Mental illness detection and analysis via social media. *Proceedings of the 2016 IEEE/ACM International Conference on Advances in Social Networks Analysis and Mining, ASONAM 2016*, pages 1418–1421.

Ivan Sekulić, Matej Gjurković, and Jan Šnajder. 2018. Not Just Depressed: Bipolar Disorder Prediction on Reddit. In *Proceedings of the 9th Workshop on Computational Approaches to Subjectivity, Sentiment Social Media Analysis (WASSA)*, pages 72–78.

Statista.com. 2020. Regional distribution of desktop traffic to Reddit.com as of September 2020, by country.

Anna Tigunova, Paramita Mirza, Andrew Yates, and Gerhard Weikum. 2019. Listening between the lines: Learning personal attributes from conversations. *The Web Conference 2019 - Proceedings of the World Wide Web Conference, WWW 2019*, 2:1818–1828.

Anna Tigunova, Andrew Yates, Paramita Mirza, and Gerhard Weikum. 2020. RedDust: a Large Reusable Dataset of Reddit User Traits. In *Proceedings ofthe 12th Conference on Language Resources and Evaluation (LREC 2020)*, pages 6118–6126.

T. Treuer and M. Tohen. 2010. Predicting the course and outcome of bipolar disorder: A review. *European Psychiatry*, 25(6):328–333.

Agnès Vayreda and Charles Antaki. 2009. Social Support and Unsolicited Advice in a Bipolar Disorder Online Forum. *Qualitative Health Research*, 19(7):931–942.

Zijian Wang and David Jurgens. 2018. It's going to be okay: Measuring Access to Support in Online Communities. In *Proceedings of the 2018 Conference on Empirical Methods in Natural Language Processing (EMNLP)*, pages 33–45.

Helen Warwick, Sara Tai, and Warren Mansell. 2019. Living the life you want following a diagnosis of bipolar disorder: A grounded theory approach. *Clinical Psychology Psychotherapy*.

Matthew L. Williams, Pete Burnap, and Luke Sloan. 2017. Towards an Ethical Framework for Publishing Twitter Data in Social Research: Taking into Account Users' Views, Online Context and Algorithmic Estimation. *Sociology*, 51(6):1149–1168.

World Health Organisation. 2018. *International classification of diseases for mortality and morbidity statistics (11th Revision)*.

Minjoo Yoo, Sangwon Lee, and Taehyun Ha. 2019. Semantic network analysis for understanding user experiences of bipolar and depressive disorders on Reddit. *Information Processing and Management*, 56(4):1565–1575.

Mark Zimmerman and Theresa A. Morgan. 2013. Problematic Boundaries in the Diagnosis of Bipolar Disorder: The Interface with Borderline Personality Disorder. *Current Psychiatry Reports*, 15(422).

## A Further method details

### A.1 Age and gender: Language use

Tigunova et al.'s (2019) $\text{HAM}_{\text{CNN-attn}}$ model predicts an age group[14] and gender for Reddit users with at least ten posts based on their up to 100 most recent posts. Separate $\text{HAM}_{\text{CNN-attn}}$ models were trained on the RedDust dataset (Tigunova et al., 2020) with the HAM open-source implementation[15] with the hyper-parameters specified by Tigunova et al. (2020) (128 CNN filters of size 2, attention layer with 150 units, 70 training epochs). Likely due to random seed variation, our trained age model had an area under the curve (AUROC) score of 0.80 compared to 0.88 in Tigunova et al. (2020). Our trained gender model had 84.9% accuracy on the RedDust test set compared to 86.0% reported by Tigunova et al. (2020).

### A.2 Age: Hybrid method

Two corrections were applied prior to the Hybrid method: The first author checked all users with a self-reported average posting age below 16 or above 60. Age at account creation predictions younger than 13 by the Language use approach were discarded as Reddit requires an age of at least 13 when signing up.

### A.3 Country

The Reddit country inference method (Harrigian, 2018) initially was a proprietary project but later the first author, Keith Harrigian, rebuilt it for the public release[16] used in this work. Therefore, the training data and model performance, provided by Keith Harrigian in personal email communication on 5th March 2021, slightly differ from the original publication. The training data consists of 56,853 automatically location-labelled users (top 5: 68.8% US, 9.4% Canada, 7.0% UK, 3.3% Australia, 1.0% Germany), of which 8.2% were identified based on self-reported locations in r/AmateurRoomPorn and the remainder by self-reported locations in reply to 'Where are you from?' questions (Harrigian, 2018). Label precision was 97.6% in a manual evaluation of 500 users[17].

The 'Global' (as opposed to US only) model was used to predict user locations, which achieves 35.6% accuracy at 100 miles in 5-fold cross validation, equal to the originally reported performance in Harrigian (2018). Overall country-level accuracy is 81.9% and is generally higher for users with more training data (95.1% US, 65.1% Canada, 82.8% UK, 44.1% Australia, 41.1% Germany).

### A.4 Gender: Username method

Wang and Jurgens (2018, p. 38) trained their long short-term memory (LSTM) gender inference model on 80% of 4,900,250 Twitter and 367,495 Reddit usernames, automatically labelled with self-reported m or f gender identity. Following them, the present work assumes usernames to perform masculine (m) gender for model predictions of 0.1 or lower, and feminine (f) for 0.9 or higher. This model and setting achieved 0.92 precision with 0.18 recall in 10% held-out Twitter and Reddit username test data (Wang and Jurgens, 2018, Figure 5 in supplementary material).

[14] younger than 14, 14-23, 24-45, 46-65, 66+, relative to the user's most recent post

[15] `https://github.com/Anna146/HiddenAttributeModels`

[16] `https://github.com/kharrigian/smgeo`

[17] `https://github.com/kharrigian/smgeo#dataset-noise`

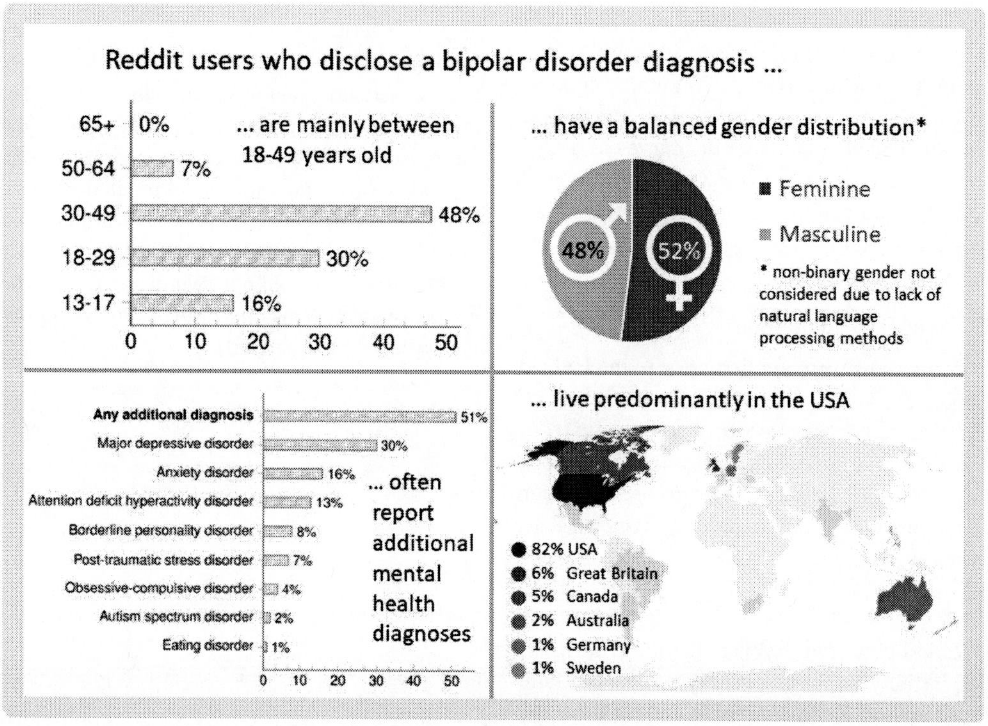


Figure 3: Visual summary of the characteristics of Reddit users who self-report a bipolar disorder diagnosis

# On the State of Social Media Data for Mental Health Research

**Keith Harrigian, Carlos Aguirre, Mark Dredze**
Johns Hopkins University
kharrigian@jhu.edu, caguirr4@jhu.edu, mdredze@cs.jhu.edu

## Abstract

Data-driven methods for mental health treatment and surveillance have become a major focus in computational science research in the last decade. However, progress in the domain remains bounded by the availability of adequate data. Prior systematic reviews have not necessarily made it possible to measure the degree to which data-related challenges have affected research progress. In this paper, we offer an analysis specifically on the state of social media data that exists for conducting mental health research. We do so by introducing an open-source directory of mental health datasets, annotated using a standardized schema to facilitate meta-analysis.[1]

## 1 Introduction

The last decade has seen exponential growth in computational research devoted to modeling mental health phenomena using non-clinical data (Bucci et al., 2019). Studies analyzing data from the web, such as social media platforms and peer-to-peer messaging services, have been particularly appealing to the research community due to their scale and deep entrenchment within contemporary culture (Perrin, 2015; Fuchs, 2015; Graham et al., 2015). Such studies have yielded novel insights into population-level mental health (De Choudhury et al., 2013; Amir et al., 2019a) and shown promising avenues for the incorporation of data-driven analyses in the treatment of psychiatric disorders (Eichstaedt et al., 2018).

These research achievements have come despite complexities specific to the mental health space often making it difficult to obtain a sufficient sample size of high-quality data. For instance, behavioral disorders are known to display variable clinical presentations amongst different populations, rendering annotations of ground truth inherently noisy (De Choudhury et al., 2017; Arseniev-Koehler et al., 2018). Scalable methods for capturing an individual's mental health status, such as using regular expressions to identify self-reported diagnoses or grouping individuals based on activity patterns, have provided opportunities to construct datasets aware of this heterogeneity (Coppersmith et al., 2015b; Kumar et al., 2015). However, they typically rely on oversimplifications that lack the same clinical validation and robustness as something like a mental health battery (Zhang et al., 2014; Ernala et al., 2019).

Ethical considerations further complicate data acquisition, with the sensitive nature of mental health data requiring tremendous care when constructing, analyzing, and sharing datasets (Benton et al., 2017). Privacy-preserving measures, such as de-identifying individuals and requiring IRB approval to access data, have made it possible to share some data across research groups. However, these mechanisms can be technically cumbersome to implement and are subject to strict governance policies when clinical information is involved due to HIPAA (Price and Cohen, 2019). Moreover, many privacy-preserving practices require that signal relevant to modeling mental health, such as an individual's demographics or their social network, are discarded (Bakken et al., 2004). This missingness has the potential to limit algorithmic fairness, statistical generalizability, and experimental reproducibility (Gorelick, 2006). Although mental health researchers may anecdotally recall difficulties acquiring quality data or reproducing prior art due to data sharing constraints, no study to our knowledge has explicitly quantified this challenge.

Indeed, prior reviews of computational research for mental health have noted several of the aforementioned challenges, but have predominantly discussed technical methods (e.g. model architectures, feature engineering) developed to surmount existing constraints (Guntuku et al., 2017; Wongkoblap

[1] https://github.com/kharrigian/mental-health-datasets



*Proceedings of the Seventh Workshop on Computational Linguistics and Clinical Psychology*, pages 15–24
June 11, 2021. 

et al., 2017). Recent work from Chancellor and De Choudhury (2020), completed concurrently with our own, was the first review to focus specifically on the shortcomings of *data* for mental health research. Our study affirms the findings of Chancellor and De Choudhury (2020), using an expanded pool of literature that more acutely focuses on *language* found in social media data. To this end, we construct a new open-source directory of mental health datasets, annotated using a standardized schema that not only enables researchers to identify *relevant* datasets, but also to identify *accessible* datasets. We draw upon this resource to offer nuanced recommendations regarding future dataset curation efforts.

## 2 Data

To generate evidence-based recommendations regarding mental health dataset curation, we require knowledge of the extant data landscape. Unlike some computational fields which have a surplus of well-defined and uniformly-adopted benchmark datasets, mental health researchers have thus far relied on a decentralized medley of resources. This fact, spurred in part by the variable presentations of psychiatric conditions and in part by the sensitive nature of mental health data, thus requires us to compile a new database of literature. In this section, we detail our literature search, establish inclusion/exclusion criteria, and define a list of dataset attributes to analyze.

### 2.1 Dataset Identification

Datasets were sourced using a breadth-focused literature search. After including data sources from the three aforementioned systematic reviews (Guntuku et al., 2017; Wongkoblap et al., 2017; Chancellor and De Choudhury, 2020), we searched for literature that lie primarily at the intersection of natural language processing (NLP) and mental health communities. We sought peer-reviewed studies published between January 2012 and December 2019 in relevant conferences (e.g. NAACL, EMNLP, ACL, COLING), workshops (e.g. CLPsych, LOUHI), and health-focused journals (e.g. JMIR, PNAS, BMJ).

We searched Google Scholar, ArXiv, and PubMed to identify additional candidate articles. We used two search term structures — 1) (mental health | `DISORDER`) + (social | electronic) + media, and 2) (machine learning | prediction | inference | detection) + (mental health | `DISORDER`). '|' indicates a logical or, and `DISORDER` was replaced by one of 13 mental health keywords.[2] Additional literature was identified using snowball sampling from the citations of these papers. To moderately restrict the scope of this work, computational research regarding neurodegenerative disorders (e.g. Dementia, Parkinson's Disease) was ignored.

### 2.2 Selection Criteria

To enhance parity amongst datasets considered in our meta-analysis, we require datasets found within the literature search to meet three additional criteria. While excluded from subsequent analysis, datasets that do not meet this criteria are maintained with complete annotations in the aforementioned digital directory. In future work, we will expand our scope of analysis to reflect the multi-faceted computational approaches used by the research community to understand mental health.

1. Datasets must contain non-clinical electronic media (e.g. social media, SMS, online forums, search query text).
2. Datasets must contain written language (i.e. text) within each unit of data .
3. Datasets must contain a dependent variable that captures or proxies a psychiatric condition listed in the DSM-5 (APA, 2013).

Our first criteria excludes research that examines electronic health records or digitally-transcribed interviews (Gratch et al., 2014; Holderness et al., 2019). Our second criteria excludes research that, for example, primarily analyzes search query volume or mobile activity traces (Ayers et al., 2013; Renn et al., 2018). It also excludes research based on speech data (Iter et al., 2018). Our third criteria excludes research in which annotations are only loosely associated with their stated mental health condition. For instance, we filter out research that seeks to identify diagnosis dates in self-disclosure statements (MacAvaney et al., 2018), in addition to research that proposes using sentiment as a proxy for mental illness (Davcheva et al., 2019). This last criteria also inherently excludes datasets that lack annotation of mental health status altogether (e.g. data dumps of online mental health support platforms and text-message counseling services) (Loveys et al., 2018; Demasi et al., 2019).

[2]Depression, Suicide, Anxiety, Mood, PTSD, Bipolar, Borderline Personality, ADHD, OCD, Panic, Addiction, Eating, Schizophrenia

## 2.3 Annotation Schema

We develop a high-level schema to code properties of each dataset. In addition to standard reference information (i.e. Title, Year Published, Authors), we note the following characteristics:

- **Platforms**: Electronic media source (e.g. Twitter, SMS)
- **Tasks**: The mental health disorders included as dependent variables (e.g. depression, suicidal ideation, PTSD)
- **Annotation Method**: Method for defining and annotating mental health variables (e.g. regular expressions, community participation/affiliation, clinical diagnosis)
- **Annotation Level**: Resolution at which ground-truth annotations are made (e.g. individual, document, conversation)
- **Size**: Number of data points at each annotation resolution for each task class
- **Language**: The primary language of text in the dataset
- **Data Availability**: Whether the dataset can be shared and, if so, the mechanism by which it may be accessed (e.g. data usage agreement, reproducible via API, distribution prohibited by collection agreement)

If a characteristic is not clear from a dataset's associated literature, we leave the characteristic blank; missing data points are denoted where applicable. While we simplify these annotations for a standardized analysis — e.g. different psychiatric batteries used to annotate depression in individuals (e.g. PHQ-9, CES-D) are simplified as "Survey (Clinical)" — we maintain specifics in the digital directory.

## 3 Analysis

Our literature search yielded 139 articles referencing 111 nominally-unique datasets. Application of exclusion criteria left us with 102 datasets. A majority of the datasets were released after 2012, with an average of 12.75 per year, a minimum of 1 (2012), and a maximum of 23 (2017). The 2015 CLPsych Shared Task (Coppersmith et al., 2015b), Reddit Self-reported Depression Diagnosis (Yates et al., 2017), and "Language of Mental Health" (Gkotsis et al., 2016) datasets were the most reused resources, serving as the basis of 7,

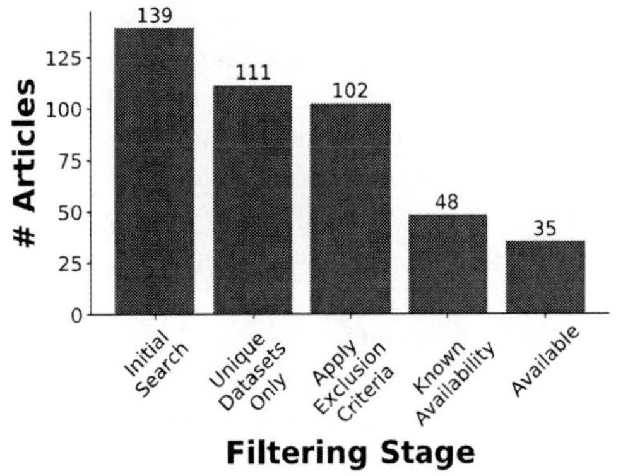


**Figure 1:** Number of articles (e.g. datasets) remaining after each stage of filtering. We were unable to readily discern the external availability of datasets for over half of the studies.

3, and 3 additional publications respectively. All datasets known to be available for distribution are available with annotations in the appendix, while remaining datasets are found our digital directory.

**Platforms**. We identified 20 unique electronic media platforms across the 102 datasets. Twitter (47 datasets) and Reddit (22 datasets) were the most widely studied platforms. YouTube, Facebook, and Instagram were relatively underutilized for mental health research — each found less than ten times in our analysis — despite being the three most-widely adopted social media platforms globally (Perrin and Anderson, 2019). We expect our focus on NLP to moderate the presence of YouTube and Instagram based datasets, though not entirely given both platforms offer expansive text fields (i.e. comments, tags) in addition to their primary content of video and images (Chancellor et al., 2016a; Choi et al., 2016). It is more likely that use of these platforms (and Facebook) for research is hindered by increasingly stringent privacy policies and ethical concerns (Panger, 2016; Benton et al., 2017).

**Tasks**. We identified 36 unique mental health related modeling tasks across the 102 datasets. While the majority of tasks were examined less than twice, a few tasks were considered quite frequently. Depression (42 datasets), suicidal ideation (26 datasets), and eating disorders (11 datasets) were the most common psychiatric conditions examined. Anxiety, PTSD, self-harm, bipolar disorder, and schizophrenia were also prominently featured conditions, each found within at least four unique datasets. A handful of studies sought to characterize finer-grained attributes associated with higher-level psychiatric conditions (e.g. symptoms of depression, stress events and stressor subjects) (Mowery et al., 2015; Lin et al., 2016). The dearth of anxiety-specific datasets was somewhat surprising given the condition's prevalence and the abundance of pyschometric batteries for assessing anxiety (Cougle et al., 2009; Antony and Barlow, 2020). That said, generalized anxiety disorder (GAD) only accounts for a small proportion of the overall prevalence of anxiety disorders (Bandelow and Michaelis, 2015) and many other types of anxiety disorders (e.g. social anxiety, PTSD, OCD, etc.) were typically treated as independent conditions (Coppersmith et al., 2015a; De Choudhury et al., 2016).

**Annotation**. We identified 24 unique annotation mechanisms. It was common for several annotation mechanisms to be used jointly to increase precision of the defined task classes and/or evaluate the reliability of distantly supervised labeling processes. For example, some form of regular expression matching was used to construct 43 of datasets, with 23 of these including manual annotations as well. Community participation/affiliation (24 datasets), clinical surveys (22 datasets), and platform activity (3 datasets) were also common annotation mechanisms. The majority of datasets contained annotations made on the individual level (63 datasets), with the rest containing annotations made on the document level (40 datasets).[3]

**Size**. Of the 63 datasets with individual-level annotations, 23 associated articles described the amount of documents and 62 noted the amount of individuals available. Of the 40 datasets with document-level annotations, 37 associated articles noted the amount of documents and 12 noted the number of unique individuals. The distribution of dataset sizes was primarily right-skewed.

One concerning trend that emerged across the datasets was the presence of a relatively low number of unique individuals. Indeed, these small sample sizes may further inhibit model generalization from platforms that are already demographically-skewed (Smith and Anderson, 2018). The largest datasets, which present the strongest opportunity to mitigate the issues presented by poorly representative online populations, tend to leverage the noisiest annotation mechanisms. For example, datasets that define a mainstream online community as a control group may expect to find approximately 1 in 20 of the labeled individuals are actually living with mental health conditions such as depression (Wolohan et al., 2018), while regular expressions may fail to distinguish between true and non-genuine disclosures of a mental health disorder up to 10% of the time (Cohan et al., 2018).

**Primary Language**. Six primary languages were found amongst the 102 datasets — English (85 datasets), Chinese (10 datasets), Japanese (4 datasets), Korean (2 datasets), Spanish (1 dataset), and Portuguese (1 dataset). This is not to say that some of the datasets do not include other languages, but rather that the predominant language found in the datasets occurs with this distribution. While an overwhelming focus on English data is a theme throughout the NLP community, it is a specific concern in this domain where culture often influences the presentation of mental health disorders (De Choudhury et al., 2017; Loveys et al., 2018).

**Availability**. We were able to identify the availability of only 48 of the 102 unique datasets in our literature search. Of these 48 datasets, 13 were known not to be available for distribution, generally due to limitations defined in the original collection agreement or removal from the public record (Park et al., 2012; Schwartz et al., 2014). The remaining 35 datasets were available via the following distribution mechanisms: 18 may be reproduced using an API and instructions provided within the associated article, 12 require a signed data usage agreement and/or IRB approval, 3 are available without restriction, and 2 may be retrieved directly from the author(s) with permission. Of the 22 datasets that used clinically-derived annotations (e.g. mental health battery, medical history), 7 were unavailable for distribution due to terms of the original data collection process and 1 was removed from the public record. The remaining 14 had unknown availability.

## 4 Discussion

In this study, we introduced and analyzed a standardized directory of social media datasets used by computational scientists to model mental health phenomena. In doing so, we have provided a valuable resource poised to help researchers quickly identify new datasets that support novel research. Moreover, we have provided evidence that affirms conclusions from Chancellor and De Choudhury (2020) and may further encourage researchers to

[3]One dataset was annotated at both a document and individual level

rectify existing gaps in the data landscape. Based on this evidence, we will now discuss potential areas of improvement within the field.

**Unifying Task Definitions**. In just 102 datasets, we identified 24 unique annotation mechanisms used to label over 35 types of mental health phenomena. This total represents a conservative estimate given that nominally equivalent annotation procedures often varied non-trivially between datasets (e.g. PHQ-9 vs. CES-D assessments, affiliations based on Twitter followers vs. engagement with a subreddit) (Faravelli et al., 1986; Pirina and Çöltekin, 2018). Minor discrepancies in task definition reflect the heterogeneity of how several mental health conditions manifest, but also introduce difficulty contextualizing results between different studies. Moreover, many of these definitions may still fall short of capturing the nuances of mental health disorders (Arseniev-Koehler et al., 2018). As researchers look to transition computational models into the clinical setting, it is imperative they have access to standardized benchmarks that inform interpretation of predictive results in a consistent manner (Norgeot et al., 2020).

**Sharing Sensitive Data**. Most existing mental health datasets rely on some form of self-reporting or distinctive behavior to assign individuals into task groups, but admittedly fail to meet ideal ground truth standards. The clinically-annotated datasets that do exist are either proprietary or do not provide a clear mechanism for inquiring about availability. The dearth of large, shareable datasets based on actual clinical diagnoses and medical ground truth is problematic given recent research that calls into question the validity of proxy-based mental health annotations (Ernala et al., 2019; Harrigian et al., 2020). By leveraging privacy-preserving technology (e.g. blockchain, differential privacy) to share patient-generated data, researchers may ultimately be able to train more robust computational models (Elmisery and Fu, 2010; Zhu et al., 2016; Dwivedi et al., 2019). In lieu of implementing complicated technical approaches to preserve the privacy of human subjects within mental health data, researchers may instead consider establishing secure computational environments that enable collaboration amongst authenticated users (Boebert et al., 1994; Rush et al., 2019).

**Addressing Bias**. There remains more to be done to ensure models trained using these datasets perform consistently irrespective of population. Several studies in our review attempted to leverage demographically-matched or activity-based control groups as a comparison to individuals living with a mental health condition (Coppersmith et al., 2015b; Cohan et al., 2018). A recent article found discrepancies between the prevalence of depression and PTSD as measured by the Centers for Disease Control and Prevention and as estimated using a model trained to detect the two conditions (Amir et al., 2019b). While the study posits reasons for the difference, it is unable to confirm any causal relationship.

More recently, Aguirre et al. (2021) found evidence of demographic (gender and racial/ethnic) bias within datasets from Coppersmith et al. (2014a, 2015c) that can create fairness issues in downstream tasks. They found poor representation and strong group imbalance in these datasets; however, simple changes in dataset size and balance alone could not fully account for performance disparities between groups. Indeed, common signs of depression recognized in prior linguistic analyses (e.g. differences in distributions for some categories of LIWC) were found not to be equally informative for all demographics. Thus, while performance disparities between demographic groups may certainly arise due to poor representation at training time, disparities may also arise due to an ill-founded assumption that mental health outcomes for all groups can be treated equivalently (Kessler et al., 2003; De Choudhury et al., 2017; Shah et al., 2019). Either way, there exists a need to rethink dataset curation and model evaluation so traditionally underrepresented groups are not further hindered from receiving adequate mental health care.

This all said, the presence of downstream bias in mental health models is admittedly difficult to define and even more difficult to fully eliminate (Gonen and Goldberg, 2019; Blodgett et al., 2020). Nonetheless, the lack of demographically-representative sampling described above would serve as a valuable starting point to address. Increasingly accurate demographic inference tools may aid in constructing datasets with demographically-representative cohorts (Huang and Carley, 2019; Wood-Doughty et al., 2020). Researchers may also consider expanding the diversity of languages in their datasets to account for variation in mental health presentation that arises due to cultural differences (De Choudhury et al., 2017; Loveys et al., 2018).

## References


Carlos Aguirre, Keith Harrigian, and Mark Dredze. 2021. Gender and racial fairness in depression research using social media. In *Proceedings of the 16th conference of the European Chapter of the Association for Computational Linguistics (EACL)*.

Silvio Amir, Mark Dredze, and John W. Ayers. 2019a. Mental health surveillance over social media with digital cohorts. In *CLPsych*.

Silvio Amir, Mark Dredze, and John W Ayers. 2019b. Mental health surveillance over social media with digital cohorts. In *CLPsych*.

Martin M Antony and David H Barlow. 2020. *Handbook of assessment and treatment planning for psychological disorders*. Guilford Publications.

American Psychiatric Association APA. 2013. *Diagnostic and statistical manual of mental disorders (DSM-5®)*. American Psychiatric Pub.

Alina Arseniev-Koehler, Sharon Mozgai, and Stefan Scherer. 2018. What type of happiness are you looking for?-a closer look at detecting mental health from language. In *CLPsych*.

John W Ayers, Benjamin M Althouse, Jon-Patrick Allem, J Niels Rosenquist, and Daniel E Ford. 2013. Seasonality in seeking mental health information on google. *American journal of preventive medicine*, 44(5):520–525.

Shrey Bagroy, Ponnurangam Kumaraguru, and Munmun De Choudhury. 2017. A social media based index of mental well-being in college campuses. *CHI*.

David E Bakken, R Rarameswaran, Douglas M Blough, Andy A Franz, and Ty J Palmer. 2004. Data obfuscation: Anonymity and desensitization of usable data sets. *IEEE Security & Privacy*.

Borwin Bandelow and Sophie Michaelis. 2015. Epidemiology of anxiety disorders in the 21st century. *Dialogues in clinical neuroscience*, 17(3):327.

Adrian Benton, Glen Coppersmith, and Mark Dredze. 2017. Ethical research protocols for social media health research. In *First ACL Workshop on Ethics in Natural Language Processing*.

Su Lin Blodgett, Solon Barocas, Hal Daumé III, and Hanna Wallach. 2020. Language (technology) is power: A critical survey of" bias" in nlp.

William E Boebert, Thomas R Markham, and Robert A Olmsted. 1994. Data enclave and trusted path system. US Patent 5,276,735.

Sandra Bucci, Matthias Schwannauer, and Natalie Berry. 2019. The digital revolution and its impact on mental health care. *Psychology and Psychotherapy: Theory, Research and Practice*.

Stevie Chancellor and Munmun De Choudhury. 2020. Methods in predictive techniques for mental health status on social media: a critical review. *NPJ digital medicine*.

Stevie Chancellor, Andrea Hu, and Munmun De Choudhury. 2018. Norms matter: contrasting social support around behavior change in online weight loss communities. In *CHI*.

Stevie Chancellor, Zhiyuan Lin, Erica L. Goodman, Stephanie Zerwas, and Munmun De Choudhury. 2016a. Quantifying and predicting mental illness severity in online pro-eating disorder communities. In *CSCW*.

Stevie Chancellor, Tanushree Mitra, and Munmun De Choudhury. 2016b. Recovery amid pro-anorexia: Analysis of recovery in social media. In *CHI*.

Dongho Choi, Ziad Matni, and Chirag Shah. 2016. What social media data should i use in my research?: A comparative analysis of twitter, youtube, reddit, and the new york times comments. In *ASIS&T*.

Arman Cohan, Bart Desmet, Andrew Yates, Luca Soldaini, Sean MacAvaney, and Nazli Goharian. 2018. Smhd: A large-scale resource for exploring online language usage for multiple mental health conditions. In *COLING*.

Glen Coppersmith, Mark Dredze, and Craig Harman. 2014a. Quantifying mental health signals in twitter. In *CLPsych*.

Glen Coppersmith, Mark Dredze, Craig Harman, and Kristy Hollingshead. 2015a. From ADHD to SAD: Analyzing the language of mental health on twitter through self-reported diagnoses. In *CLPsych*.

Glen Coppersmith, Mark Dredze, Craig Harman, Kristy Hollingshead, and Margaret Mitchell. 2015b. CLPsych 2015 shared task: Depression and PTSD on twitter. In *CLPsych*.

Glen Coppersmith, Craig Harman, and Mark Dredze. 2014b. Measuring post traumatic stress disorder in twitter. In *ICWSM*.

Glen Coppersmith, Ryan Leary, Eric Whyne, and Tony Wood. 2015c. Quantifying suicidal ideation via language usage on social media. In *Joint Statistics Meetings Proceedings, Statistical Computing Section, JSM*, volume 110.

Glen Coppersmith, Kim Ngo, Ryan Leary, and Anthony Wood. 2016. Exploratory analysis of social media prior to a suicide attempt. In *CLPsych*.

Jesse R Cougle, Meghan E Keough, Christina J Riccardi, and Natalie Sachs-Ericsson. 2009. Anxiety disorders and suicidality in the national comorbidity survey-replication. *Journal of psychiatric research*, 43(9):825–829.

Elena Davcheva, Martin Adam, and Alexander Benlian. 2019. User dynamics in mental health forums – a sentiment analysis perspective. In *Wirtschaftsinformatik*.

Munmun De Choudhury. 2015. Anorexia on tumblr: A characterization study. In *5th international conference on digital health 2015*.

Munmun De Choudhury, Michael Gamon, Scott Counts, and Eric Horvitz. 2013. Predicting depression via social media. *ICWSM*.

Munmun De Choudhury and Emre Kiciman. 2017. The language of social support in social media and its effect on suicidal ideation risk. In *ICWSM*.

Munmun De Choudhury, Emre Kıcıman, Mark Dredze, Glen Coppersmith, and Mrinal Kumar. 2016. Discovering shifts to suicidal ideation from mental health content in social media. *CHI*.

Munmun De Choudhury, Sanket S. Sharma, Tomaz Logar, Wouter Eekhout, and René Clausen Nielsen. 2017. Gender and cross-cultural differences in social media disclosures of mental illness. In *CSCW*.

Orianna Demasi, Marti A. Hearst, and Benjamin Recht. 2019. Towards augmenting crisis counselor training by improving message retrieval. In *CLPsych*.

Sarmistha Dutta, Jennifer Ma, and Munmun De Choudhury. 2018. Measuring the impact of anxiety on online social interactions. In *ICWSM*, pages 584–587.

Ashutosh Dhar Dwivedi, Gautam Srivastava, Shalini Dhar, and Rajani Singh. 2019. A decentralized privacy-preserving healthcare blockchain for iot. *Sensors*.

Johannes C Eichstaedt, Robert J Smith, Raina M Merchant, Lyle H Ungar, Patrick Crutchley, Daniel Preoţiuc-Pietro, David A Asch, and H Andrew Schwartz. 2018. Facebook language predicts depression in medical records. *Proceedings of the National Academy of Sciences*.

Ahmed M Elmisery and Huaiguo Fu. 2010. Privacy preserving distributed learning clustering of healthcare data using cryptography protocols. In *2010 IEEE 34th Annual Computer Software and Applications Conference Workshops*.

Sindhu Kiranmai Ernala, Michael L Birnbaum, Kristin A Candan, Asra F Rizvi, William A Sterling, John M Kane, and Munmun De Choudhury. 2019. Methodological gaps in predicting mental health states from social media: Triangulating diagnostic signals. In *CHI*.

Carlo Faravelli, Giorgio Albanesi, and Enrico Poli. 1986. Assessment of depression: a comparison of rating scales. *Journal of affective disorders*.

Christian Fuchs. 2015. *Culture and economy in the age of social media*. Routledge.

George Gkotsis, Anika Oellrich, Tim Hubbard, Richard Dobson, Maria Liakata, Sumithra Velupillai, and Rina Dutta. 2016. The language of mental health problems in social media. In *CLPsych*.

Hila Gonen and Yoav Goldberg. 2019. Lipstick on a pig: Debiasing methods cover up systematic gender biases in word embeddings but do not remove them.

Marc H Gorelick. 2006. Bias arising from missing data in predictive models. *Journal of clinical epidemiology*.

Melissa W Graham, Elizabeth J Avery, and Sejin Park. 2015. The role of social media in local government crisis communications. *Public Relations Review*.

Jonathan Gratch, Ron Artstein, Gale M. Lucas, Giota Stratou, Stefan Scherer, Angela Nazarian, Rachel Wood, Jill Boberg, David DeVault, Stacy Marsella, David R. Traum, Albert A. Rizzo, and Louis-Philippe Morency. 2014. The distress analysis interview corpus of human and computer interviews. In *LREC*.

Sharath Chandra Guntuku, David B Yaden, Margaret L Kern, Lyle H Ungar, and Johannes C Eichstaedt. 2017. Detecting depression and mental illness on social media: an integrative review. *Current Opinion in Behavioral Sciences*, 18:43–49.

Keith Harrigian, Carlos Aguirre, and Mark Dredze. 2020. Do models of mental health based on social media data generalize? In *"Findings of ACL: EMNLP"*.

Eben Holderness, Philip Cawkwell, Kirsten Bolton, James Pustejovsky, and Mei-Hua Hall. 2019. Distinguishing clinical sentiment: The importance of domain adaptation in psychiatric patient health records. In *ClinicalNLP*.

Binxuan Huang and Kathleen M Carley. 2019. A hierarchical location prediction neural network for twitter user geolocation.

Molly Ireland and Micah Iserman. 2018. Within and between-person differences in language used across anxiety support and neutral reddit communities. In *CLPsych*.

Dan Iter, Jong Yoon, and Dan Jurafsky. 2018. Automatic detection of incoherent speech for diagnosing schizophrenia. In *CLPsych*.

Jared Jashinsky, Scott H. Burton, Carl Lee Hanson, Joshua H. West, Christophe G. Giraud-Carrier, Michael D Barnes, and Trenton Argyle. 2014. Tracking suicide risk factors through twitter in the us. *Crisis*, 35 1:51–9.

Ronald C Kessler, Patricia Berglund, Olga Demler, Robert Jin, Doreen Koretz, Kathleen R Merikangas, A John Rush, Ellen E Walters, and Philip S Wang. 2003. The epidemiology of major depressive disorder: results from the national comorbidity survey replication (ncs-r). *Jama*, 289(23):3095–3105.

Mrinal Kumar, Mark Dredze, Glen Coppersmith, and Munmun De Choudhury. 2015. Detecting changes in suicide content manifested in social media following celebrity suicides. *HT*.

Yaoyiran Li, Rada Mihalcea, and Steven R. Wilson. 2018. Text-based detection and understanding of changes in mental health. In *SocInfo*.

Huijie Lin, Jia Jia, Quan Guo, Yuanyuan Xue, Qi Li, Jie Huang, Lianhong Cai, and Ling Feng. 2014. User-level psychological stress detection from social media using deep neural network. In *22nd ACM international conference on Multimedia*.

Huijie Lin, Jia Jia, Liqiang Nie, Guangyao Shen, and Tat-Seng Chua. 2016. What does social media say about your stress?. In *IJCAI*, pages 3775–3781.

David E Losada, Fabio Crestani, and Javier Parapar. 2017. erisk 2017: Clef lab on early risk prediction on the internet: experimental foundations. In *CLEF*.

David E Losada, Fabio Crestani, and Javier Parapar. 2018. Overview of erisk: early risk prediction on the internet. In *CLEF*.

Kate Loveys, Jonathan Torrez, Alex Fine, Glen Moriarty, and Glen Coppersmith. 2018. Cross-cultural differences in language markers of depression online. In *CLPsych*.

Sean MacAvaney, Bart Desmet, Arman Cohan, Luca Soldaini, Andrew Yates, Ayah Zirikly, and Nazli Goharian. 2018. Rsdd-time: Temporal annotation of self-reported mental health diagnoses. In *CLPsych*.

David N. Milne, Glen Pink, Ben Hachey, and Rafael A. Calvo. 2016. CLPsych 2016 shared task: Triaging content in online peer-support forums. In *CLPsych*.

Danielle Mowery, Craig Bryan, and Mike Conway. 2015. Towards developing an annotation scheme for depressive disorder symptoms: A preliminary study using twitter data. In *CLPsych*.

Danielle L. Mowery, Albert Park, Craig J Bryan, and Mike Conway. 2016. Towards automatically classifying depressive symptoms from twitter data for population health. In *PEOPLES*.

Beau Norgeot, Giorgio Quer, Brett K Beaulieu-Jones, Ali Torkamani, Raquel Dias, Milena Gianfrancesco, Rima Arnaout, Isaac S Kohane, Suchi Saria, Eric Topol, et al. 2020. Minimum information about clinical artificial intelligence modeling: the mi-claim checklist. *Nature medicine*.

Galen Panger. 2016. Reassessing the facebook experiment: critical thinking about the validity of big data research. *Information, Communication & Society*, 19(8):1108–1126.

Minsu Park, Chiyoung Cha, and Meeyoung Cha. 2012. Depressive moods of users portrayed in twitter.

A Perrin and M Anderson. 2019. Share of us adults using social media, including facebook, is mostly unchanged since 2018. pew research center.

Andrew Perrin. 2015. Social media usage. *Pew research center*, pages 52–68.

Inna Pirina and Çağrı Çöltekin. 2018. Identifying depression on reddit: The effect of training data. In *SMM4H*.

W Nicholson Price and I Glenn Cohen. 2019. Privacy in the age of medical big data. *Nature medicine*.

Brenna N Renn, Abhishek Pratap, David C Atkins, Sean D Mooney, and Patricia A Areán. 2018. Smartphone-based passive assessment of mobility in depression: Challenges and opportunities. *Mental health and physical activity*, 14:136–139.

Sarah Rush, Sara Britt, and John Marcotte. 2019. Icpsr virtual data enclave as a collaboratory for team science.

Koustuv Saha and Munmun De Choudhury. 2017. Modeling stress with social media around incidents of gun violence on college campuses. *CSCW*.

H. Andrew Schwartz, Johannes Eichstaedt, Margaret L. Kern, Gregory Park, Maarten Sap, David Stillwell, Michal Kosinski, and Lyle Ungar. 2014. Towards assessing changes in degree of depression through Facebook. In *CLPsych*.

Ivan Sekulic, Matej Gjurković, and Jan Šnajder. 2018. Not just depressed: Bipolar disorder prediction on reddit. In *9th Workshop on Computational Approaches to Subjectivity, Sentiment and Social Media Analysis*.

Deven Shah, H Andrew Schwartz, and Dirk Hovy. 2019. Predictive biases in natural language processing models: A conceptual framework and overview. *arXiv preprint arXiv:1912.11078*.

Guangyao Shen, Jia Jia, Liqiang Nie, Fuli Feng, Cunjun Zhang, Tianrui Hu, Tat-Seng Chua, and Wenwu Zhu. 2017. Depression detection via harvesting social media: A multimodal dictionary learning solution. In *IJCAI*.

Judy Hanwen Shen and Frank Rudzicz. 2017. Detecting anxiety through reddit. In *CLPsych*.

Han-Chin Shing, Suraj Nair, Ayah Zirikly, Meir Friedenberg, Hal Daumé, and Philip Resnik. 2018. Expert, crowdsourced, and machine assessment of suicide risk via online postings. In *CLPsych*.

Aaron Smith and Monica Anderson. 2018. Social media use in 2018. *Pew*.

Elsbeth Turcan and Kathy McKeown. 2019. Dreaddit: A Reddit dataset for stress analysis in social media. In *LOUHI*.

JT Wolohan, Misato Hiraga, Atreyee Mukherjee, Zeeshan Ali Sayyed, and Matthew Millard. 2018. Detecting linguistic traces of depression in topic-restricted text: Attending to self-stigmatized depression with NLP. In *LCCM Workshop*.

Akkapon Wongkoblap, Miguel A Vadillo, and Vasa Curcin. 2017. Researching mental health disorders in the era of social media: systematic review. *JMIR*, 19(6):e228.

Zach Wood-Doughty, Paiheng Xu, Xiao Liu, and Mark Dredze. 2020. Using noisy self-reports to predict twitter user demographics.

Andrew Yates, Arman Cohan, and Nazli Goharian. 2017. Depression and self-harm risk assessment in online forums. In *EMNLP*.

Lei Zhang, Xiaolei Huang, Tianli Liu, Zhenxiang Chen, and Tingshao Zhu. 2014. Using linguistic features to estimate suicide probability of chinese microblog users. In *HCC*.

Haining Zhu, Joanna Colgan, Madhu Reddy, and Eun Kyoung Choe. 2016. Sharing patient-generated data in clinical practices: an interview study. In *AMIA*.

Ayah Zirikly, Philip Resnik, Özlem Uzuner, and Kristy Hollingshead. 2019. CLPsych 2019 shared task: Predicting the degree of suicide risk in Reddit posts. In *CLPsych*.

## A Available Datasets

Ultimately, we identified 35 unique mental health datasets that were available for distribution. A subset of annotations for these datasets, along with original reference information, can be found in Table 1 (see next page).

We categorize dataset availability using four distinct distribution mechanisms.

- `DUA`: The dataset requires researchers to sign a data usage agreement that outlines the terms and conditions by which the dataset may be analyzed; in some cases, this also requires institutional authorization and oversight (e.g. IRB approval)
- `API`: The dataset may be reproduced (with a reasonable degree of effort) using instructions provided in the dataset's primary article and access to a public-facing application programming interface (API)
- `AUTH`: The dataset may be accessed by directly contacting the original author(s)
- `FREE`: The dataset is hosted on a public-facing server, accessible by all without any additional restrictions

Of the datasets that were available for distribution via one of the above mechanisms, we noted the following 27 unique mental health conditions/predictive tasks:

- Attention Deficit Hyperactivity Disorder (ADHD)
- Alcoholism (ALC)
- Anxiety (ANX)
- Social Anxiety (ANXS)
- Asperger's (ASP)
- Autism (AUT)
- Bipolar Disorder (BI)
- Borderline Personality Disorder (BPD)
- Depression (DEP)
- Eating Disorder (EAT)
- Recovery from Eating Disorder (EATR)
- General Mental Health Disorder (MHGEN)
- Obsessive Compulsive Disorder (OCD)
- Opiate Addiction (OPAD)
- Opiate Usage (OPUS)
- Post Traumatic Stress Disorder (PTSD)
- Panic Disorder (PAN)
- Psychosis (PSY)
- Trauma from Rape (RS)
- Schizophrenia (SCHZ)
- Seasonal Affective Disorder (SAD)
- Self Harm (SH)
- Stress (STR)
- Stressor Subjects (STRS)
- Suicide Attempt (SA)
- Suicidal Ideation (SI)
- Trauma (TRA)

| Reference | Platform(s) | Task(s) | Level | Individuals | Documents | Availability |
|---|---|---|---|---|---|---|
| Coppersmith et al. (2014a) | Twitter | BI, PTSD, SAD, DEP | Ind. | 7k | 16.7M | DUA |
| Coppersmith et al. (2014b) | Twitter | PTSD | Ind. | 6.3k | - | DUA |
| Jashinsky et al. (2014) | Twitter | SI | Doc. | 594k | 733k | API |
| Lin et al. (2014) | Twitter, Sina Weibo, Tencent Weibo | STR, STRS | Ind. | 23.3k | 490k | API |
| Coppersmith et al. (2015a) | Twitter | ANX, EAT, OCD, SCHZ, SAD, BI, PTSD, DEP, ADHD | Ind. | 4k | 7M | DUA |
| Coppersmith et al. (2015b) | Twitter | PTSD, DEP | Ind. | 1.7k | - | DUA |
| De Choudhury (2015) | Tumblr | EAT, EATR | Ind. | 28k | 87k | API |
| Kumar et al. (2015) | Reddit, Wikipedia | SI | Ind. | 66k | 19.1k | API |
| Mowery et al. (2015) | Twitter | DEP | Doc. | - | 129 | AUTH |
| Chancellor et al. (2016b) | Tumblr | EATR | Ind. | 13.3k | 67M | API |
| Coppersmith et al. (2016) | Twitter | SA | Ind. | 250 | - | DUA |
| De Choudhury et al. (2016) | Reddit | PSY, EAT, ANXS, SH, BI, PTSD, RS, DEP, PAN, SI, TRA | Ind. | 880 | - | API |
| Gkotsis et al. (2016) | Reddit | ANX, BPD, SCHZ, SH, ALC, BI, OPAD, ASP, SI, AUT, OPUS | Ind. | - | - | API |
| Lin et al. (2016) | Sina Weibo | STR | Doc. | - | 2.6k | FREE |
| Milne et al. (2016) | Reach Out | SH | Doc. | 1.2k | - | DUA |
| Mowery et al. (2016) | Twitter | DEP | Doc. | - | 9.3k | AUTH |
| Bagroy et al. (2017) | Reddit | MHGEN | Doc. | 30k | 43.5k | API |
| De Choudhury and Kiciman (2017) | Reddit | SI | Ind. | 51k | 103k | API |
| Losada et al. (2017) | Reddit | DEP | Ind. | 887 | 530k | DUA |
| Saha and De Choudhury (2017) | Reddit | STR | Doc. | - | 2k | API |
| Shen et al. (2017) | Twitter | DEP | Ind. | 300M | 10B | FREE |
| Shen and Rudzicz (2017) | Reddit | ANX | Doc. | - | 22.8k | API |
| Yates et al. (2017) | Reddit | DEP | Ind. | 116k | - | DUA |
| Chancellor et al. (2018) | Reddit | EAT | Doc. | - | 2.4M | API |
| Cohan et al. (2018) | Reddit | ANX, EAT, OCD, SCHZ, BI, PTSD, DEP, ADHD, AUT | Ind. | 350k | - | DUA |
| Dutta et al. (2018) | Twitter | ANX | Ind. | 200 | 209k | API |
| Ireland and Iserman (2018) | Reddit | ANX | Ind. | - | - | API |
| Li et al. (2018) | Reddit | MHGEN | Ind. | 1.8k | - | API |
| Losada et al. (2018) | Reddit | EAT, DEP | Ind. | 1.5k | 1.2M | DUA |
| Pirina and Çöltekin (2018) | Reddit | DEP | Doc. | - | 1.2k | API |
| Shing et al. (2018) | Reddit | SI | Ind. | 1.9k | - | DUA |
| Sekulic et al. (2018) | Reddit | BI | Ind. | 7.4k | - | API |
| Wolohan et al. (2018) | Reddit | DEP | Ind. | 12.1k | - | API |
| Turcan and McKeown (2019) | Reddit | STR | Doc. | - | 2.9k | FREE |
| Zirikly et al. (2019) | Reddit | SI | Ind. | 496 | 32k | DUA |

**Table 1:** Characteristics of datasets that meet our inclusion criteria and are known to be accessible. The full set of annotations may be found in our digital directory (`https://github.com/kharrigian/mental-health-datasets`).

# Individual differences in the Movement-Mood Relationship in Digital Life Data

**Glen Coppersmith**
Qntfy
glen.coppersmith@qntfy.com

**Alex B. Fine**
Qntfy
alex.fine@qntfy.com

**Patrick Crutchley**
Qntfy
patrick.crutchley@qntfy.com

**Joshua Carroll**
Qntfy
josh.carroll@qntfy.com

## Abstract

Our increasingly digitized lives generate troves of data that reflect our behavior, beliefs, mood, and wellbeing. Such "digital life data" provides crucial insight into the lives of patients outside the healthcare setting that has long been lacking, from a better understanding of mundane patterns of exercise and sleep routines to harbingers of emotional crisis. Moreover, information about individual differences and personalities is encoded in digital life data. In this paper we examine the relationship between mood and movement using linguistic and biometric data, respectively. Does increased physical activity (movement) have an effect on a person's mood (or vice-versa)? We find that weak group-level relationships between movement and mood mask interesting and often strong relationships between the two for individuals within the group. We describe these individual differences, and argue that individual variability in the relationship between movement and mood is one of many such factors that ought be taken into account in wellbeing-focused apps and AI systems.

## 1 Introduction

Health and wellbeing research generally seeks to find patterns that hold for all members of a population. A familiar example is the claim that those who exercise more are happier (Stubbe et al., 2007). While this claim has intuitive appeal for most people, there are many individuals for whom this relationship does not seem to hold (e.g., someone who is challenged with chronic pain that is exacerbated by exercise). Where chronic pain is an extreme example, there are many more subtle ways that a person's individual circumstances might cause them to deviate from expected population norms.

Generally speaking, whether this relationship holds across the population or varies across individuals is an empirical question, and one with profound implications for delivering effective clinical guidance and for the design of mental health and wellness technology (e.g., Menke, 2018). This may be one of the contributing factors to the difficulty that mental health interventions face with retention and attrition over the course of treatment: what was designed for the population does not necessarily adapt to a particular individual's life. Preventing attrition is considered a longstanding and core challenge in the design and execution of studies and interventions alike (Eysenbach, 2005; Christensen and Mackinnon, 2006). This is more pronounced in digital mental health apps, many of which are designed to support long term behavior change, yet face significant difficulty retaining users, with a recent study indicating a median retention rate of just 3.3% of users retained after 30 days of usage (Baumel et al., 2019). This strongly suggests a need to understand the individual differences between users that might have an effect on retention and attrition and use that information to augment intervention approaches or suggest novel ones.

Collecting the data necessary to quantify these individual differences has been a challenge historically, especially with traditional behavioral methods (e.g., questionnaires). With the increasing ubiquity of mobile devices, the relevant data can now be captured and recorded to support large-scale, fine-grained analysis and intervention. Recent work shows that indices of mood, mental health, and wellbeing can be estimated from social media behavior (De Choudhury et al., 2013a,b; Coppersmith et al., 2017, 2016, 2015; Schwartz et al., 2016; Resnik et al., 2015; Cohan et al., 2016; Wang et al., 2014; Park et al., 2015; Eichstaedt et al., 2015). Here, we explore the relationships between mood, emotion, and mental health conditions derived from machine classifiers and Fitbit metrics.



*Proceedings of the Seventh Workshop on Computational Linguistics and Clinical Psychology*, pages 25–31
June 11, 2021. 

## 2 Data

Users come from the `OurDataHelps.org` program, which enables participants to donate social media and wearable data to support mental health research. For each of the users ($n = 160$) included in this analysis, we analyzed historic data from at least one source of language (Twitter or Facebook) and subsequent actigraphic data collected via a Fitbit device. All users had at least 30 days in which their wearable recorded data and in which posted at least once on social media. All data analyzed was from before the COVID-19 pandemic, associated lockdown, and changes in pattern of life that it induced. Users opted-in to data collection via oAuth, which was subjected to deidentification and stored following the ethical protocols of (Benton et al., 2017). Due to differences in models of wearable devices, users had different aspects of their movement recorded, so we analyzed data elements common across at least 20 users.

## 3 Methods

We analyzed language data using previously-trained models of mood, emotion, and mental health. Each model examines the text of social media posts using a simple lexicon or character $n$-gram language model (CLM), and produces a score relevant to a psychological variable.

We use models created by Coppersmith et al. to score for ADHD, anxiety, bipolar disorder, borderline personality disorder, depression, eating disorders, PTSD, and schizophrenia (Coppersmith et al., 2015). Briefly, these models estimate the relative likelihood that a given text was generated by a user at risk for a specific condition (e.g., PTSD) or a matched control, with one model created per condition. The data used to compare language was derived from users who made self-statements of diagnosis (e.g., "I was diagnosed with PTSD") publicly on social media. For each user, we estimated age and gender via a classifier similar in spirit to (Sap et al., 2014). An age- and gender-matched control user was identified from a large English-speaking sample.

For each string of characters (i.e., character $n$-gram) the model measured how likely it was to occur in the population with the condition and in the matched controls. This forms the basis of the scoring for the model, optimized to provide a score even from short texts. While many machine learning open vocabulary approaches are tuned to look at all the language that a person generates to estimate risk, the models used here are tuned to work for small amounts of text, given the present task. We refer the reader to Coppersmith et al. (2015) for further details on the pre-processing steps.

For scoring emotion, we used a CLM trained from messages that contain hashtagged emotions (e.g., *#joy*), from Coppersmith et al. (2016). For scoring sentiment, we used VADER, a closed-vocabulary and rule-based tool specifically tuned for social media data (Hutto and Gilbert, 2014). We report each individual sentiment separately (positive, neutral, negative) as well as the compound sentiment, meant to give a single overall score of the sentiment expressed in the text. We used DepecheMood to estimate mood, another closed-vocabulary approach, with high-coverage and high-precision (Staiano and Guerini, 2014).

All data of each type recorded from midnight to midnight in each user's local timezone is collapsed into a single number capturing the value for that day. For language data this is the average score for each model across all messages. For wearables, we use the most straightforwardly interpretable version of the data (e.g., hours of sleep) as retrieved from the API. The movement and physical data recorded from the user's wearable (steps, average heart rate) is similarly accumulated from midnight to midnight, with the exception of sleep data which, following Fitbit's reporting feature, is recorded on the morning the user wakes up (e.g., the two hours of sleep from 10pm until midnight is included in the next day's sleep total). Since we were primarily concerned with the relationship between movement and psychological variables measured by language, we excluded any day for which we did not have both movement and language data. Note that the unit of analysis of language here is the language generated in a single day, models tuned for relatively small amounts of text, like closed-vocabulary lexica and machine learning models trained to predict on short texts were ideal.

We calculated Pearson's $r$ for each person between each pair of variables, treating each day as a separate observation. Because this is an exploratory analysis and we wish to focus our discussion on effects that are most likely to hold promise for future work, we artificially set the $r$ value to 0 for all subsequent analysis for any correlation where the $p$-value associated with Pearson's $r$ is greater than 0.01. This $p$-value was selected such that for any pair of variables we compare, we would expect 1-2 of the 160 users to be spuriously identified as having a significant when no relationship existed. We opted for a more conservative cutoff here than the traditional $p < 0.05$ since analyses at that $p$-value would allow for an expected 8 spurious correlations to be falsely indicated, which could significantly influence the subsequent analytic step. Furthermore, the exploratory nature of the work obviates the need to address multiple comparisons using a technique such as a Bonferroni correction.

## 4 Results

Figure 1 shows the correlation matrix with Pearson's $r$ computed across all users. The models described above are shown in the same order on both axes. The color of each cell captures the Pearson's $r$ between the variables, averaged across users, with white indicating a lack of correlation (an $r$ near 0 or correlations with high $p$ values which were treated as $r = 0$, as noted). Blue indicates positive correlation and red indicates negative correlation – the solid dark blue diagonal reflects the fact that each variable correlates perfectly with itself. The variables are grouped by the construct measured, separated by black lines: emotion, mental health conditions, mood, sentiment, and movement. While some significant relationship can be seen between various language and movement measures, the vast majority seem to be near $r = 0$. The notable exception is sleep onset latency (i.e., the amount of time it takes to fall asleep) which has generally negative relationships with positive emotions and moods and a positive relationship with negative emotions and moods. This finding is in line with other work examining the link between aspects of sleep and wellbeing (Short et al., 2013).

However, this picture shows nuance when we examine correlation matrices computed for each *individual*. Figure 1 shows exemplar correlation matrices for individual users. Note that significant relationships exist for individuals that were not observed for the group. This suggests that the relationships between psychological phenomena and aspects of movement are not uniform in direction or magnitude.

Figure 2 illustrates this point in more detail with a histogram of the distribution of correlations between a few measures of movement and mood. All correlations that were not significant were excluded from these histograms. Note that there are users for whom there are statistically significant correlations, in both the positive and negative directions, of both large and small magnitudes. Many of the other histograms for other pairwise comparisons, excluded for brevity, show similar patterns. Taken with the previous results, this demonstrates that relationships between movement- and mood-related constructs exhibit sufficient individual-level variability in both direction and magnitude that inferences about these relationships must explicitly and quantitatively account for this variability.

These results highlight the need for personalized approaches to improving mental health and wellbeing through movement- or activity-based interventions.

## 5 Anecdotes

A subset of the users opted in to allow us to discuss the results and data with them in order to allow for validation of the findings.

For one user, many aspects of their sleep are more strongly correlated with emotions than for the population. The amount of time spent in bed was correlated with negative emotions and negatively correlated with joy. Similarly, the number of times they were awakened during the night was positively correlated with posts classified as angry or annoyed the following day. This suggests that this user's mood is particularly sensitive to sleep, relative to the general population. This aligned with the user's subjective impressions of their experience.

For another pair of users, we found significant correlations involving the time spent sedentary throughout the course of the day. For one user, the time spent sedentary during the day was positively correlated with positive mood outcomes, while the second user demonstrated a negative correlation between these two measures. Subjective reports from these users was consistent with these findings: the first indicated that if they were sitting still it meant that their children were being well-behaved, and thus was indicative of a pleasant day. The second reported, by contrast, that if they were sitting still throughout the day, that indicated a long day of meetings, which tended to increase their frustration and negative mood.

## 6 Discussion

We replicated previous work finding some significant relationships between movement and mood at a population level (i.e., sleep latency's relation-

Figure 1: Pearson's $r$ values for correlations between model outputs, averaged across the population (Top) and exemplar individuals (Middle). Strength of correlation is indicated by color (Bottom) Note the strong correlations between language and movement measures, and that they are between different measures for different users.

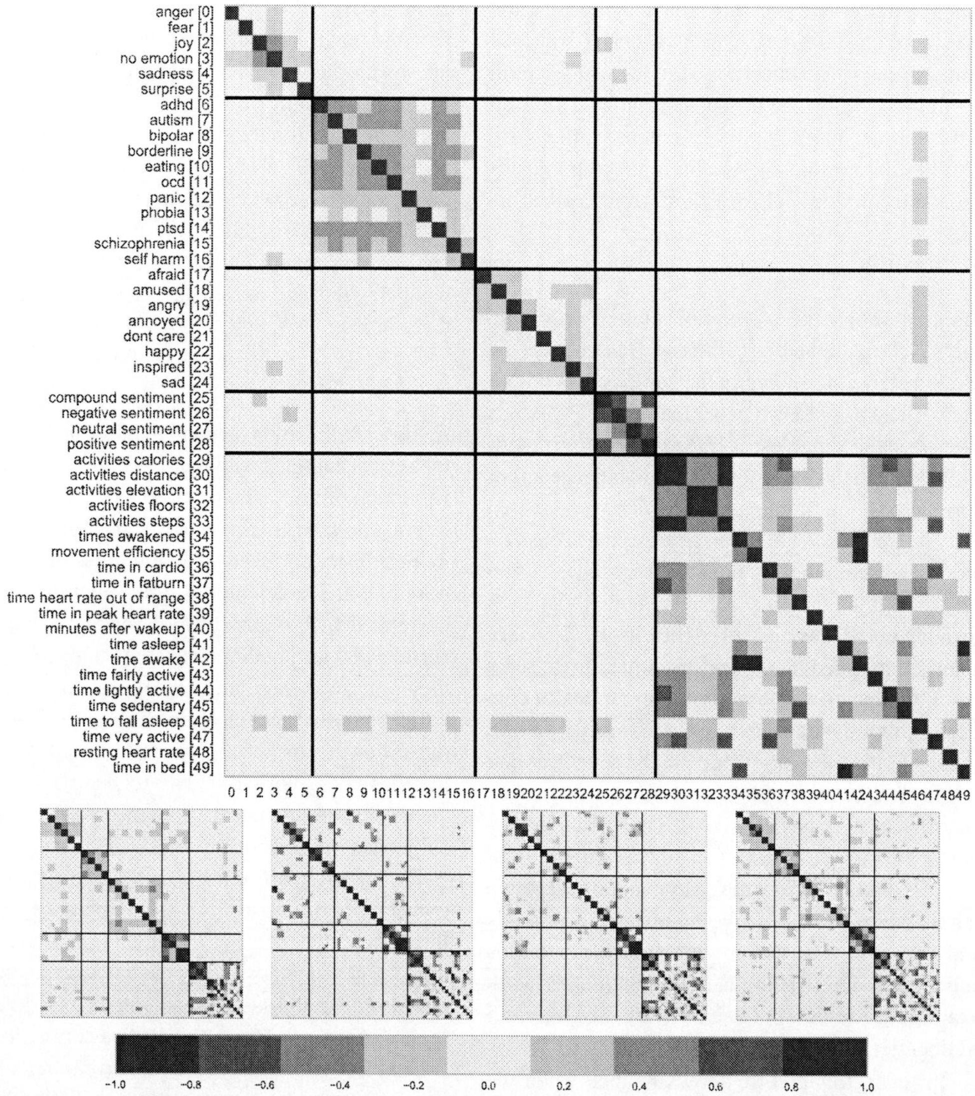


Figure 2: Histogram of Pearson's $r$ values for users with statistically significant correlations between (left) number of steps per day and posts with positive sentiment and (right) time asleep and posts with the blasé "don't care" mood.

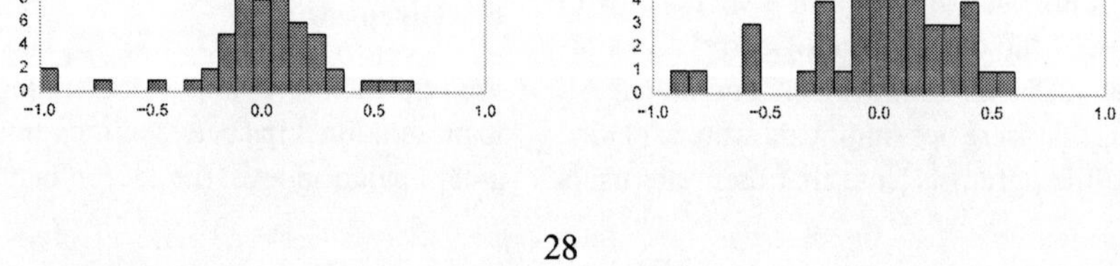

ship to a range of psychological factors), while also demonstrating that significant relationships between movement and mood exist for individuals that do not hold across the population. This supports anecdotal and observational experience where, for population-level findings, there are individuals who seem to defy the expected trend.

The results reported here hold promise for future work, both theoretical and applied. Further study, with a larger subject pool, will allow us to examine *structured variability*, i.e., subpopulations with homogeneous relationships between movement and mood. There are well-established statistical techniques for characterizing and simultaneously modeling individual- and group-level relationships like those under discussion here, including multi-level modeling (Gelman and Hill, 2007), as well as numerous clustering techniques for inferring homogeneous subsets of users in a principled way (e.g., hierarchical clustering; Johnson, 1967). Without a strong *a priori* hypothesis for how many such homogeneous subsets of users exist, techniques with an inherent measure of cluster quality to suggest the number of clusters would be worthwhile. With the inherent relational nature of the data, it may be prudent to approach this clustering problem via techniques that take advantage of this information explicitly in the form of a (dis)similarity matrix (e.g., spectral clustering; Ng et al., 2001). Moreover, for developers of mental health and wellness technology that hinges on providing users with guidance related to movement and sleep, these results point the way forward for user testing that may enhance the quality and efficacy of these tools.

## 7 Caveats

Because the results reported here are based on donated digital life data, we expect this sample is biased in certain ways, assuming that the propensity to (1) share data without compensation, (2) actively contribute to mental health research, and (3) come across the donation opportunity at `OurDataHelps.org` are not uniformly distributed throughout the population. For example, in a project similar to `OurDataHelps.org`, we solicited data donation from veterans of the US Armed Forces. To date, 22% of individuals that donated their data to this project identify as female. By contrast, according the Department of Veterans Affairs, roughly 9% of US veterans are women (of Veterans Affairs et al., 2017). Thus, women are over-represented in our sample. It is difficult to say exactly why this is, but the bias is most likely due to a confluence of factors, including gender-based differences in the propensity to participate in research that is considered altruistic or pro-social (e.g., Bani and Giussani, 2010) as well as idiosyncrasies in the way the study was promoted. However, we expect that this sort of bias would work *against* the observed pattern (i.e., since the population is more homogeneous than the general population, the relationship between movement and mood is less likely to vary significantly across individuals).

One underlying assumption of this work is that posts on social media have some reflection of the emotional state, mood, or other transient psychological phenomena that a person is experiencing. There is some controversy about the extent and strength of this relationship, with some finding significant reflections of emotion and mood in daily language (e.g., posts on social media Chen et al., 2020) while others fail to find these relations (e.g., in everyday speech Sun et al., 2020).

## 8 Conclusion

We empirically explored the relationship between a variety of movement and mood measures using social media posts and wearable data from 160 users. The relationships uncovered are more nuanced than the population-level conclusions that are generally popularized by the press and highlight the need for individualized approaches to movement-based wellbeing interventions.

Ultimately, understanding the relationship between movement and mood for a particular individual will allow for tailoring of wellbeing and mental health interventions to their specific needs, and thus increase our collective ability to tailor mental health and wellbeing interventions to the user. At minimum, this lays the foundation to provide some predictive ability for how a user may be willing to accept and engage with a suggested exercise-based intervention. The results reported here serve as a particularly strong indication of the promise held in personalized wellbeing interventions, and are consonant with a rich body of recent work highlighting the need for personalized medicine in general.

## Acknowledgments

We would like to thank the people who generously donated their data to `OurDataHelps.org`, without whom this research would not be possible.

## References


Marco Bani and Barbara Giussani. 2010. Gender differences in giving blood: a review of the literature. *Blood Transfusion*, 8(4):278.

Amit Baumel, Frederick Muench, Stav Edan, and John M Kane. 2019. Objective user engagement with mental health apps: systematic search and panel-based usage analysis. *Journal of medical Internet research*, 21(9):e14567.

Adrian Benton, Glen Coppersmith, and Mark Dredze. 2017. Ethical research protocols for social media health research. In *Proceedings of the First ACL Workshop on Ethics in Natural Language Processing*, pages 94–102.

Lushi Chen, Walid Magdy, Heather Whalley, and Maria Klara Wolters. 2020. Examining the role of mood patterns in predicting self-reported depressive symptoms. In *12th ACM Conference on Web Science*, pages 164–173.

Helen Christensen and Andrew Mackinnon. 2006. The law of attrition revisited. *Journal of medical Internet research*, 8(3):e20.

Arman Cohan, Sydney Young, and Nazli Goharian. 2016. Triaging mental health forum posts. In *Proceedings of the Third Workshop on Computational Linguistics and Clinical Psychology*, pages 143–147.

Glen Coppersmith, Mark Dredze, Craig Harman, and Kristy Hollingshead. 2015. From ADHD to SAD: Analyzing the language of mental health on Twitter through self-reported diagnoses. In *Proceedings of the Workshop on Computational Linguistics and Clinical Psychology: From Linguistic Signal to Clinical Reality*, Denver, Colorado, USA. North American Chapter of the Association for Computational Linguistics.

Glen Coppersmith, Casey Hilland, Ophir Frieder, and Ryan Leary. 2017. Scalable mental health analysis in the clinical whitespace via natural language processing. In *Biomedical & Health Informatics (BHI), 2017 IEEE EMBS International Conference on*, pages 393–396. IEEE.

Glen Coppersmith, Kim Ngo, Ryan Leary, and Tony Wood. 2016. Exploratory data analysis of social media prior to a suicide attempt. In *Proceedings of the Workshop on Computational Linguistics and Clinical Psychology: From Linguistic Signal to Clinical Reality*, San Diego, California, USA. North American Chapter of the Association for Computational Linguistics.

Munmun De Choudhury, Scott Counts, and Eric Horvitz. 2013a. Predicting postpartum changes in emotion and behavior via social media. In *Proceedings of the ACM Annual Conference on Human Factors in Computing Systems (CHI)*, pages 3267–3276. ACM.

Munmun De Choudhury, Michael Gamon, Scott Counts, and Eric Horvitz. 2013b. Predicting depression via social media. In *Proceedings of the 7th International AAAI Conference on Weblogs and Social Media (ICWSM)*.

Johannes C. Eichstaedt, Hansen Andrew Schwartz, Margaret L. Kern, Gregory Park, Darwin R. Labarthe, Raina M. Merchant, Sneha Jha, Megha Agrawal, Lukasz A. Dziurzynski, Maarten Sap, Christopher Weeg, Emily E. Larson, Lyle H. Ungar, and Martin E. P. Seligman. 2015. Psychological language on Twitter predicts county-level heart disease mortality. *Psychological Science*, 26(2):159–169.

Gunther Eysenbach. 2005. The law of attrition. *Journal of medical Internet research*, 7(1):e11.

Andrew Gelman and Jennifer Hill. 2007. *Data analysis using regression and multilevel/hierarchical models*, volume Analytical methods for social research. Cambridge University Press, New York.

Clayton J Hutto and Eric Gilbert. 2014. Vader: A parsimonious rule-based model for sentiment analysis of social media text. In *Eighth International AAAI Conference on Weblogs and Social Media*.

Stephen C Johnson. 1967. Hierarchical clustering schemes. *Psychometrika*, 32(3):241–254.

Andreas Menke. 2018. Precision pharmacotherapy: psychiatry's future direction in preventing, diagnosing, and treating mental disorders. *Pharmacogenomics and personalized medicine*, 11:211.

Andrew Ng, Michael Jordan, and Yair Weiss. 2001. On spectral clustering: Analysis and an algorithm. *Advances in neural information processing systems*, 14:849–856.

Gregory Park, H Andrew Schwartz, Johannes C Eichstaedt, Margaret L Kern, Michal Kosinski, David J Stillwell, Lyle H Ungar, and Martin EP Seligman. 2015. Automatic personality assessment through social media language. *Journal of personality and social psychology*, 108(6):934.

Philip Resnik, William Armstrong, Leonardo Claudino, Thang Nguyen, Viet-An Nguyen, and Jordan Boyd-Graber. 2015. Beyond lda: exploring supervised topic modeling for depression-related language in twitter. In *Proceedings of the 2nd Workshop on Computational Linguistics and Clinical Psychology: From Linguistic Signal to Clinical Reality*, pages 99–107.

Maarten Sap, Greg Park, Johannes C. Eichstaedt, Margaret L. Kern, David J. Stillwell, Michal Kosinski, Lyle H. Ungar, and H. Andrew Schwartz. 2014. Developing age and gender predictive lexica over social media. In *Proceedings of the Conference on Empirical Methods in Natural Language Processing (EMNLP)*, pages 1146–1151.

H Andrew Schwartz, Maarten Sap, Margaret L Kern, Johannes C Eichstaedt, Adam Kapelner, Megha Agrawal, Eduardo Blanco, Lukasz Dziurzynski, Gregory Park, David Stillwell, et al. 2016. Predicting individual well-being through the language of social media. In *Biocomputing 2016: Proceedings of the Pacific Symposium*, pages 516–527. World Scientific.

Michelle A Short, Michael Gradisar, Leon C Lack, Helen R Wright, and Hayley Dohnt. 2013. The sleep patterns and well-being of australian adolescents. *Journal of adolescence*, 36(1):103–110.

Jacopo Staiano and Marco Guerini. 2014. Depechemood: a lexicon for emotion analysis from crowd-annotated news. *arXiv preprint arXiv:1405.1605*.

JH Stubbe, MHM De Moor, DI Boomsma, and Eco J C de Geus. 2007. The association between exercise participation and well-being: a co-twin study. *Preventive medicine*, 44(2):148–152.

Jessie Sun, H Andrew Schwartz, Youngseo Son, Margaret L Kern, and Simine Vazire. 2020. The language of well-being: Tracking fluctuations in emotion experience through everyday speech. *Journal of Personality and Social Psychology*, 118(2):364.

Department of Veterans Affairs et al. 2017. Women veterans report: The past, present, and future of women veterans. *Retrieved from Washington, DC: https://www. va. gov/vetdata/docs/specialreports/women_veterans_2015_final. pdf*.

Rui Wang, Fanglin Chen, Zhenyu Chen, Tianxing Li, Gabriella Harari, Stefanie Tignor, Xia Zhou, Dror Ben-Zeev, and Andrew T Campbell. 2014. Studentlife: assessing mental health, academic performance and behavioral trends of college students using smartphones. In *Proceedings of the 2014 ACM International Joint Conference on Pervasive and Ubiquitous Computing*, pages 3–14. ACM.

# Dissociating Semantic and Phonemic Search Strategies in the Phonemic Verbal Fluency Task in early Dementia

**Hali Lindsay** and **Philipp Mueller**
German Research Center for Artificial Intelligence (DFKI), Saarbrücken, Germany
{hali.lindsay, philipp.mueller}@dfki.de

**Nicklas Linz** and **Mario Mina**
ki elements, Saarbrücken, Germany
{nicklas, mario.mina}@ki-elements.de

**Radia Zeghari**
The CoBTeK, Université Cote d'Azur (UCA), Nice, France
radia.zeghari@gmail.com

**Alexandra König**
Stars Team, Institut National de Recherche en Informatique et en Automatique (INRIA), Valbonne, France
alexandra.konig@inria.fr

**Johannes Tröger**
German Research Center for Artificial Intelligence (DFKI), Saarbrücken, Germany
johannes.troeger@dfki.de

## Abstract

Effective management of dementia hinges on timely detection and precise diagnosis of the underlying cause of the syndrome at an early mild cognitive impairment (MCI) stage. Verbal fluency tasks are among the most often applied tests for early dementia detection due to their efficiency and ease of use. In these tasks, participants are asked to produce as many words as possible belonging to either a semantic category (SVF task) or a phonemic category (PVF task). Even though both SVF and PVF share neurocognitive function profiles, the PVF is typically believed to be less sensitive to measure MCI-related cognitive impairment and recent research on fine-grained automatic evaluation of VF tasks has mainly focused on the SVF. Contrary to this belief, we show that by applying state-of-the-art semantic and phonemic distance metrics in automatic analysis of PVF word productions, in-depth conclusions about production strategy of MCI patients are possible. Our results reveal a dissociation between semantically- and phonemically-guided search processes in the PVF. Specifically, we show that subjects with MCI rely less on semantic- and more on phonemic processes to guide their word production as compared to healthy controls (HC). We further show that semantic similarity-based features improve automatic MCI versus HC classification by 29% over previous approaches for the PVF. As such, these results point towards the yet underexplored utility of the PVF for in-depth assessment of cognition in MCI.

## 1 Introduction

Dementia is a syndrome primarily presenting with broad cognitive impairments. There are multiple underlying causes that result in dementia such as Alzheimer's Disease (AD) or fronto-temporal lobar degeneration or focal lesions (MacPherson et al., 2016). These sub-forms have different neurocognitive profiles. The most-common Alzheimer's Disease (AD)-related dementia is typically driven by an amnestic cognitive impairment (Kidd, 2008) whereas the fronto-temporal dementia is often associated with executive function impairment (Huey et al., 2009).

Early identification of dementia as well as precise differentiation between dementia sub-forms is crucial for effective management of the syndrome (Thyrian et al., 2016). Pairing high diagnostic sensitivity with ease of use, verbal fluency tests (VF) are amongst the most-applied tests in cognitive assessment of dementia (Troyer et al., 1997). In these tests, participants are asked to produce as many words from a specific category as they can in a fixed time. The two main variants of VF tests are the semantic verbal fluency (SVF) and the phonemic verbal fluency (PVF). In the SVF,



*Proceedings of the Seventh Workshop on Computational Linguistics and Clinical Psychology*, pages 32–44

the word category is defined by semantics (e.g. all animal words), whereas in the PVF participants need to produce words starting with a specific letter (e.g. "S"). Traditionally, test scores are computed by counting the number of correctly named words within the given time (Gomez and White, 2006). Although both VF variants are quite similar in the way they engage different neurocognitive functions, the cognitive strategies of the task can indicate different patterns of the underlying neuropathology. For instance, an SVF impairment is often only regarded as evidence for amnestic dementia (Vaughan et al., 2016; Teng et al., 2013) whereas a PVF impairment is almost exclusively regarded as evidence for fronto-temporal dementias (Dubois et al., 2000).

Recently, advanced Natural Language Processing (NLP) techniques have been applied to allow for in-depth analysis of the produced word sequence in VF tasks, particularly for the SVF (Linz et al., 2017a; Kim et al., 2019; Diaz-Orueta et al., 2020; Zemla et al., 2020). By extracting clusters from the produced word sequence and by modelling the semantic relationships between- and within these clusters, it is possible to disentangle the effects of memory impairment from effects of executive function impairment (Tröger et al., 2019). Despite the success of these qualitative features in the SVF, their utility for automatic analysis of the PVF remains underexplored.

In this paper, we investigate both phonemic and semantic motivations for the underlying strategy of the phonemic verbal fluency task, and thereby reduce the gap between clinical theory and computational approaches to evaluating cognitive speech tasks. By contrasting semantic and phonemic distance measure in an analysis based on time bins, we show a dissociation between semantically- and phonemically-guided search processes: Subjects with mild cognitive impairment (MCI) exhibit significantly less semantic similarity in their productions as compared to healthy controls (HC). Finally, in experiments on automatic classification of MCI vs. HC from PVF word productions, we show that semantic features improve over previous approaches by 29%. Taken together, our results pave the way towards more fine-grained analysis of the PVF task that can help to improve clinical decision processes.

## 2 Clinical Background

### 2.1 Cognitive Processes in VF

Verbal Fluency tasks (VF) require a network of cognitive processes activating—-a region associated with language (Vigneau et al., 2006)—-the frontal lobe (Coslett et al., 1991; Miller, 1984), specifically the left hemisphere (Birn et al., 2010; Troyer et al., 1998; Mueller et al., 2015), as well as the temporal lobe (Newcombe, 1969; Cerhan et al., 2002).

VF are used to assess semantic memory and executive functions as a good VF performance hinges on intact semantic memory stores as well as the ability to access these memory stores (Chertkow and Bub, 1990; Hodges et al., 1992; Mueller et al., 2015). Executive functioning, specifically, working memory is thought to allow a person to effectively search through phonological and semantic stores while regulating and adapting the search strategy to produce more words over the task (Faust, 2012; Rende et al., 2002; Troyer et al., 1997; Rosen, 1980).Both PVF and SVF are hypothesised to span multiple overlapping cognitive abilities; executive, verbal, and attention abilities (Mueller et al., 2015; Li et al., 2017; Shao et al., 2014; Schmidt et al., 2017). However, there is evidence that each task measures a set of distinct cognitive processes.

PVF burdens executive resources whereas the SVF demands linguistic-conceptual knowledge (Thompson-Schill et al., 1997; Vigneau et al., 2006; Shao et al., 2014; Mueller et al., 2015; Schmidt et al., 2017; Birn et al., 2010). SVF is theorized to engage the temporal lobe for lexical-semantic access and retrieval from semantic store (Newcombe, 1969; Mueller et al., 2015; Cerhan et al., 2002) where as the PVF is thought to rely on executive functioning and prefrontal lobe processes (Mueller et al., 2015) as well as phonological and orthographic cues for word retrieval (Li et al., 2017; Clark et al., 2013). Generally, it is hypothesised that SVF requires both semantic and retrieval processes whereas PVF relies only on retrieval processes (Fisher et al., 2004). However, there is conflicting research that PVF taps into the semantic network, although to a lesser extent than semantic fluency (Lezak et al., 2004; Mueller et al., 2015; Schmidt et al., 2017; Clark et al., 2013).

Bizzozero et al. (2013) investigated the extent to which SVF and PVF were related to semantic and attention processes and found evidence of semantic processes in both SVF and PVF. Nutter-Upham et al. (2008) observed a larger effect size for the amnestic MCI (aMCI) group's deficit on semantic verbal fluency (Cohen's d=0.98) than for their deficit on phonemic verbal fluency (Cohen's d=0.66), due to greater variability in phonemic verbal fluency performance. Therefore, an alternative interpretation is that their findings actually do reflect a preferential deficit on semantic verbal fluency in aMCI. Supporting these findings, imaging studies combined with factor analysis have also suggested that the PVF task is relies on both semantic and phonemic processes (Schmidt et al., 2017; Clark et al., 2013).

### 2.2 VF for Diagnosis

Both the Phonemic and Semantic varieties of verbal fluency are commonly used to diagnosis and monitor cognitive decline such as mild cognitive impairment (MCI) and Alzheimer's Disease and Related Dementias (ADRD) (Marra et al., 2011; Clark et al., 2009; Gomez and White, 2006; Troyer et al., 1998).

SVF has been found to be more impaired than PVF in ADRD (Cerhan et al., 2002; Barr and Brandt, 1996; Zhao et al., 2013) and deficits in both semantic and phonemic memory have been reported. However there is conflicting research for PVF and SVF in the MCI group. For aMCI, only the SVF shows impairment (Hodges, 2006; Murphy et al., 2006; Teng et al., 2013). While other studies show decline on both the PVF and SVF task for MCI (Mueller et al., 2015; Vita et al., 2014; Nutter-Upham et al., 2008). Rinehardt et al. (2014) compared controls with aMCI, non-aMCI and AD and found that both MCI groups were less impaired on the SVF than the PVF, behaving more like controls than the AD group.

Clark et al. (2013) considered computationally-based phonemic and semantic measures when analyzing the PVF and SVF tasks in relation to gray matter correlates for HC, MCI and AD. They concluded that both tasks showed greater semantic motivations than phonemic motivation, even in the PVF task.

PVF may be a sensitive test for investigating phonemic and semantic processes but a global word count does not provide the in-depth information needed to understand the underlying cognitive processes (Gomez and White, 2006; Becker and Salles, 2016). In this paper, we apply recently developed automatic analysis techniques from computational linguistics to the PVF to obtain a better insight into the degradation of semantic and phonemic processes.

## 3 Previous Work

### 3.1 Analyzing Semantic and Phonemic Strategy for VF

Several modes of analysis have been proposed with the goal of observing the role that different cognitive strategies play throughout VF tasks.

Much work has been done on the semantic variety of verbal fluency, specifically for the animal category. Troyer et al. (1997) introduced a semantically-motivated hierarchical list of animals for determining semantic clusters. To overcome this time-intensive and subjective annotation process, previous research worked on automatically producing semantic clusters over SVF productions (Ryan, 2013; Pakhomov et al., 2015b, 2016; Linz et al., 2017b; König et al., 2018; Kim et al., 2019). For example, Pakhomov et al. (2015a) compared traditional and novel computational methods of evaluating SVF using medical imaging techniques between healthy and cognitively impaired individuals. The semantic relatedness of words was determined using latent semantic analysis of word co-occurrences from a large online corpora. This study showed that computational methods of evaluating the SVF were beneficial in understanding the relationships between the different cognitive processes.

Building off of this, Linz et al. (2017a) used neural word embeddings as a data-driven way to model semantic clustering in the SVF task. König et al. (2018) showed high correlations ($r = 0.9$) between automatically extracted clustering and switching features and clinical methods. From these clusters, several features including cluster size or number of switches between clusters were calculated to reflect cognitive processes (Linz et al., 2017a; König et al., 2018).

In addition to the SVF, Troyer et al. (1997) proposed a rule-based method for finding phonemically-related clusters of words in PVF productions. Lindsay et al. (2019) automated this rule-based method for determining phonemic clusters, and proposed three additional phonemic similarity metrics for evaluating the PVF task on healthy German students, namely the Levenshtein distance (LD), phonemically-weighted Levenshtein distance (PHON-LD), as well as position-weighted Levenshtein distance (POS-LD). Clark et al. (2013) pro-

| | HC | MCI | $p$ |
|---|---|---|---|
| N (#Female) | 34(6) | 48(22) | - |
| Age | 73.56(6.74) | 75.02(7.68) | 0.40 |
| Education | 12.65(1.82) | 10.71(4.01) | 0.08 |
| MMSE | 28.76(1.28) | 25.79(2.74) | <0.01 |

Table 1: Demographic information for the French population used. Age and Education are given in years. The Mini-Mental State Exam (MMSE) is a test to measure cognitive function (Max score 30). Means are given for the populations with standard deviation in parentheses. Significance testing between groups is reported in $p$ column.

posed another phonemic distance measure using an English pronouncing dictionary and a formula for measuring string overlap to estimate phonemic-relatedness of adjacent words over the task.

Recently, (Linz et al., 2019) considered a binning-based approach (Fernaeus et al., 2008) for the automatic analysis of the SVF. In this approach, features were calculated separately on non-overlapping, 10-second time bins, which alloweda deeper investigation into the evolution of a participant's production strategy over time. Linz et al. (2019) used temporal binning to analyse at what points in time during SVF word production HC differed from MCI and AD patients with respect to word count, transition length, and word frequency.

To conclude, while previous works introduced metrics for quantifying semantic as well as phonemic similarity in VF word productions, no comprehensive comparison of these metrics was performed on the PVF in a clinical setting. This leaves a gap between clinical theory of motivating cognitive strategies and computational methods as to how to automatically evaluate both phonemic and semantic strategy for the PVF task. To allow for a fine-grained analysis of production strategy over the course of the PVF task, we analyze semantic and phonemic distance metrics in the temporal binning framework.

### 3.2 PVF-based MCI Classification

Compared to the amount of work on HC versus MCI classification from the SVF (Linz et al., 2017a; König et al., 2018), considerably less studies have investigated this classification task using the PVF (Ryan, 2013; Lindsay et al., 2020). Ryan (2013) used logistic regression to classify between HC and MCI using only repetitions (AUC=0.53) and word count (AUC=0.5) from the PVF. Lindsay et al. (2020) reported a baseline PVF experiment between HC and MCI and reported an AUC of 0.75 using only word count on a very small dataset (8HC/19MCI). Additional temporal features lowered the classification (AUC=0.55). To the best of our knowledge, no study at the present time has investigated HC versus MCI classification with the PVF using phonemic and semantic measures.

## 4 Methods

### 4.1 Data

The data used in this research was collected during the Dem@Care (Karakostas et al., 2017) and ELEMENT (Tröger et al., 2017) projects. Participants were recruited through the Memory Clinic located in Nice University Hospital at the Institute Claude Pompidou in Nice, France. The study was approved by the Nice Ethics Committee. All participants were native speakers of French and asked to give informed consent before participating in the study. The French data was collected in the form of speech recordings via an automated recording application installed on a tablet computer. The recordings were manually transcribed in PRAAT (Boersma and Weenink, 2009) according to the CHAT protocol (MacWhinney, 1991). Participants were asked to complete a battery of cognitive tests, including a 60 second phonemic verbal fluency task for the letter category *F*. Demographics for the data used are displayed in Table 1. A Mann-Whitney U test was conducted between the HC and MCI populations to check for significant differences between age (W = 1106, p-value = 0.40) and education (W = 1492, p-value = 0.08) but none were found.

### 4.2 Binning, Clustering & Global Resolutions of VF Analysis

We look at three resolutions of the verbal fluency task that have been applied to the SVF task and consider them for the PVF task; temporal binning, clustering and switching and global features. Each method provides a different resolution for looking word retrieval strategy. Temporal binning (Linz et al., 2019; Fernaeus et al., 2008) gives the finest resolution of strategy. The clustering is motivated by clinical theory to investigate the different cognitive processes (Troyer et al., 1998). Global features are what are the current norm in clinical practice (Troyer et al., 1998; Gomez and White, 2006).

#### 4.2.1 Binning Methods

To produce temporal bins for the PVF, we follow the methodology in (Linz et al., 2019) that was previously used for SVF. The complete 60-second PVF response is split into into six 10-seconds bins. This produces a new resolution of the task from which we can then compute features. As done in (Linz et al., 2019), we include the word count as well as the average temporal distance(TD) between consecutive words. In addition, we include the average semantic distance between consecutive words as well as the averages of the three phonemic distance measures LD, PHON-LD, and POS-LD. This allows for a separate investigation of the phonemic, semantic and temporal measures that guide search processes during the span of the word production in the PVF task.

**Semantic Distance (SD)** We follow Linz et al. (2017a) who computed semantic similarity between two words as the cosine distance between their embedding vectors. To construct word embeddings, FastText models (Bojanowski et al., 2016) are used. For this paper, the cosine distance is used, where $Cosine_{distance} = 1 - Cosine_{similarity}$.

**Levenshtein Distance (LD)** Lindsay et al. (2019) used the Levenshtein distance as a measure of phonetic distance when evaluating the PVF task. They first phonetically transliterate the word using the python package epitran (Mortensen et al., 2018). They then proposed using the traditional levenshtein distance to measures the number of edits (insertions, substitutions and deletions) between consecutive words (Levenshtein, 1966). They also proposed two weighted measures of LD as described below.

**Phonemic-weighted Levenshtein Distance (PHON-LD)** In addition to LD, Lindsay et al. (2019) proposed a phonemically weighted version of levenshtein distance. Using the epitran package, each phoneme has a corresponding 21-length phonological vector to represents the characteristics of the sound (e.g. voice/unvoiced, front/back). When computing the levenshtein distance, they weighted substitutions as the cosine between the to phonological vectors. Insertions and deletion are still valued at 1.

**Position-weighted Levenshtein Distance (POS-LD)** Lindsay et al. (2019) also investigated a position weighted levenshtein distance as the distance between phonetic representations of consecutive words, weighted for position in the word. Deletions, insertions and substitutions are set weighted by exponential distribution (with $\lambda = 0.5$) at the position of the phoneme in the word.

**Temporal Distance (TD)** The temporal distance is defined as the time in seconds between the boundaries of consecutive words in the PVF production.

#### 4.2.2 Clustering Methods

Clustering-based approaches for VF evaluation consist of two steps. First, the produced word sequence is partitioned into a set of clusters. Second, features (e.g. mean cluster size) are computed from the automatically produced clusters. In this study, we consider a rule-based phonemic clustering as well as an automated version of semantic clustering, and temporal clustering to investigate production. For each both phonemic and semantic clustering types, the mean cluster size and number of switches are computed.

**Phonemic Clustering** In the case of phonemic clustering features, we determine clusters in the word sequence following the phonemically-motivated, clinical approach from Troyer et al. (1997) that was automated by Lindsay et al. (2019). This approach uses phonemic similarity rules to determine whether subsequent words belong to the same cluster or not.

**Semantic Clustering** Semantic Clusters are determined as in Linz et al. (2017a). Using the semantic distance method described previously, a semantic threshold is determined for each participant by averaging the semantic distance between all words in the production. If the semantic distance between consecutive words is lower than the threshold, the words are said to be in a cluster. If the semantic distance between consecutive words is greater than the threshold, this introduces a cluster boundary.

To obtain semantic word embeddings, the pre-trained French fastText model is used. This model is trained on Common Crawl and Wikipedia corpora using the continuous bag of words (CBOW) algorithm with a negative sampling loss function. FastText models are trained at the character level using a character n-gram model. The 300-dimension

| | HC | | MCI | | HC v. MCI | |
|---|---|---|---|---|---|---|
| | **Mean** | **SE** | **Mean** | **SE** | *W* | *p* |
| *Average Over Bins* | | | | | | |
| Word Count | 2.70 | 0.17 | 2.00 | 0.11 | 1145 | **0.002** |
| Semantic Distance | 0.54 | 0.12 | 0.57 | 0.12 | 584 | **0.040** |
| Temporal Distance | 4.25 | 0.29 | 5.96 | 0.36 | 496 | **0.002** |
| LD | 3.09 | 0.13 | 2.57 | 0.11 | 1125 | **0.004** |
| PHON-LD | 1.92 | 0.08 | 1.70 | 0.06 | 1016 | 0.060 |
| POS-LD | 1.66 | 0.05 | 1.49 | 0.04 | 1096 | **0.008** |
| *Rule-Based Phonemic Clustering* | | | | | | |
| Mean Cluster Size | 4.63 | 1.74 | 4.02 | 1.57 | 1042 | **0.033** |
| Number of Switches | 2.51 | 1.17 | 2.19 | 0.98 | 947.5 | 0.195 |
| *Automatic Semantic Clustering* | | | | | | |
| Mean Cluster Size | 2.81 | 0.79 | 2.63 | 0.83 | 928.5 | 0.287 |
| Number of Switches | 9.09 | 4.15 | 7.04 | 3.27 | 1077 | **0.014** |

Table 2: Significance testing results between HC and MCI for the binning and clustering methods with a Mann-Whitney U test. The p-value is reported and a significance level is set at 0.05. Significant values are shown in bold type face. Standard Error (SE). Means and SE are provided to understand relationship between the groups. The top half of the table reports values for the binning analysis. The bottom half of the table reports significance results for the clustering analysis.

model is used for this analysis. For specific numerical parameter values, or to download the models used in this research, please see the link in the footnote[1].

### 4.3 Global Features

In addition to the binning features and clustering features in (Section 4.2.2), we include the traditional way of evaluating verbal fluency tasks, which computes aggregate features for the whole 60 second long word production. For an overview of all features used, please see Appendix A. The most general and widely adopted measures of verbal fluency are the word count and repetition count (Spreen et al., 1991; Tombaugh et al., 1999). The word count is the count of all relevant words produced in (e.g. all words said start with the letter *F*), excluding repeated words. The repetition count is the number of words produce more than once.

### 4.4 Experiments

Statistical Analysis was done in R Studio (R Core Team, 2017). All coding experiments are implemented using python 3.7. For significance testing, a non-parametric Mann-Whitney U test for significance is always reported.

#### 4.4.1 Comparing Strategic Processes With Binning Methods

To visualize what the strategic process over the duration of the PVF task, we plot the group averages of each feature across the bins. For overall performance, we plot the average word count and transition time by bin. To investigate semantic processes we plot the semantic distance between the words in each bin. To investigate the phonemic measures, we plot the LD, PHON-LD, and POS-LD.

In addition, we compute the bin average and standard error (se) for each group over all distance measures. A non-parametric Mann-Whitney U test for significance is reported to see if the bin averages differ between groups.

#### 4.4.2 Classification Experiments

The classification models are created using the scikit-learn library[2] (Pedregosa et al., 2011).

For the classification application of these features, we focused on an early diagnostic scenario; distinguishing between healthy controls and mild cognitive impairment. To observe how age and education bias our classifier, we trained individual models on each potential bias (Nogueira et al., 2016; Petti et al., 2020). For the clinical baseline, a model was produced by training on only word count (word count) (Lindsay et al., 2020). To compare to previous work, a model was trained on number of repetitions (Ryan, 2013).

In addition to the baseline comparison experiments, we investigated individual and combined models. Four individual models were built using the features for semantic clustering, semantic binning, phonemic clustering or phonemic binning.

[1]https://fasttext.cc/docs/en/crawl-vectors.html

[2]sklearn version==0.24.0 for python 3.7

To investigate the proposed analysis modes and cognitive strategies, we built four combined models; all binning features (binning), all clustering features (clustering), all semantic features (semantic), and all phonemic features (phonemic).

Finally, we investigate a model using all features (All) and compare the models performance to the proposed baselines.

**Classification Specifications** To compare these methods, the extremely randomized trees (also known as extra trees) algorithm is used to train a classifier for each experimental scenario. This algorithm was chosen due to its ability to reduce variance and lesser likelihood of overfitting on a relatively small dataset with high dimensionality. Due to the limited amount of data available (34HC/48MCI), training-testing data splits were created using leave one out cross validation to maximize the amount of training data available, while still testing on every available data point. Due to the extreme randomness of the algorithm chosen, performance metrics can fluctuate between runs. To nullify the potential of the bias effects of random initialization, the experiment is repeated 50 times. For each model, the Area Under the Receiver Operator Curve (AUC) is averaged of the 50 iterations and reported.

## 5 Results

Results from the experiments to investigate strategic process as described in Section 4.4.1 are visualized in Figure 1. Significance testing between the HC and MCI groups are given in Table 2

### 5.1 Strategic Processes

For all binning features, excluding word count, a lower average bin distance represents a higher similarity between adjacent words. Compared to the HC group, the MCI group has a lower average word count, is less semantically motivated and more phonemically related. They also have longer transition times. The MCI group also show significantly smaller phonemic cluster (p=0.03) and lower number of semantic switches (p=0.01).

### 5.2 Classification results

To reduce the complexity of Figure 2, baseline and combined classifications are visualized with ROC-AUC curves and additional classification experiments are reported in the text of this section.

Both the age (AUC=0.41) and education (AUC=0.24) models perform below chance. The most common clinical evaluation, word count, performs at chance (AUC=0.50). The model trained using all features (AUC=0.71) proposed in this study improves over all baselines including the previous Ryan (2013) model (AUC=0.42) by 29 points.

Not shown in Figure 2, we compare each of the semantic and phonemic process in combination with the binning and clustering methods. Semantic clustering methods (AUC=0.61) achieve similar performance when used for binning (AUC=0.64) where as phonemic features are best when combined with the binning methods (AUC=0.70) but perform poorly for clustering (AUC=0.45).

As shown in Figure 2, the combined binning methods (AUC=0.67) perform similarly to the combined clustering methods (AUC=0.64). The combined phonemic features (AUC=0.76) perform the best overall for the early diagnostic classification scenario.

## 6 Discussion

The phonemic verbal fluency task remains underexplored in its use for clinical assessment as well as research of MCI.

However, in this paper we show, that with state-of-the-art semantic as well as phonemic distance metrics, the PVF can reveal neurocognitive function involvement that is crucial to better assess MCI. Our data shows that with recent semantic and phonemic similarity metrics, we can capture MCI-related impairments, such as a general semantic impairment, that have also been reported in the SVF (Verma and Howard, 2012; Taler and Phillips, 2008) but not on the PVF. Our results show significantly lower semantic distance for HC responses when compared to the MCI group in the PVF task which is, by nature, phonemically motivated. In return, MCI patients show significantly lower phonemic distance. This could possibly be explained by the MCI group relying heavily on a phonemic strategy to guide their search rather than a utilizing a semantic strategy. The higher semantic distance for the MCI group could be interpreted as a structural deficit to access semantic memory efficiently as has been shown to be very prominent at all stages of AD-related dementia (Verma and Howard, 2012).

This is especially striking as one would expect

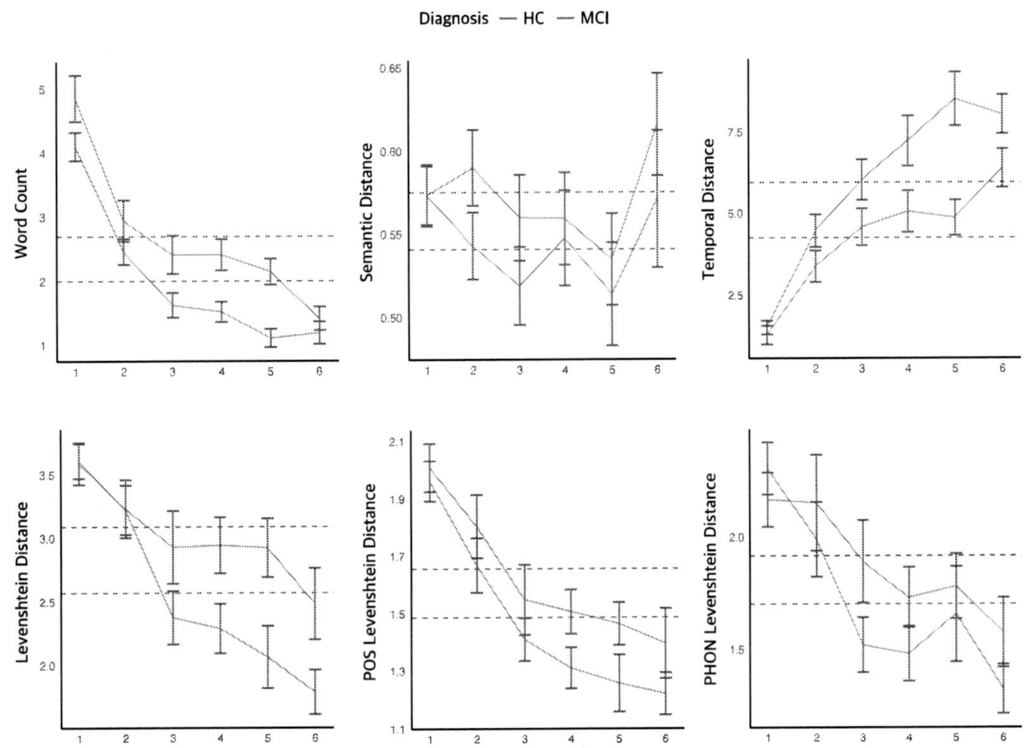


Figure 1: Graphical representation of binning results for each distance measure. Standard error bars are given for the HC and MCI groups at each bin. The dashed line represents the group average overall bins. For interpreting semantic and phonemic (LD, POS-LD, PHON-LD) distance metrics, a lower distance is interpreted as indicating a higher similarity.

the phonemic distance to increase as more words are produced (with a larger number of words per bin, the mean distance of adjacent words should be higher). Such an increase is the case for the phonemic distance where MCIs produce fewer words overall and are more phonemically related in comparison to HC, who produce more words and have a larger average phonemic distance over the bins. However, the exact opposite is the case for the semantic distance where MCIs produce fewer words while generating a list of less semantically related words in comparison to the HC group. This strongly points towards the conclusion that MCI patients struggle to exploit the associative network of their semantic memory.

By making neurocognitive processes visible in the PVF that are traditionally reserved for the SVF in clinical practice, the PVF becomes significantly more relevant to real-world MCI and dementia assessment. In order to support the diagnostic usage of the PVF for MCI assessment, we simulate a diagnostic decision scenario through downstream machine learning classification using the semantic as well as phonemic features in the PVF. Our results show that by using semantic and phonemic features we can improve classification results over previous clinical and automatic baselines. The all features model (AUC=0.71) out performs both the word count (AUC=0.50) and previous work of Ryan (2013) (AUC=0.42).

Both clustering (AUC=0.64) and binning (AUC=0.67) methods of analysis perform comparatively. Both the semantic (AUC=0.65) and phonemic (AUC=0.76) measures outperform the clinical baselines (0.50). The classification results support that while the task is overall a phonemic task, semantic investigation of the PVF is relevant for future research and capable of discriminating between HC and MCI better than the clinical baseline.

As an additional finding, the machine learning task benefits from a combined binning and cluster-

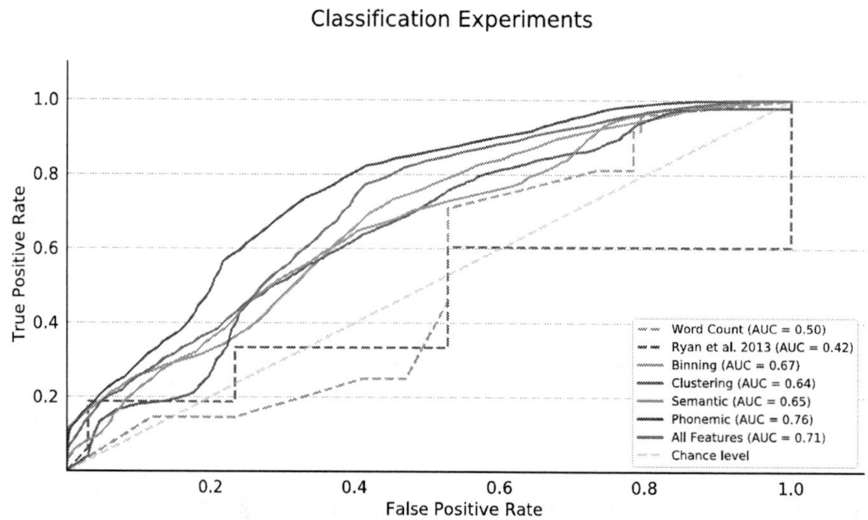


Figure 2: Visualzation of the ROC curve for the binary classification results between HC and MCI. Baseline methods are dashed in shades of gray. Ryan et al. 2013 is a previously published approach for comparison. Resolution modes are given in red. Strategy classifications are given in blue. The over all experiment is in green. AUC scores are given in the legend in the lower right corner. A perfect classification is 1.0. Chance is illustrated at 0.50.

ing approach when modelling the phonemic processes (AUC=0.76), increasing over only phonemic clustering (AUC=0.45) or phonemic binning methods (AUC=0.70) for classification.

## 7 Conclusion

This paper set out to investigate the ability of computational linguistic techniques for understanding phonemic and semantic cognitive processes of the under-explored phonemic verbal fluency task. Utilizing three resolutions of analysis, temporal binning, clustering and global measures, combined with semantic and phonemic distance measures, we found semantic impairment in a phonemic task as has been hypothesized in previous clinical research. In addition to giving a finer-resolution for understanding the PVF task, the additional phonemic and semantic features improved classification over previous clinical and automatic baselines for early dementia detection with the PVF task. Future work should investigate these measures in additional languages and possibly combine the features presented in this paper with medical imaging techniques to see if the findings can be replicated.

## Acknowledgements

This research was funded by MEPHESTO project Q10 (BMBF Grant Number 01IS20075).

## References

Amy Barr and Jason Brandt. 1996. Word-list generation deficits in dementia. *Journal of clinical and experimental neuropsychology*, 18(6):810–822.

Natalia Becker and Jerusa Salles. 2016. Methodological criteria for scoring clustering and switching in verbal fluency tasks. *Psico-USF*, 21:445–457.

Rasmus M Birn, Lauren Kenworthy, Laura Case, Rachel Caravella, Tyler B Jones, Peter A Bandettini, and Alex Martin. 2010. Neural systems supporting lexical search guided by letter and semantic category cues: a self-paced overt response fmri study of verbal fluency. *Neuroimage*, 49(1):1099–1107.

Ilaria Bizzozero, Stefania Scotti, Francesca Clerici, Simone Pomati, Marcella Laiacona, and Erminio Capitani. 2013. On which abilities are category fluency and letter fluency grounded a confirmatory factor analysis of 53 alzheimer's dementia patients. *Dementia and geriatric cognitive disorders extra*, 3(1):179–191.

Paul Boersma and David Weenink. 2009. Praat: doing phonetics by computer (version 5.1.13).

Piotr Bojanowski, Edouard Grave, Armand Joulin, and Tomas Mikolov. 2016. Enriching word vectors with subword information. *arXiv preprint arXiv:1607.04606*.

Jane Cerhan, Robert Ivnik, Glenn Smith, Eric Tangalos, Ronald Petersen, and Brad Boeve. 2002. Diagnostic utility of letter fluency, category fluency, and fluency difference scores in alzheimer's disease. *The Clinical neuropsychologist*, 16:35–42.

Howard Chertkow and Daniel Bub. 1990. Semantic memory loss in dementia of alzheimer's type. *Brain : a journal of neurology*, 113 ( Pt 2):397–417.

David Clark, Virginia Wadley, P. Kapur, Thomas DeRamus, Brandon Singletary, Anthony Nicholas, P.D. Blanton, K. Lokken, Hrishikesh Deshpande, D. Marson, and Georg Deutsch. 2013. Lexical factors and cerebral regions influencing verbal fluency performance in mci. *Neuropsychologia*, 54.

L. J. Clark, M. Gatz, L. Zheng, Y. L. Chen, C. McCleary, and W. J. Mack. 2009. Longitudinal Verbal Fluency in normal Aging, Preclinical, and Prevalent Alzheimer's Disease. *Am J Alzheimers Dis Other Demen*, 24(6):461–468.

H. Coslett, Dawn Bowers, Mieke Verfaellie, and K Heilman. 1991. Frontal verbal amnesia. phonological amnesia. *Archives of neurology*, 48:949–55.

Unai Diaz-Orueta, Alberto Blanco-Campal, Melissa Lamar, David J. Libon, and Teresa Burke. 2020. Marrying past and present neuropsychology: Is the future of the process-based approach technology-based? *Frontiers in Psychology*, 11:361.

B. Dubois, A. Slachevsky, I. Litvan, and B. Pillon. 2000. The fab. *Neurology*, 55(11):1621–1626.

Miriam Faust, editor. 2012. *The handbook of neuropsychology of language. Volume 1: Language processing in the brain: basic science. Volume 2: Language processing in the brain: clinical populations*. Wiley-Blackwell, Chichester. Bibtex: faust_handbook_2012.

Sven-Erik Fernaeus, Per Östberg, Åke Hellström, and Lars-Olof Wahlund. 2008. Cut the coda: Early fluency intervals predict diagnoses. *Cortex*, 44(2):161 – 169.

Nancy J. Fisher, Mary C. Tierney, Byron P. Rourke, and John P. Szalai. 2004. Verbal fluency patterns in two subgroups of patients with alzheimer's disease. *The Clinical Neuropsychologist*, 18(1):122–131. PMID: 15595364.

Rowena G. Gomez and Desirée A. White. 2006. Using verbal fluency to detect very mild dementia of the Alzheimer type. *Archives of Clinical Neuropsychology*, 21(8):771–775.

John R Hodges. 2006. Alzheimer's centennial legacy: origins, landmarks and the current status of knowledge concerning cognitive aspects. *Brain*, 129(11):2811–2822.

John R. Hodges, David P. Salmon, and Nelson Butters. 1992. Semantic memory impairment in alzheimer's disease: Failure of access or degraded knowledge? *Neuropsychologia*, 30(4):301–314.

E. Huey, E. Goveia, S. Paviol, M. Pardini, F. Krueger, G. Zamboni, M. Tierney, E. Wassermann, and J. Grafman. 2009. Executive dysfunction in frontotemporal dementia and corticobasal syndrome. *Neurology*, 72:453 – 459.

Anastasios Karakostas, Alexia Briassouli, Konstantinos Avgerinakis, Ioannis Kompatsiaris, and Magda Tsolaki. 2017. The dem@care experiments and datasets: a technical report. *CoRR*, abs/1701.01142.

Parris Kidd. 2008. Alzheimer's disease, amnestic mild cognitive impairment, and age-associated memory impairment: Current understanding and progress toward integrative prevention. *Alternative medicine review : a journal of clinical therapeutic*, 13:85–115.

Najoung Kim, Jung-Ho Kim, Maria K. Wolters, Sarah E. MacPherson, and Jong C. Park. 2019. Automatic scoring of semantic fluency. *Frontiers in Psychology*, 10:1020.

A. König, N. Linz, J. Töger, M. Wolters, J. Alexandersson, and P. Robert. 2018. Fully automatic analysis of semantic verbal fluency performance for the assessment of cognitive decline. *Dementia and Geriatric Cognitive Disorders*. Accepted.

Vladimir I Levenshtein. 1966. Binary codes capable of correcting deletions, insertions, and reversals.

Muriel Deutsch Lezak, Diane B Howieson, David W Loring, Jill S Fischer, et al. 2004. *Neuropsychological assessment*. Oxford University Press, USA.

Yunqing Li, Ping Li, Qing X. Yang, Paul J. Eslinger, Chris T. Sica, and Prasanna Karunanayaka. 2017. Lexical-semantic search under different covert verbal fluency tasks: An fmri study. *Frontiers in Behavioral Neuroscience*, 11:131.

Hali Lindsay, Nicklas Linz, Johannes Tröger, and Jan Alexandersson. 2019. Automatic data-driven approaches for evaluating the phonemic verbal fluency task with healthy adults. In *ICNLSP*.

Hali Lindsay, Johannes Tröger, Jan Alexandersson, and Alexandra König. 2020. What difference does it make? early dementia detection using the semantic and phonemic verbal fluency task. In *LREC 2020 Workshop RaPID-3: Resources and ProcessIng of Linguistic, Para-Linguistic and Extra-Linguistic Data from People with Various Forms of Cognitive/Psychiatric Impairments*.

Nicklas Linz, Kristina Lundholm Fors, Hali Lindsay, Marie Eckerström, Jan Alexandersson, and Dimitrios Kokkinakis. 2019. Temporal analysis of the semantic verbal fluency task in persons with subjective and mild cognitive impairment. In *Proceedings of the Sixth Workshop on Computational Linguistics and Clinical Psychology*, pages 103–113, Minneapolis, Minnesota. Association for Computational Linguistics.

Nicklas Linz, Johannes Tröger, Jan Alexandersson, and Alexandra König. 2017a. Using Neural Word Embeddings in the Analysis of the Clinical Semantic Verbal Fluency Task. In *Proceedings of the 12th International Conference on Computational Semantics (IWCS)*.

Nicklas Linz, Johannes Tröger, Jan Alexandersson, Maria Wolters, Alexandra König, and Philippe Robert. 2017b. Predicting dementia screening and staging scores from semantic verbal fluency performance. In *2017 IEEE International Conference on Data Mining Workshops (ICDMW)*, pages 719–728.

Sarah MacPherson, Colm Healy, Michael Allerhand, Barbara Spano, Carina Tudor-Sfetea, Mark White, Daniela Smirni, Tim Shallice, Edgar Chan, Marco Bozzali, and Lisa Cipolotti. 2016. Cognitive reserve and cognitive performance of patients with focal frontal lesions. *Neuropsychologia*, 96.

Brian MacWhinney. 1991. *The CHILDES project: Tools for analyzing talk.* Lawrence Erlbaum Associates, Inc.

Camillo Marra, Monica Ferraccioli, Maria Vita, Davide Quaranta, and Guido Gainotti. 2011. Patterns of cognitive decline and rates of conversion to dementia in patients with degenerative and vascular forms of mci. *Current Alzheimer research*, 8:24–31.

Edgar Miller. 1984. Verbal fluency as a function of a measure of verbal intelligence and in relation to different types of cerebral pathology. *British Journal of Clinical Psychology*, 23(1):53–57.

David R. Mortensen, Siddharth Dalmia, and Patrick Littell. 2018. Epitran: Precision G2P for many languages. In *Proceedings of the Eleventh International Conference on Language Resources and Evaluation (LREC 2018)*, Paris, France. European Language Resources Association (ELRA).

Kimberly Diggle Mueller, Rebecca L Koscik, Asenath LaRue, Lindsay R Clark, Bruce Hermann, Sterling C Johnson, and Mark A Sager. 2015. Verbal fluency and early memory decline: results from the wisconsin registry for alzheimer's prevention. *Archives of Clinical Neuropsychology*, 30(5):448–457.

Kelly J. Murphy, Jill B. Rich, and Angela K. Troyer. 2006. Verbal fluency patterns in amnestic mild cognitive impairment are characteristic of alzheimer's type dementia. *Journal of the International Neuropsychological Society*, 12(4):570–574.

Freda Newcombe. 1969. Missile wounds of the brain: A study of psychological deficits.

Dalia Santos Nogueira, Elizabeth Azevedo Reis, and Ana Vieira. 2016. Verbal fluency tasks: Effects of age, gender, and education. *Folia Phoniatrica et Logopaedica*, 68(3):124–133.

Katherine E Nutter-Upham, Andrew Saykin, Laura Rabin, Robert Roth, Heather Wishart, Nadia Pare, and Laura Flashman. 2008. Verbal fluency performance in amnestic mci and older adults with cognitive complaints. *Arch Clin Neuropsychol*, 23(3):229–41.

Serguei V.S. Pakhomov, Lynn Eberly, and David Knopman. 2016. Characterizing Cognitive Performance in a Large Longitudinal study of Aging with Computerized Semantic Indices of Verbal Fluency. *Neuropsychologia*, 89:42–56.

Serguei V.S. Pakhomov, David T. Jones, and David S. Knopman. 2015a. Language networks associated with computerized semantic indices. *NeuroImage*, 104:125–137.

Serguei V.S. Pakhomov, Susan E. Marino, Sarah Banks, and Charles Bernick. 2015b. Using Automatic Speech Recognition to Assess Spoken Responses to Cognitive Tests of Semantic Verbal Fluency. *Speech Communication*, 75:14–26.

F. Pedregosa, G. Varoquaux, A. Gramfort, V. Michel, B. Thirion, O. Grisel, M. Blondel, P. Prettenhofer, R. Weiss, V. Dubourg, J. Vanderplas, A. Passos, D. Cournapeau, M. Brucher, M. Perrot, and E. Duchesnay. 2011. Scikit-learn: Machine Learning in Python. *Journal of Machine Learning Research*, 12:2825–2830.

Ulla Petti, Simon Baker, and Anna Korhonen. 2020. A systematic literature review of automatic alzheimer's disease detection from speech and language. *Journal of the American Medical Informatics Association*, 27(11):1784–1797.

R Core Team. 2017. *R: A Language and Environment for Statistical Computing*. R Foundation for Statistical Computing, Vienna, Austria.

Barbara Rende, Gail Ramsberger, and Akira Miyake. 2002. Commonalities and differences in the working memory components underlying letter and category fluency tasks: A dual-task investigation. *Neuropsychology*, 16:309–21.

Eric Rinehardt, Katie Eichstaedt, John A Schinka, David A Loewenstein, Michelle Mattingly, Jean Fils, Ranjan Duara, and Mike R Schoenberg. 2014. Verbal fluency patterns in mild cognitive impairment and alzheimer's disease. *Dementia and geriatric cognitive disorders*, 38(1-2):1–9.

Wilma G Rosen. 1980. Verbal fluency in aging and dementia. *Journal of clinical and experimental neuropsychology*, 2(2):135–146.

James Ryan. 2013. *A System for Computerized Analysis of Verbal Fluency Tests.* Ph.D. thesis.

Charlotte SM Schmidt, Lena V Schumacher, Pia Römer, Rainer Leonhart, Lena Beume, Markus Martin, Andrea Dressing, Cornelius Weiller, and Christoph P Kaller. 2017. Are semantic and phonological fluency based on the same or distinct sets of cognitive processes? insights from factor analyses in healthy adults and stroke patients. *Neuropsychologia*, 99:148–155.

Zeshu Shao, Esther Janse, Karina Visser, and Antje S. Meyer. 2014. What do verbal fluency tasks measure? Predictors of verbal fluency performance in older adults. *Frontiers in Psychology*, 5:772.

O. Spreen, P.P.O. Spreen, E. Strauss, and P.P.E. Strauss. 1991. *A Compendium of Neuropsychological Tests: Administration, Norms, and Commentary*. Maestría de neuropsicología. Oxford University Press.

Vanessa Taler and Natalie A Phillips. 2008. Language performance in alzheimer's disease and mild cognitive impairment: a comparative review. *Journal of clinical and experimental neuropsychology*, 30(5):501–556.

Edmond Teng, Judith Leone-Friedman, Grace J. Lee, Stephanie Woo, Liana G. Apostolova, Shelly Harrell, John M. Ringman, and Po H. Lu. 2013. Similar Verbal Fluency Patterns in Amnestic Mild Cognitive Impairment and Alzheimer's Disease. *Archives of Clinical Neuropsychology*, 28(5):400–410.

Sharon L Thompson-Schill, Mark D'Esposito, Geoffrey K Aguirre, and Martha J Farah. 1997. Role of left inferior prefrontal cortex in retrieval of semantic knowledge: a reevaluation. *Proceedings of the National Academy of Sciences*, 94(26):14792–14797.

Jochen René Thyrian, Tilly Eichler, Andrea Pooch, Kerstin Albuerne, Adina Dreier, Bernhard Michalowsky, Wolfgang Hoffmann, and Diana Wucherer. 2016. Systematic, early identification of dementia and dementia care management are highly appreciated by general physicians in primary care - results within a cluster-randomized-controlled trial (delphi). *Journal of Multidisciplinary Healthcare*, 9:183.

Tom N Tombaugh, Jean Kozak, and Laura Rees. 1999. Normative data stratified by age and education for two measures of verbal fluency: Fas and animal naming. *Archives of Clinical Neuropsychology*, 14(2):167–177.

Johannes Tröger, Nicklas Linz, Jan Alexandersson, Alexandra König, and Philippe Robert. 2017. Automated Speech-based Screening for Alzheimer's Disease in a Care Service Scenario. In *Proceedings of the 11th EAI International Conference on Pervasive Computing Technologies for Healthcare*.

Angela K Troyer, Morris Moscovitch, and Gordon Winocur. 1997. Clustering and Switching as Two Components of Verbal Fluency: Evidence From Younger and Older Healthy Adults. *Neuropsychology*, 11(1):138–146.

Angela K Troyer, Morris Moscovitch, Gordon Winocur, Michael P Alexander, and Don Stuss. 1998. Clustering and switching on verbal fluency: the effects of focal frontal- and temporal-lobe lesions. *Neuropsychologia*, 36(6):499 – 504.

Johannes Tröger, Nicklas Linz, Alexandra König, P. Robert, Jan Alexandersson, Jessica Peter, and Jutta Kray. 2019. Exploitation vs. exploration—computational temporal and semantic analysis explains semantic verbal fluency impairment in alzheimer's disease. *Neuropsychologia*, 131.

Roisin M. Vaughan, Robert F. Coen, RoseAnne Kenny, and Brian A. Lawlor. 2016. Preservation of the semantic verbal fluency advantage in a large population-based sample: Normative data from the tilda study. *Journal of the International Neuropsychological Society*, 22(5):570–576.

Malvika Verma and Robert J Howard. 2012. Semantic memory and language dysfunction in early alzheimer's disease: a review. *International journal of geriatric psychiatry*, 27(12):1209–1217.

Mathieu Vigneau, V Beaucousin, Pierre-Yves Hervé, Hugues Duffau, Fabrice Crivello, O Houdé, Bernard Mazoyer, and N Tzourio-Mazoyer. 2006. Meta-analyzing left hemisphere language areas: Phonology, semantics, and sentence processing. *NeuroImage*, 30:1414–32.

Maria Vita, Camillo Marra, Pietro Spinelli, Alessia Caprara, Eugenia Scaricamazza, Diana Castelli, Serena Canulli, Guido Gainotti, and Davide Quaranta. 2014. Typicality of words produced on a semantic fluency task in amnesic mild cognitive impairment: Linguistic analysis and risk of conversion to dementia. *Journal of Alzheimer's disease : JAD*, 42.

Jeffrey C Zemla, Kesong Cao, Kimberly D Mueller, and Joseph L Austerweil. 2020. Snafu: The semantic network and fluency utility. *Behavior research methods*, pages 1–19.

Qianhua Zhao, Qihao Guo, and Zhen Hong. 2013. Clustering and switching during a semantic verbal fluency test contribute to differential diagnosis of cognitive impairment. *Neuroscience bulletin*, 29(1):75–82.

# A Appendix

| Category | Feature Name | Description |
|---|---|---|
| **Global Features** | *Measures that span over the task as a whole* | |
| | Word Count | The total number of words excluding repetitions. Scoring system used in clinical practice |
| | Number of Repetitions | Number of repetitions said during the task. Previously suggested in Ryan (2013). |
| **Phonemic Features** | *Rule-based measures for phonemic clustering strategies proposed by Troyer et al. (1997) and automated by Lindsay et al. (2019)* | |
| | Mean Cluster Size | Average number of words in clinical phonemic clusters |
| | Number of Switches | Total number of switches between clinical phonemic clusters |
| **Semantic Features** | *Automatic data-driven methods for determining semantically motivated clusters as proposed in Linz et al. (2017a)* ... | |
| | Mean Cluster Size | Average number of words in a semantic cluster |
| | Number of Switches | Total number of switches between semantic clusters |
| **Binning Features** | *10-second binning approach for finer resolution of task proposed by Linz et al. (2019); The following features are computed for each of the six, 10-second bins.* | |
| | Word Count by Bin | The number of words per 10 second bin |
| | LD by Bin | Levenshtein distance per 10 second bin |
| | POS-LD by Bin | Position-weighted Levenshtein distance per 10 second bin |
| | PHON-LD by Bin | Phonemic-weighted Levenshtein distance per 10 second bin |
| | Semantic Distance by Bin | Semantic Distance between consecutive words per 10 second bin |
| | Mean Temporal Distance by Bin | The average transition time in seconds between the end of one word and the onset of the next word by 10 second bin |

Table 3: The following features were extracted from the PVF task produced by the participants.

# Demonstrating the Reliability of Self-Annotated Emotion Data

**Anton Malko**[1], **Cecile Paris**[1,2], **Andreas Duenser**[1], **Maria Kangas**[3], **Diego Mollá**[1,2], **Ross Sparks**[1], **Stephen Wan**[1]

[1]Data61, CSIRO, Australia
[2]Department of Computing, Macquarie University, Sydney, Australia
[3]Department of Psychology, Macquarie University, Sydney, Australia

anton.malko@csiro.au, cecile.paris@csiro.au, andreas.duenser@csiro.au, maria.kangas@mq.edu.au, diego.molla-aliod@mq.edu.au, ross.sparks@csiro.au, stephen.wan@csiro.au

## Abstract

Vent is a specialised iOS/Android social media platform with the stated goal to encourage people to post about their feelings and explicitly label them. In this paper, we study a snapshot of more than 100 million messages obtained from the developers of Vent, together with the labels assigned by the authors of the messages. We establish the quality of the self-annotated data by conducting a qualitative analysis, a vocabulary-based analysis, and by training and testing an emotion classifier. We conclude that the self-annotated labels of our corpus are indeed indicative of the emotional contents expressed in the text and thus can support more detailed analyses of emotion expression on social media, such as emotion trajectories and factors influencing them.

## 1 Introduction

Social media platforms are being widely used by people to express their feelings. While some such platforms are generic in their purpose (e.g., Twitter), others have specific goals, such as connecting with people with similar health issues (e.g., PatientsLikeMe[1]). Vent[2] belongs to the latter class of platforms: its stated goal is to encourage people to express and share their feelings. Vent enables people to post messages expressing their own feelings and to react to posts from others. Interestingly, Vent requires people to label their posts with the emotion they feel at the time of posting. The platform thus provides us with an opportunity to study, at scale, how people express emotions, to what emotions they react, how emotions change over time, and what factors influence their trajectory.

Vent data is self-annotated for emotion, which is of particular interest to us. Studies on emotions in social media often derive labels from texts, either with the help of annotators, or using sentiment analysis techniques (see, for example, reviews of annotated datasets by Bostan and Klinger (2018); Mohammad (2020)). We note, however, that information that external observers (annotators or algorithms) can extract from a text may not be sufficient to reliably identify the affective state of the text's author at the time of posting. This could be because the texts are too short to provide enough context, are ambiguous, or require extra-textual context to interpret. Even when richer context is available, external observers may not necessarily assign a definitive affective label to a text. For example, psychological construction theory (Barrett, 2006) states that emotion labels are a result of *categorisation* of the current state of the organism, in the current context; consequently, the same episode may be categorised differently by the person who experiences it and by an outside observer. Given this, self-assigned affective labels may provide a more direct access to a person's emotional state than labels attributed after the fact.

Our ultimate goal is to study emotion trajectories (on social media) and the factors that affect them, potentially leading to the automatic identification of mental health issues. However, before we can employ data such as that provided in Vent to study emotion sharing and changes in emotions, we must establish whether the self-annotated labels are reasonable indicators of emotional states. This is because, even with self-assigned labels, there are concerns that may arise: for example, the label choice may be a byproduct of poor user interface design. Establishing that the labels are reasonable is thus our central aim in this paper. We conduct a multi-step analysis of the Vent data, showing that we can use this kind of data to study how people express their feelings and how people react to them.

[1]https://www.patientslikeme.com/
[2]https://www.vent.co/



*Proceedings of the Seventh Workshop on Computational Linguistics and Clinical Psychology*, pages 45–54
June 11, 2021. ©2021 Association for Computational Linguistics

The rest of the paper is structured as follows. We begin with a short summary of related research in Section 2. This is followed by a description of the Vent platform and the data we have from it in Section 3. We describe the data selection steps in Section 4. We then present the analysis steps we have taken to ascertain that the labels adequately reflect the affective states expressed in the texts in Section 5. Section 6 concludes this paper and outlines future research directions.

## 2 Related Work

There is a growing number of datasets annotated with affect information. Many of these are annotated by experts or via crowdsourcing and fall out of the scope of our work. Instead, we refer the reader to the surveys by Bostan and Klinger (2018); Mohammad (2020).

To the best of our knowledge, self-annotated affective datasets are rare; the reviews by Bostan and Klinger (2018); Mohammad (2020) mention only one such dataset. ISEAR ("International Survey on Emotion Antecedents and Reactions") is a self-labelled affective dataset created by Scherer and Wallbott (1994). It was collected by administering a questionnaire, in which people were asked to describe recent experiences of one of the seven emotions (Anger, Fear, Joy, Sadness, Disgust, Shame, Guilt) and to answer questions about their physiological and psychological state during these emotion episodes. Overall, roughly 3,000 people from 37 countries completed the questionnaire, providing 7,666 textual descriptions. In comparison, our dataset contains considerably more data.

A more widely used approach to produce emotion annotation without using experts is to rely on distant supervision — for example, treating Twitter hashtags like `#happy` or `#sad` as self-assigned emotion labels. Examples of datasets constructed with distant supervision include those by Mohammad (2012); Roberts et al. (2012); Wang et al. (2012); Qadir and Riloff (2013); Mohammad and Kiritchenko (2015); Volkova and Bachrach (2016); Abdul-Mageed and Ungar (2017). Emotion classifiers using these datasets are reported to perform well: the best results thus far were produced by Abdul-Mageed and Ungar (2017), who used a Gated Recurrent Neural Network (GRNN) classifier on 1.6 million tweets labelled with emotions from Plutchik's categorisation (Plutchik, 1980) and reached an averaged F1-score[3] of 0.9568.

Lykousas et al. (2019) used web-scraping techniques to collect 33 million messages from the Vent platform, from around 1 million users with public profiles (meaning that anybody on the platform could see these posts). They presented a broad descriptive exploration of these data, along with an analysis of emotions in texts and user networks, but they did not investigate the quality of the annotations. In comparison, our dataset is directly provided via a 2019 data science partnership with Vent.

Our data includes all posts (anonymised for this research). Our goal here is to assess the alignment between affect in self-assigned affective labels and texts.

## 3 Vent and its Dataset

Vent advertises itself as a platform to "Express your feelings and connect with people who care". Vent is thus specifically geared towards sharing one's emotions, unlike Twitter or Facebook, which support many other activities. This makes Vent particularly interesting for investigating emotion expression on social media. Users (*venters*) register anonymously, with only an email address. Once registered, they can create short text messages (*vents*), read messages by other venters and react to them, using comments or *interactions* (short predefined reactions, for example, "HUG", "LOL", or an emoji).

Vent's creators have given us access to the data from the platform over a 5-year period, from the late 2013 until the end of 2019, as part of a collaborative project to study mental health.[4] Overall, the raw dataset contains over 107 million vents, from close to 1.5 million users, including both public and private posts, along with additional types of information, namely comments, interactions, follower/followee links and the information on discussion groups. Due to ethical and privacy concerns, the dataset is not publicly available.

Vent's labels[5] are arranged in a two-level hier-

[3] A classification performance metric, which takes into account both the classifier's accuracy on the target class (Recall), and its ability to avoid classifying non-target examples as target (Precision). It is defined as a harmonic mean of Precision and Recall; its worst value is 0 and its best value is 1 (Chicco and Jurman, 2020).

[4] The project was approved by the CSIRO ethics committee; reference number 165/19.

[5] For clarity, we will use different fonts to refer to Vent's label categories (e.g., `Sadness`) and real affective states (e.g., sadness).

archy. At the top level, there are 85 emotion categories, which we can categorise into these 5 groups:

**Affective states.** This group contains the following 9 categories: `Affection`, `Anger`, `Creativity`, `Fear`, `Feelings`, `Happiness`, `Positivity`, `Sadness`, `Surprise`.

**Dates.** There are 46 categories linked to dates and seasonal events, such as `Autumn`, `Ramadan`, `Paralympics`, etc.

**Groups of people.** There are 13 categories in this group, e.g., `Women HM`, `Pride'18`, etc.

**Character/Role/Imaginary content.** This group contains 7 categories related to fictional and imaginary topics such as `Vampire`, `Star Wars`.

**Miscellanea.** There are 10 categories of miscellaneous nature, e.g., `Candy`, `Gaming`.

The nine categories related to affective states are always available to the users. All other categories generally have to be paid for individually, although they can become temporarily available for free on special occasions (e.g., on Halloween).[6]

At the second level of the hierarchy, there are 1,187 labels. Figure 1 shows examples of labels within a subset of the 9 always available categories.

When users want to create a message, they first go through a labelling interface: all labels from a given category are presented on a single screen, and swiping the screen to the left or right switches between label categories. The name of the current category is *not* shown to the users by default and is only indicated by the background colour of the screen. It becomes visible if one taps on the scrolling control.

In the current version of Vent, when users create a new vent, their label choice screen starts with the label category from their most recent vent. This might introduce biases to the data: for example, users may just proceed with the first choice they see (e.g., if they need to share some intense emotion experience and accurate labelling is not important to them at the moment). We also note that people *have to* select a single label. Finally, the inventory of labels is pre-defined. In some situations, this may cause people to choose a label that does not exactly match their current dominant state.

[6]This has changed in the most recent versions of Vent: currently, one has to pay a monthly subscription fee to unlock all additional label categories.

## 4 Data selection

For our analyses, we restrict our data to the vents that correspond to the following six high-level categories, which we call "core categories": `Affection`, `Anger`, `Fear`, `Happiness`, `Sadness`, and `Surprise`. These labels are always available to the users. Importantly, out of all Vent's categories, they are most easily interpretable in terms of affective states. Many psychological accounts of human emotion repertoire include some or all of these categories (see, e.g., Table 1 of Ortony and Turner's (1990) publication); and they map one-to-one onto Shaver et al.'s (1987) classification. Vents with these labels account for 45.4% of the total number of vents.

In addition, we exclude the following categories of users:

1. **Official Vent account.** Vent has an official account, which consists mostly of a) questionnaires about experiences on Vent; b) technical information (e.g., planned maintenance) and c) discussion of possible/existing label categories.

2. **Robots.** The following heuristic was used: a user is a robot if (1) they created at least 100 messages within a day, (2) they posted vents on no more than 10 distinct dates, and (3) at least 99% of the vents were posted within a single day. Using this rule, we discovered 258 users, who created 187,063 messages. A manual analysis suggested that our heuristic is satisfactory: only 1 of 30 randomly selected users in this subset was not a robot. One additional robot account with 10,219 vents not satisfying the heuristic criterion was further excluded during manual exploration.

3. **Users with fewer than 20 vents.** The purpose of this filter was to ensure that the users we include have at least some experience in using the app.

The resulting dataset contains 45,194,018 vents from 372,662 users. It is used for the qualitative analysis in Section 5.1.

For the more detailed automated analyses in Sections 5.2–5.4, we further subset these data in the following way. Most categories contain labels which

| Surprise | Feelings | Positivity | Anger | Affection |
|---|---|---|---|---|
| AR Libre AR | AU TRUE BLUE AU | BR Independente BR | CA PROUD CA | FR SUPPORTIVE FR |
| COOL | CR Pura Vida CR | CO Berraco CO | HR Nezavisan HR | JP Supportive EC |
| TIDY | FR REVOLUTIONARY FR | DE Einheitlich DE | LB Independent LB | TR Supportive TR |
| MESSY | KR Gwangbok KR | Eclipsing | MX Viva MX | BLESSED |
| EXPLORATIVE | MY Merdeka MY | THANKFUL | NO Uavhengig NO | LOVING |
| RICH | SY PEACEFUL SY | STUFFED | PL Independent PL | ANGELIC |
| Amazed | US INDEPENDENT US | HYPED | Angry | Kind |
| Astonished | US Patriotic US | GENEROUS | Annoyed | PRIDEFUL |
| Concerned | Safe | PURRFECT | Bitter | APPRECIATIVE |
| Conflicted | Pawsome | Relaxed | Disgusted | Adoring |
| Confused | HONEST | DETERMINED | Done | Affectionate |
| Curious | Intelligent | Reflective | Exasperated | Caring |
| Dazed | FESTIVE | WISHFUL | Frustrated | Cuddly |
| Embarrassed | Observant | ∞ STIMTASTIC ∞ | Furious | Devoted |

Figure 1: A selection of label categories and labels.

are less clearly connected to affect or only used rarely (e.g., "Independent" or "Viva" in `Anger` in Figure 1) — we exclude them from consideration. Next, we sample 1.8 million vents per core label category, filtering out (a) vents only containing words "null", "test" or "testing"; (b) tag memes. Tag memes are explained in Section 5.1 and are identified with a regular expression.[7] Finally, we exclude non-English vents, as identified by the `langid`[8] tool, which removes approximately 7% of the messages. The resulting subset contains 1.5–1.6 million messages per label category; we will refer to it as "the reduced dataset".

## 5 Assessing the alignment of the labels and the texts

To ascertain the alignment between the text and the labels, we conduct the following analyses:

1. A qualitative analysis, conducted manually on a subset of the data in order to identify potentially non-affective uses of the labels;

2. A vocabulary-based analysis, in which we gather statistics on the presence of emotionally loaded words in the vents using word-emotion associations;

3. An emoji-based analysis, in which we examine the top 10 emojis in each label category of interest; and

4. A text-to-label machine learning classifier analysis, in which we train a BERT model to establish whether textual information beyond simple keywords helps to differentiate between individual label categories.

We use these four methods to establish that the self-annotated labels do indeed reflect emotional state. The methods are complementary. The qualitative analysis attempts to capture idiosyncratic uses of labels, which may be hard to anticipate and thus hard to analyse automatically. The vocabulary-based and emoji-based analyses establish whether individual emotion-loaded tokens in the texts are congruent with the labels. Finally, the classification approach allows the exploration of the connection between entire texts and their labels, capitalising on context beyond individual tokens. These analyses are described below.

### 5.1 Qualitative analysis

During an initial data exploration, we found the following cases of non-affective uses of the labels:

1. **Vents with "default" labels**. Some people choose default labels for their vents, occasionally stating reasons for doing this: for example, liking the colour of a specific category or being too lazy to chose a label for every vent.

2. **Vents from bio accounts**. Vent allows users to add biographical information to their accounts; however, some users create separate dedicated accounts just to post messages containing such information. Posts in these accounts include not only demographic facts, but also topics of interest, and guidelines for followers (describing who should or should not follow).

[7] `(.*tagged by.*) | (.*i tag.*) | (.*tagging.*)`

[8] `https://github.com/saffsd/langid.py`

3. **Tag memes**: We observed the occurrence of user-generated questionnaires on a wide variety of topics (e.g., "What kind of vent user are you?", "common fears"). Vent users refer to them as "tag memes". Such questionnaires often follow a specific template, so we could identify them based on a regular expression. A manual analysis of 100 messages identified using the regular expression we employed showed that 18 of them were not tag memes.

To assess the relative presence of the above non-affective uses of the labels, we inspected 1,000 randomly selected vents from the dataset (after applying the filters described in Section 4). The sample did not contain instances of people mentioning default emotions. The sample contained 4 tag memes (0.4% of the sample), and only 1 vent from a bio account. We therefore conclude that clearly non-affective uses of labels are rare.

### 5.2 Vocabulary-based analysis

After performing the qualitative analysis, we consider emotionally loaded words present in the texts. The data used for this analysis is a sample of 1.5 million vents per category from the reduced dataset, to have a balanced distribution across categories.

The emotionally loaded words are obtained from the the NRC Emotion Lexicon (henceforth, EmoLex) (Mohammad and Turney, 2012). EmoLex is one of the largest emotion lexicons. It contains 14,182 words and indicates whether they are associated with one of 10 affective states: Plutchik's eight (Plutchik, 1980), plus "positive" and "negative". Each word can be associated with any number of affective states.

For this analysis, we only consider EmoLex words associated with at least one specific emotion, excluding words which only have generic associations with positive and/or negative affect. This results in 4,463 unique words out of 14,182 and 8,265 word-affect association pairs. Around 70% of the vents have words from this set.

Table 1 shows the lexicon coverage per label category. Within all label categories, except `Surprise`, words related to a corresponding emotion[9] are found in the largest proportion of vents. For example, if we consider vents labelled with `Anger` (second column of Table 1), EmoLex words related to anger are found in the largest proportion of these vents. Such associations also hold at the more general level of emotional valence (positive vs. negative affect): within a given label category, EmoLex words associated with emotions of matching valence are generally found in a larger proportion of vents: e.g., within `Sadness`, more vents contain words related to anger, fear and sadness than to anticipation, joy and trust.

Conversely, if we examine what Vent category has the largest percentage of words from a determined EmoLex emotion category (by analysing Table 1 row by row, instead of column by column), we observe that closely related categories are most likely. For example, if we know that a vent has sadness-related words, it is most likely to be labelled with `Sadness`. This pattern holds in virtually all cases when there exists a one-to-one mapping from an emotion to a Vent category, with only two exceptions: EmoLex words associated with joy are most likely to be found in `Affection` vents, and EmoLex words associated with surprise are most likely to be found in `Affection` and `Happiness` vents. A similar pattern is observed for emotion valence: for example, words associated with anger are most likely to be found in vents labelled with any category with negative valence: `Anger`, `Fear`, `Sadness`. These results suggest

Table 1: Percentage of vents having at least one word associated with a given emotion. 'Any' – proportion of vents with at least one word associated with any emotion. Af – `Affection`, An – `Anger`, Fe – `Fear`, Ha – `Happiness`, Sa – `Sadness`, Su – `Surprise`. Maximum values in each column (excluding the 'Any' row) are highlighted in bold.

| EmoLex emotion category | Vent label category: Af | An | Fe | Ha | Sa | Su |
|---|---|---|---|---|---|---|
| anger | 24 | **42** | 34 | 23 | 34 | 27 |
| anticipation | 37 | 33 | 38 | 40 | 32 | 33 |
| disgust | 21 | 38 | 30 | 21 | 30 | 24 |
| fear | 23 | 38 | **39** | 23 | 36 | 27 |
| joy | **47** | 30 | 30 | **42** | 30 | 32 |
| sadness | 25 | 40 | 38 | 23 | **42** | 28 |
| surprise | 24 | 21 | 22 | 24 | 21 | 20 |
| trust | 39 | 35 | 35 | 40 | 32 | **35** |
| any | 69 | 72 | 71 | 69 | 68 | 66 |

[9]Vent category of `Affection` does not have a corresponding emotion in EmoLex, but arguably, joy is the closest option. Plutchik considered love to be a combination of joy and trust (e.g., see (Plutchik, 1980, p.21); "trust" is called "acceptance" in the reference), and interestingly, high proportion of `Affection` vents have words related to these emotions.

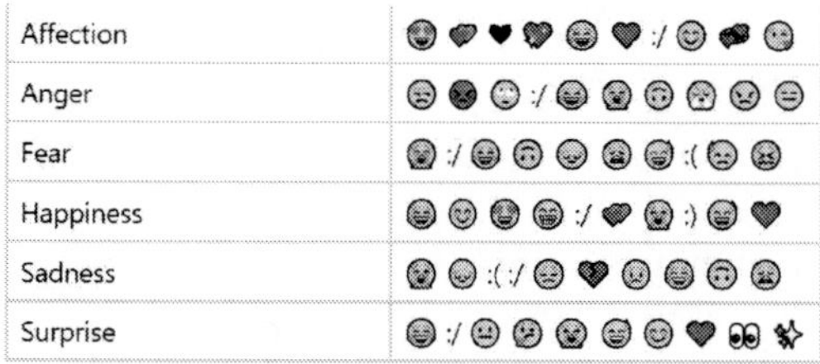


Figure 2: Top 10 emojis per category. Emojis are ordered from most to least used.

that when people use words associated with a given emotion, they are more likely to choose the corresponding Vent label.

### 5.3 Emoji based analysis

The previous section showed that the data was consistent with the EmoLex resource. We perform a similar analysis with emojis, which are not included in EmoLex, checking to see that these distant supervision labels are generally consistent with the self-annotated labels. We carry out a separate analysis of the most used emojis, using the same dataset of 1.5 million vents per label category. Emojis were identified using the `emoji`[10] and `emot`[11] Python libraries.

Figure 2 shows that the use of emojis is congruent with the category. For example, the top 10 emojis in `Affection` contain more hearts than any other category; and emojis indicating angry faces only appear in the top 10 list for `Anger`. We can observe the same at the level of affect valence as well. For example, the "`:(`" emoticon does not appear in the top 10 list for `Affection` and `Happiness`; hearts do not appear in the top 10 list for `Anger`, `Fear` and `Sadness` (with the exception of the *broken* heart in `Sadness`).

This analysis of the use of emojis per Vent category is consistent with the vocabulary-based analysis of the previous section.

### 5.4 Emotion classification

Our final analysis to assess the alignment of the labels and the texts has been conducted by training a neural emotion classifier with the Vent data, and observing the results on a separate test data, also drawn from the Vent data. The rationale is that, if the classifier can identify the labels, then these labels are used in a consistent way. Of course this does not indicate *per se* that the self-annotated labels are correct, because there might have been a bias in the labelling process which has been captured as a pattern picked by the classifier. But combined with the analysis described in the previous sections, good classification results would give additional evidence for the validity of the self-annotated data.

[10]`https://github.com/carpedm20/emoji/`
[11]`https://github.com/NeelShah18/emot`

Table 2: EmoLex-based models. F1-score by class: mean value (stddev) across the five runs

| Label category | Precision | Recall | F1 |
|---|---|---|---|
| `Affection` | – | – | – |
| `Anger` | 0.26 (0.002) | 0.16 (0.002) | 0.20 (0.002) |
| `Fear` | 0.25 (0.004) | 0.15 (0.002) | 0.19 (0.003) |
| `Happiness` | 0.29 (0.002) | 0.19 (0.002) | 0.23 (0.002) |
| `Sadness` | 0.27 (0.003) | 0.19 (0.003) | 0.22 (0.003) |
| `Surprise` | 0.22 (0.003) | 0.08 (0.001) | 0.12 (0.002) |

For the classification data, we create 5 random subsets with 500,000 + 50,000 + 50,000 vents (train-dev-test) per category, each time sampling from the reduced dataset. All texts are lowercased.

We use two simple classifiers as baselines. In the first one, labels are simply chosen at random from Vent's core categories. This classifier produces Precision of 0.17, Recall of 0.17 and F1-score of 0.17 for all classes. The second classifier is based on EmoLex. For each vent in our sample, we predict the EmoLex emotion associated with the largest number of words in this vent. Ties (including cases where vents contained no words from EmoLex) are broken at random. As Vent's `Affection` does not map directly onto EmoLex emotions, we exclude it from consideration in this particular analysis. The classification results generally improve over the random baseline, but the gains are small: the macro F1-score ranged from 0.189 to 0.192, with a mean of 0.190 and a standard deviation of 0.001. The F1 scores by class averaged across all five runs are given in Table 2.

Finally, we use a BERT-based model (De-

Table 3: BERT-based models. F1-score by class: mean value (stddev) across the five runs.

| Label category | Precision | Recall | F1 |
|---|---|---|---|
| Affection | 0.62 (0.005) | 0.65 (0.005) | 0.63 (0.005) |
| Anger | 0.57 (0.005) | 0.57 (0.004) | 0.57 (0.000) |
| Fear | 0.54 (0.005) | 0.49 (0.005) | 0.52 (0.005) |
| Happiness | 0.58 (0.000) | 0.59 (0.004) | 0.58 (0.005) |
| Sadness | 0.54 (0.005) | 0.60 (0.000) | 0.56 (0.004) |
| Surprise | 0.52 (0.004) | 0.47 (0.004) | 0.49 (0.004) |

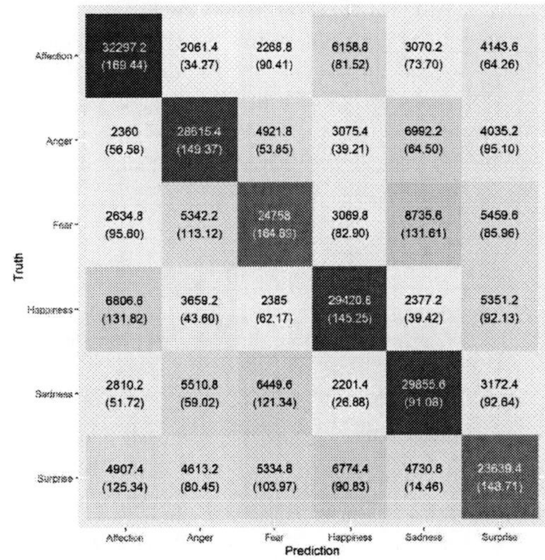


Figure 3: Confusion matrix for BERT model's predictions. Numbers correspond to mean values (stddev) across the five runs.

vlin et al., 2019).[12] The model's standard lexicon is manually augmented with emojis, using `emoji2vec` pre-trained embeddings (Eisner et al., 2016). We use the following hyperparameters. The maximum sequence length for the BERT tokeniser is set at 128. The learning rate is $3 \cdot 10^{-5}$. The batch size is 512 (spread over 4 GPUs). The number of epochs is 2, with checkpoints every 150 batches. The best checkpoint (as measured by macro F1) is saved.

We train a separate model on each random subset. Macro F1 score ranges from 0.560 to 0.562, with a mean of 0.561 and a standard deviation of 0.001. Table 3 shows F1-score by class, and Figure 3 shows the confusion matrix for the model's predictions; in both cases the values are averaged across the five runs.

The BERT-based classifier has improved performance, indicating that context over and above emotionally loaded keywords contains considerable amount of information benefiting classification. With respect to the alignment between labels and texts, the results are consistent with the results of the vocabulary-based and emoji-based analyses (Figure 3). The correct label is predicted most frequently. Incorrectly predicting a label referring to the emotion of similar valence is more likely than predicting a label of the opposite valence: e.g., when the true label is `Happiness`, `Affection` is a more likely incorrect prediction than `Anger`, `Fear` or `Sadness`. As before, the category of `Surprise` appears to be less clearly connected with the texts properties: the classifier made the biggest number of mistakes on it, and these mistakes were relatively evenly spread across the other 5 categories.

To better understand the classifier's performance, we visually inspect 60 random sentences (10 per label category) in which the classifier made a wrong prediction. Given that the variability between the models in the five runs is small, we only examine predictions from a single model with the best macro F1 score. Table 4 shows the results. As recommended by Benton et al. (2017), all specific examples are rephrased to protect users privacy. In the majority of the vents (45), the label assigned by the classifier is consistent with the text. Common reasons for the mistakes include lack of context which would allow to clearly differentiate between several possible affective states (e.g., `Affection` and `Happiness`, or `Anger` and `Sadness`); multiple emotions clearly expressed in the text (in some cases the classifier did capture one of the emotions, while the label reflected another). In a minority of cases, it is not immediately clear whether the labels fit the text (8 cases). In two such cases, the orthography is quite severely affected. In four cases, the Vent label hierarchy is to blame: the lower level label matched the sentence, but the category it be-

[12] `bert-base-uncased` from the HuggingFace Transformers library (Wolf et al., 2019).

Table 4: Analysis of 60 random examples in which there is a mismatch between the gold label existing in Vent and the automated label assigned by the classifier. "No context" — not enough context to assign a label, given just text. "Both" — both the gold and the automated label fit the sentence, and (a) "Both conceivable" — it is hard to choose between them; (b) "Gold better" — the gold label appears a better fit; (c) "Automated better" — the automated label appears to be a better fit. "Gold only" — only the gold label fits. "Automated only" — only the automated label fits. "Neither" — neither the gold nor the automated labels fit. Examples are accompanied by the gold label (in bold) and the automated label.

| Type | Count | Example |
|---|---|---|
| No context | 3 | (1) "*Ahaa*" (**Anger**; Affection) |
| Both | 45 | |
| Both conceivable | 26 | (2) "*It seems I am always the problem*" (**Anger**; Sadness) |
| Gold better | 10 | (3) "*Nowadays movies are very strange*" (**Surprise**, Happiness) |
| Automated better | 9 | (4) "*Why can't I fall asleep. It's always this way, I want to sleep and not be stressed. Everything is going to be even worse tomorrow. I just wanna f***ing sleep... [several more similar sentences]*" (**Fear**; Anger) |
| Gold only | 4 | (5) "*This crazy woman told me to stop watching animes and study instead. My animes have more culture than you.*" (**Anger**; Happiness) |
| Automated only | 2 | (6) "*I hate friends who do what you ask them not to. If I tell not to look at me, f***ing don't. F***ING LISTEN TO ME*" (**Fear**; Anger) |
| Neither | 6 | (7) "*Can't wait until the evening, I do need some time for myself*" (**Fear**; Happiness) |

longed to did not. One example is the vent "*I am leaving tomorrow, this is sad, but also a relief, as I am tired and want to be home.*" — the lower level label is "Stressed", which is congruent with the text; however this label falls under `Fear` category, which is a worse fit for the message.

The model performance, and consistency with the vocabulary and emoji analysis performed in Sections 5.2–5.3, gives further evidence that the affective information contained in vents is congruent with the assigned labels.

## 6 Conclusions

In this paper, we have presented an analysis of the quality of self-annotated emotion data from the Vent platform, which is specifically focused on emotion sharing. Our results suggest that self-assigned labels in Vent have a reasonable degree of connection to the affective states expressed in the texts. A qualitative analysis of the vents and their labels indicates that labels which are not meant to communicate affect are rare. A vocabulary-based analysis based on EmoLex shows that Vent labels align with affect polarity of the texts, and that words associated with a certain EmoLex emotion are most frequently encountered in vents in the corresponding Vent category. The top 10 emojis in each category are consistent with the category label. Finally, a BERT classification model can predict correct labels most often, and the classification mistakes often preserve emotion valence. Overall, we conclude that self-assigned labels produced in a non-controlled naturalistic setting can be used as a reasonably accurate representation of the author's affective state, and thus can support more complex analyses of emotions in social media.

Our analyses focused on the assumption that each text conveys one dominant emotion which may or may not be congruent with the assigned label. We adopted this approach as a first step, allowing us to explore simple models matching the structure of the data (one message – one label). This is an oversimplification, as suggested by examples such as (4) in Table 4 or the earlier example about going home (*"this is sad, but also a relief"*). Several emotions may be expressed in a single text, either because the emotional state of the author evolved during the writing of the message, or because the author had mixed emotions (e.g., Larsen and McGraw (2014)). As Vent only allows one label per message, the presence of vents containing mixed emotions could lower the observed alignment between the labels and the texts.[13] Thus, understanding whether and how mixed emotions are expressed in naturalistic data such as those from Vent would be important in this line of research, and we may explore it in our future work.

One particular research direction we are currently exploring is tracking the changes in reported emotion over time, the factors influencing these changes, and the connection of these properties with mental health well-being.

## Acknowledgements

We would like to thank Dean Serroni and Albert Jou from Vent for granting us access to the data and for providing additional information about the data and the app. We would also like to thank the two anonymous reviewers for providing feedback on an earlier version of this paper.

## References

Muhammad Abdul-Mageed and Lyle Ungar. 2017. EmoNet: Fine-grained emotion detection with gated recurrent neural networks. In *Proceedings of the 55th Annual Meeting of the Association for Computational Linguistics (Volume 1: Long Papers)*, pages 718–728, Vancouver, Canada. Association for Computational Linguistics.

Lisa Feldman Barrett. 2006. Solving the emotion paradox: Categorization and the experience of emotion. *Personality and Social Psychology Review*, 10(1):20–46.

Adrian Benton, Glen Coppersmith, and Mark Dredze. 2017. Ethical research protocols for social media health research. In *Proceedings of the First ACL Workshop on Ethics in Natural Language Processing*, pages 94–102, Valencia, Spain. Association for Computational Linguistics.

Laura-Ana-Maria Bostan and Roman Klinger. 2018. An analysis of annotated corpora for emotion classification in text. In *Proceedings of the 27th International Conference on Computational Linguistics*, pages 2104–2119, Santa Fe, New Mexico, USA. Association for Computational Linguistics.

[13]For example, an anonymous reviewer raised a possibility that the BERT classifier might have showed lower results on `Fear` and `Surprise`, because these emotions are more commonly associated with mixed emotional experiences.

Davide Chicco and Giuseppe Jurman. 2020. The advantages of the Matthews correlation coefficient (MCC) over F1 score and accuracy in binary classification evaluation. *BMC Genomics*, 21(1).

Jacob Devlin, Ming-Wei Chang, Kenton Lee, and Kristina Toutanova. 2019. BERT: Pre-training of deep bidirectional transformers for language understanding. In *Proceedings of the 2019 Conference of the North American Chapter of the Association for Computational Linguistics: Human Language Technologies, Volume 1 (Long and Short Papers)*, pages 4171–4186, Minneapolis, Minnesota. Association for Computational Linguistics.

Ben Eisner, Tim Rocktäschel, Isabelle Augenstein, Matko Bošnjak, and Sebastian Riedel. 2016. emoji2vec: Learning emoji representations from their description. In *Proceedings of The Fourth International Workshop on Natural Language Processing for Social Media*, pages 48–54, Austin, TX, USA. Association for Computational Linguistics.

Jeff T. Larsen and A. Peter McGraw. 2014. The case for mixed emotions. *Social and Personality Psychology Compass*, 8(6):263–274.

Nikolaos Lykousas, Costantinos Patsakis, Andreas Kaltenbrunner, and Vicenç Gómez. 2019. Sharing emotions at scale: The vent dataset. In *Proceedings of the Thirteenth International AAAI Conference on Web and Social Media (ICWSM 2019)*.

Saif Mohammad. 2012. #emotional tweets. In **SEM 2012: The First Joint Conference on Lexical and Computational Semantics – Volume 1: Proceedings of the main conference and the shared task, and Volume 2: Proceedings of the Sixth International Workshop on Semantic Evaluation (SemEval 2012)*, pages 246–255, Montréal, Canada. Association for Computational Linguistics.

Saif M. Mohammad. 2020. Sentiment analysis: Detecting valence, emotions, and other affectual states from text. *Emotion measurement*.

Saif M Mohammad and Svetlana Kiritchenko. 2015. Using hashtags to capture fine emotion categories from tweets. *Computational Intelligence*, 31(2):301–326.

Saif M. Mohammad and Peter D. Turney. 2012. Crowdsourcing a word-emotion association lexicon. *Computational Intelligence*, 29(3):436–465.

Andrew Ortony and Terence J. Turner. 1990. What's basic about basic emotions? *Psychological Review*, 97(3):315–331.

Robert Plutchik. 1980. A general psychoevolutionary theory of emotion. In *Theories of Emotion*, pages 3–33. Elsevier.

Ashequl Qadir and Ellen Riloff. 2013. Bootstrapped learning of emotion hashtags #hashtags4you. In *Proceedings of the 4th Workshop on Computational Approaches to Subjectivity, Sentiment and Social Media*

*Analysis*, pages 2–11, Atlanta, Georgia. Association for Computational Linguistics.

Kirk Roberts, Michael A. Roach, Joseph Johnson, Josh Guthrie, and Sanda M. Harabagiu. 2012. EmpaTweet: Annotating and detecting emotions on Twitter. In *Proceedings of the Eighth International Conference on Language Resources and Evaluation (LREC'12)*, pages 3806–3813, Istanbul, Turkey. European Language Resources Association (ELRA).

Klaus R. Scherer and Harald G. Wallbott. 1994. Evidence for universality and cultural variation of differential emotion response patterning. *Journal of Personality and Social Psychology*, 66(2):310–328.

Phillip Shaver, Judith Schwartz, Donald Kirson, and Cary O'Connor. 1987. Emotion knowledge: Further exploration of a prototype approach. *Journal of Personality and Social Psychology*, 52(6):1061–1086.

Svitlana Volkova and Yoram Bachrach. 2016. Inferring perceived demographics from user emotional tone and user-environment emotional contrast. In *Proceedings of the 54th Annual Meeting of the Association for Computational Linguistics (Volume 1: Long Papers)*, pages 1567–1578, Berlin, Germany. Association for Computational Linguistics.

Wenbo Wang, Lu Chen, Krishnaprasad Thirunarayan, and Amit P Sheth. 2012. Harnessing twitter" big data" for automatic emotion identification. In *2012 International Conference on Privacy, Security, Risk and Trust and 2012 International Confernece on Social Computing*, pages 587–592. IEEE.

Thomas Wolf, Lysandre Debut, Victor Sanh, Julien Chaumond, Clement Delangue, Anthony Moi, Pierric Cistac, Tim Rault, Rémi Louf, Morgan Funtowicz, and Jamie Brew. 2019. Huggingface's transformers: State-of-the-art natural language processing. *ArXiv preprint*.

# Hebrew Psychological Lexicons

**Natalie Shapira, Dana Atzil-Slonim, Daniel Juravski, Moran Baruch, Adar Paz, Dana Stolowicz-Melman, Tal Alfi-Yogev, Roy Azoulay, Adi Singer, Maayan Revivo, Chen Dahbash, Limor Dayan, Tamar Naim, Lidar Gez, Boaz Yanai, Adva Maman, Adam Nadaf, Elinor Sarfati, Amna Baloum, Tal Naor, Ephraim Mosenkis, Matan Kenigsbuch, Badreya Sarsour, Yarden Elias, Liat Braun, Moria Rubin, Jany Gelfand Morgenshteyn, Noa Bergwerk, Noam Yosef, Sivan Peled, Coral Avigdor, Rahav Obercyger, Rachel Mann, Tomer Alper, Inbal Beka, Ori Shapira, Yoav Goldberg**
**Bar-Ilan University, Israel**

## Abstract

We introduce a large set of Hebrew lexicons pertaining to psychological aspects. These lexicons are useful for various psychology applications such as detecting emotional state, well being, relationship quality in conversation, identifying topics (e.g., family, work) and many more. We discuss the challenges in creating and validating lexicons in a new language, and highlight our methodological considerations in the data-driven lexicon construction process. Most of the lexicons are publicly available, which will facilitate further research on Hebrew clinical psychology text analysis. The lexicons were developed through data driven means, and verified by domain experts, clinical psychologists and psychology students, in a process of reconciliation with three judges. Development and verification relied on a dataset of a total of 872 psychotherapy session transcripts. We describe the construction process of each collection, the final resource and initial results of research studies employing this resource.

## 1 Introduction

A lexicon is the vocabulary of a domain of knowledge, and can be a valuable tool in the analysis of many psychological tasks. For example, in detecting clients' mental states, emotions and symptoms (Guntuku et al., 2017; Trotzek et al., 2018).

Lexicons are especially advantageous when data is scarce. Often in psychotherapy research, few samples are available in clinical trials, and confidentiality limits sharing of data. Scarcity of data is particularly challenging in less common languages like Hebrew. Recent data-hungry models are not practical in such cases where data is small, while other approaches, applying the use of lexicons, are more effective for predictive abilities. Moreover, lexicons can be shared across studies and serve as *clinical markers* (e.g., Al-Mosaiwi and Johnstone, 2018).

Additionally, through their simplicity, lexicons enable easy interpretation of results. They can be elaborate for indicating psychological states within text, e.g., in accordance to the frequency of specified terms within a passage (Tausczik and Pennebaker, 2010).

Lexicons are widely used in research and industry due to their proven effectiveness and ease of use. There are several psycho-linguistic lexicons, amongst them the Linguistic Inquiry and Word Count (LIWC) (Pennebaker et al., 2015), Vader-lexicon (Hutto and Gilbert, 2014), NRC-Sentiment-Emotion-Lexicon (Mohammad and Turney, 2013), MRC (Coltheart, 1981), and DLATK (Schwartz et al., 2017), however *no valid psycho-linguistic lexicon for Hebrew exists*. [1]

Several approaches are generally employed for developing lexicons. One prevalent method involves judging collected words with domain experts (Pennebaker et al., 2015) or with crowdsourcing (Tanana et al., 2016). There are also various methods for translating existing lexicons from other languages (e.g., triangulation-based, machine translation and then manual fine-tuning). However lexicon translation tends to be impractical since direct translation leads to incomplete or wrong results (Massó et al., 2013) . In particular, the Hebrew language poses many word-level translation obstacles due to its morphologically-rich form and ambiguous orthography (as outlined in Section 2).

We describe the development of a collection of Hebrew psychological lexicons that were created between the years 2018 and 2021. We utilize a base dataset of 872 psychotherapy sessions, described in Section 3, to either validate or extract words for the lexicons. The first set of lexicon collections (Section 4) are devised by domain experts, and verified using the base dataset. The word lists in the second set (Section 5) are fully automatically generated

[1] A large collection of Hebrew NLP resources are available at `https://github.com/NLPH/NLPH`.



*Proceedings of the Seventh Workshop on Computational Linguistics and Clinical Psychology*, pages 55–69
June 11, 2021. 

| | Expert Knowledge Based Lexicons | | | | Data-Driven Lists | | Expert Knowledge + Automatic Methods | |
|---|---|---|---|---|---|---|---|---|
| | | | | | Supervised | Unsupervised | Translation | Expansion |
| **Collection name** | **Valence (Positive-Negative)** | **Emotional Variety** | **Paralinguistics** | **Depressive Characteristics** | **Well-Being** | **Conversation Topics** | **Hebrew LIWC** | **Extended Emotional Variety** |
| **Number of lexicons/lists** | 2 | 42 | 11 | 14 | 2 | 200 | ∼40 out of 125 | 44 |
| **Total number of words** | 200 | 7313 | 154 | 194 | 40 | 4000 | under construction | under construction |
| **Coverage** | 2000 most frequent word types in dataset | 5000 most frequent word types in dataset | 31,067 tokens 1022 word types | several hundred most important word types | 139 non-clinical sessions 38 clinical sessions | the whole dataset ∼5 million tokens | - | - |
| **Verified by at least three domain experts** | yes | yes | yes | yes | - | - | yes | under construction |
| **Initial research use case** | yes | work in progress | yes | yes | - | yes data-dependent | - | - |
| **Freely available** | yes | yes | yes | yes | yes | yes | internal use only | will be released |

Table 1: A summary of the presented lexicons and word lists.

based on the dataset, and mainly serve for textual analysis of psychotherapy sessions. Section 6 combines domain experts and automatic methods for the preparation of lexicons. For each of the lexicon collections and methods, we provide a use-case in the clinical psychotherapy domain, illustrating their usefulness and effectiveness. See Table 1 for a description and statistics on the lexicons.

While many of the lexicon types described are common in the psychology domain, we additionally introduce two new lexicon types. The first is an *emotional-variety* lexicon type with *complementary-emotions*, i.e., each emotion lexicon has a complementing-emotion lexicon, valuable for reducing noise when analyzing emotion. The second type is for *paralinguistic* categorization, which enables the classification of different nonverbal vocal behavioral events within psychotherapy sessions.

Most of the lexicons freely available,[2] which will facilitate further research on Hebrew clinical psychology text analysis. The methods described may also aid in the establishment of additional lexicons in Hebrew and in other languages.

## 2 Challenges with Lexicon Translation

While methods for translating existing lexicons from other languages have been exploited before, lexicon translation yields wrong categorization of words (Massó et al., 2013). This is particularly the case when involving morphologically rich languages, and is also due to word ambiguity and cultural influence on languages.

In Hebrew, like in other Semitic (e.g., Arabic) and Indo-European languages (e.g., Spanish, Dutch), there are inflections and verb conjugations that have no direct conversion in English. Van Wissen and Boot (2017) address the problem by converting each word in a lexicon to its lemma (i.e., canonical form) and then using an existing list to expand to the various linguistic conjugations. In Hebrew it is possible to retrieve all the different inflections and verb conjugations for many words using specialized linguistic lexicons, such as the MILA lexicon (Itai and Wintner, 2008).[3] Even so, it is not always the case that all forms of a word should be included in the same lexicon. For example, in the *emotion variety* lexicon collection (Section 4.2), the word רגוע 'ragua' (relaxed) appears in the *not-nervous* lexicon and תרגיע 'targia' (calm down) appears in the *not-guilty* lexicon, sharing the same root form but having different semantic emotional classification.

In addition, there may be situations of ambiguity in which words with completely different meanings are mapped to the same lemma, e.g., the words (1) חימה 'chema' (anger) and חמה 'chama' (sun) have the same orthographic lemma חמה; (2) עדשות 'adashot' (contact lenses) and עדשים 'adashim' (lentils) have the same orthographic lemma עדשה 'adasha', thus adding noise to the directly-translated lexicon.

Furthermore, when expanding a lexicon around a word, ignoring diacritics often yields ambiguous forms. For example, while the word אחלה 'achla' (cool) is in the *positive emotion* lexicon (Section 4.1), without diacritics the optional base forms are איחל 'ichel' (wish), חילה 'chila' (to make ill), אחלה 'achla' (cool) and חלה 'chala' (to become ill), having different emotional polarity. Then, each of these words is also expanded with all their inflections, e.g., חליתי 'chaliti' (I became ill), adding up to hundreds of words to the wrong lexicon.

[2]https://github.com/natalieShapira/HebrewPsychologicalLexicons. As LIWC is commercial, we cannot publicly release the translated lexicons described in Section 6.1

[3]We use the BGU-version of the lexicon, which is bundled with the YAP Hebrew parser (More and Tsarfaty, 2016) as the file bgulex.utf8.hr.

Another problem is that there are lexicon types whose translation is not straightforward. For example, the *I words* lexicon in LIWC is a small set of 12 distinct words (e.g., *I, me, mine*) (Tausczik and Pennebaker, 2010) and can be used to count the frequency of all the occurrences of first-person mentions in a given text passage in English. However, Hebrew's morphological system preclude such word-counting method for seeking "I words" in the text passage, as the first-person status is often realized morphologically, and may appear on many word forms. Hebrew words follow a complex morphological structure, with both derivational and inflectional elements, that can encode gender, number, tense, person, possessive and noun-compounding. For example, **אהבתי** 'ahavti' (I loved), **אוהב** 'ohav' (I will love), **אוהבת** 'ohevet' (I-feminine love/she loves), **אהובי** 'ahuvi' (my love), Therefore, preprocessing of syntactic and morphological parsing is a critical phase for extracting the relevant details (e.g., the first person singular counts).

Lastly, the ambiguous interpretation in different languages makes out-of-context translation impossible. For example, the word 'dear' will be translated in Hebrew to the word יקר 'yakar' , but יקר 'yakar' also means 'expensive'. While 'dear' in LIWC is a word with positive polarity, 'expensive' is not. We cannot assume that if a resource is valid in language A, then its translation into language B will necessarily give us a valid resource in language B.

Relatedly, language is strongly culturally influenced, and a word may be categorized differently across languages and cultural context in terms of human psychology, especially around emotion or sentiment (Wierzbicka, 1985). For example, the color green, will refer to jealousy and envy in some cultures: "green-eyed monster" was first used by William Shakespeare about jealousy. There are proverbs in Hebrew that associate envy to the green color: "green with envy". In addition, in Hebrew ירוק ('yarok' green) can be used as a mockery of a person with no experience in his or her field, like an unripe fruit, especially used in the military context—a recruit. In contrast, green serves as a religious/sacred symbol in Islam as Muhammad's favorite color. (See also cultural differences in a study that examined the relationship between colors and emotions by Hupka et al., 1997.)

## 3 Base Dataset Description

All our lexicons rely on a dataset[4] of a total of 872 psychotherapy session transcripts from 74 different client-therapist dyads (pairs) consisting of a total of about 5 million tokens—100 thousand word types (unique words). All sessions are labeled with psychological analysis information that assists in generating a lexicon and/or verifying one. We infer relevant session-level labels from questionnaires filled by the participants at each session: (1) clients self-reported their well-being, measured using the ORS questionnaire (Miller et al., 2003), which is considered to be an indicator for progress in treatment; (2) therapists and clients reported on interpersonal relational events that occurred during a session, corresponding to tensions or breakdowns in their collaborative relationship (alliance ruptures), measured by the PSQ questionnaire (Muran et al., 2004); (3) therapists and clients reported emotional states measured by the POMS questionnaire (McNair, 1992).

## 4 Lexicons Based on Expert Knowledge

The approach employed for creating the following lexicons is inspired by that of Pennebaker et al. (2015), specifically via a three-judge (domain experts) reconciliation procedure for admitting words into a lexicon.

### 4.1 Valence (Positive and Negative)

A fundamental aspect to consider in psychological analysis is detecting positive and negative emotion. With regards to clinical text analysis, words identified as emotionally positive or negative have been shown to correlate to clinical conditions (Morales et al., 2017).

To create the positive and negative emotion lexicons, we collected the 2000 most frequent words (including stop words) from our base data as candidates. We found that these 2000 most frequent words cover 86% of all tokens in all transcripts. Three judges independently rated whether each word should be categorized as generally having a positive and/or negative emotion, after which a reconciliation process was conducted to resolve conflicting decisions. Initial Fleiss' Kappa (Fleiss, 1971) for interrater agreement was 0.54 (moderate

[4]See the appendix for more details about the participants, demographics information, treatment, transcriptions, questionnaires and ethical concerns.

agreement) and the final was 0.95, indicating almost perfect agreement (Landis and Koch, 1977). The main changes following the reconciliation process was (1) the addition of words with low polarity/confidence e.g., the word אבל 'aval' (but) was added in the second phase to the negative list; (2) the correction of errors and mistakes e.g., the word אוקי 'okay' (OK), was included in the positive list while the word אוקיי which is the same meaning 'okay' (OK), was not included; (3) better agreement on 'mixed emotion words' that evoked both positive *and* negative emotions (8.7% e.g., mother, feeling, power) compared to words evoking any emotion (73% e.g., also, like, type). There were no words with hard disagreement, i.e., where at least one of the judges marked the word as positive only and another judge marked it as negative only. In total, the lexicons contain 200 positive and negative emotion word types. To avoid ambiguities and encourage uniformity between future studies, we released only one version of lexicons (majority of two judges excluding mixed emotion words).[5]

Based on the two lexicons, we calculated the number of positive and negative emotion words within each session transcript (an hour of conversation) in the dataset. On average, there were 185 positive emotion words and 327 negative emotion words per session. 15% of the all tokens in the transcripts were emotion words.

**Usage** In one study conducted in our lab, we found correlations between a client's and therapist's positive/negative emotion words and client's and therapist's positive/negative emotions as reported in the POMS questionnaire. In another study, that uses our positive-negative emotion lexicons, Shapira et al. (2020) examined the relationship between the number of emotion words spoken in a session and the client's self-reported questionnaire regarding her well-being. The findings are consistent with the literature and in line with theoretical views highlighting the role of positive emotions and negative emotions and the association to well-being (e.g., Blatt (1995); Shahar et al. (2020); Morales et al. (2017)). Finally, Juravski (2020) also shows a correlation between the use of positive and negative emotion lexicons to predicted emojis by a pretrained model based on Twitter data,[6] contributing to the mutual validation of the tools. The above studies show that positive and negative emotion lexicons can be leveraged for automatic detection of emotional state and well-being within texts.

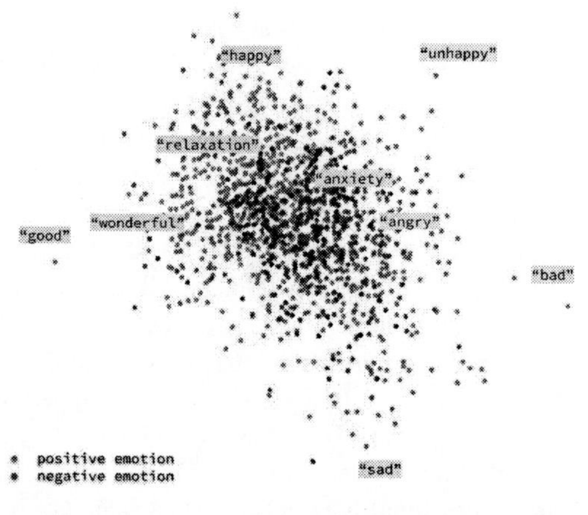


Figure 1: 2D-Projection of emotion word embeddings.

## 4.2 Emotional Variety

A great and diverse variety of emotional states exist, and in this section we describe the process of developing lexicons that relate to this variety. Our motive for developing these emotional lexicons stems from a basic notion in psychotherapy research: the ability to be in touch with emotional experiences, to portray them in words and to give them meaning, as a result of treatment, has been found to effectively predict improvement in mental well-being. This is consistent across various therapeutic models and types of mental disorders (Greenberg et al., 2012).

The development of the emotion lexicon was carried out in several stages. We first compiled a list of emotions on the basis of the POMS emotion questionnaire (see Appendix A.2.2), Robert plutchik's "wheel of emotions" (Plutchik, 2000) and those described by Ong et al. (2018). The list includes: *enthusiastic, amused, proud, interested, calm, sad, ashamed, guilty, hostile, nervous, anger, contentment, anxiety, vigor, joy, disgust, surprise, trust, anticipation, confusion, fatigue*.

For each emotion we created another category that is the complement of that emotion (e.g. *not_sad* as the complement of *sad*), hence resulting in a total of 42 categories.

The main purpose for categorizing complementing emotions is to enable more precise word categorization when requiring emotional analysis of

[5] Other versions (e.g. consensual words, words with low polarity, mixed emotions words) can be obtained upon request.

[6] https://hub.docker.com/r/danieljuravski/hemoji

text. An additional important motive is the long-term thought for allowing automatic expansion of these lexicon seeds (Section 6.2) using semantic-based methods.[7] Having a complementing-emotion word list can assist in the expansion process of the corresponding emotion lexicon by providing indicators for what might *not* categorize to that emotion. Figure 1 shows the projection of a list of positive and negative (complementing) emotion word embeddings.[8] While most words indeed separate to two different clusters, the clusters intersect considerably. This illustrates that it is not enough to assume that words will semantically cluster together by their emotional category. Having an emotion's complementary lexicon can be advantageous for finding new words for that emotion.[9] To the best pf our knowledge, we are the first to propose complementary-emotion lexicons.

In the second stage of the lexicons' development, 19 advanced undergraduate psychology students were given the list of emotional categories and were asked to suggest at least five appropriate words for each. Words could be produced either associatively or through active search (e.g., by using an online Hebrew thesaurus [10]). We additionally conducted a similar classification annotation procedure as described in Section 4.1, whereas in this case the 5000 most frequent words, covering 90% of all tokens in all transcripts, were tagged with one of the 28 emotion categories (not every word evoked an emotion). These were merged with the freely-suggested words from above.

The final collection of emotional variety lexicon seeds consists of a total of 7313 emotion words. The percentages of judges' agreement for the rating phase ranged from 98% to 100% agreement. This lexicon collection is available as a ready-to-use version. An expanded version of this lexicon is currently in the works (with the algorithm mentioned above, in Appendix A.3).

[7]Such as with the *word-similarity* package, pretrained on Hebrew Twitter word embeddings. `https://github.com/Ronshm/hebrew-word2vec`

[8]Using the *Tensorflow Embedding Projector* tool. `https://projector.tensorflow.org`

[9]See Appendix A.3 for a potential algorithm that could be used to expand emotion lexicons, using the complementing lexicon.

[10]such as `https://synonyms.reverso.net/synonym/he/`

**Therapist:** Shall I get you a glass of water? *<In a whisper>*

**Client:** *<Sounds of silent crying. Pulling the nose>* yes *<Like clearing throat>*, yes.

Figure 2: An example of paralinguistic event annotations (in *italics*) within the transcription, described in free text by the transcriber.

### 4.3 Paralinguistics Events

Paralinguistic events refer to non-verbal vocal elements of interpersonal language communication that accompany the verbal message. This component of communication may change meaning, create nuance or convey emotion, through the use of various techniques such as pitch and volume, weight, intonation, silences, laughter, etc. (Valstar et al., 2013), and may be expressed consciously or unconsciously (Harris and Rubinstein, 1975) by participants. Sometimes these elements are considered aphonemic, i.e., they cannot even be spelled out (Trager, 1961). All of these phenomena are inherent in the speech sequence, and are often processed as words in automatic speech processing – a *high tone* in speech as an indication of anxiety or a *breathy voice* as an indication of attractiveness – are already processed into the voice message.

Paralinguistic elements are of great importance in the therapeutic context. To date, much credible evidence has accumulated in research that confirms that characteristics of voice significantly influence the formation and development of the therapeutic relationship (Sikorski, 2012). In the clinical setting, paralinguistic communication is of fundamental importance to therapist-client dynamics. For example, through unconscious perception of change in the client's paralinguistic events, the therapist (while noticing the overt meaning conveyed through semantic channels) can adjust his or her own paralinguistics, and with a good understanding of the client's inner state, he or she can encourage expansion of the client's awareness (Rocco et al., 2013). Moreover, a strong association between vocal characteristics and certain psychopathological states has been documented, e.g., depression accompanied by slow, long, and intertwined speech in breaks (Ellgring and Scherer, 1996).

The paralinguistic events were labeled (as comments) in our transcripts dataset by the transcribers as free text (see examples in Figure 2). A total of 31,067 tokens occur in the transcriber comments, of which 2147 are unique and 1022 appear at least twice. The most frequent tokens are: "laughing"

**LOW_TONE** = {quiet} שקט, {mumble} ממלמל, {with mumble} במלמול, {whisper} בלחש, ...
**HIGH_TONE** = {loud} גבוה, {shouting} צועק, {loud} חזק, {loud} רם, {roaring} שאגה, ...
**IMITATIONS_TONE** = {imitation} חיקוי, {theatrical} תיאטרלית, {fake} מזויף, {childish} ילדותי, ...
**CRYING** = {crying} בוכה, {choking} חנוק, {shivering} רועד, {sobbing} מתייפחת, {tears} מדמעות, ...
**SMIRK** = {smirk} מגחכת, {smirk} גיחוך, {smirk} מגחך, {smirk} בגיחוך, {smirk} מגחכות, ...
**TUT-TUT** = {tut-tut} צקצוק, {tut-tut} מצקצק, {tut-tut} מצקצקת, {tut-tut} צקצקו, ...
**SIGH** = {sigh} נאנחת, {sigh} נאנח, {sigh} אנחה, {sigh} באנחה, ...
**BODY** = {coughing} משתעלת, {yawning} מפהק, {breathing} נושמת, {sipping} לוגם, ...
**HUMMING** = {nodding} מהנהנת, {humming} מהמהם, {aha} אהא, {ahm} אהמ, ...
**JOY** = {laughs} צוחקת, {amused} משועשע, {with humor} בהומור, {giggling} בצחקוק, ...
**SARCASM** = {cynically} בציניות, {cynically} ציני

Figure 3: Paralinguistic categories (lexicons) and examples of words within them.

(feminine singular) at a frequency of 22%, "laughing" (masculine singular) at 5.3%, "tut-tut" (3.5%), "sigh" (2.5%), "laugh" (feminine plural; 2.3%), "giggle" (feminine singular; 1.8%), "of" (1.2%), "tongue" (1.2%), "cry" (referred to in masculine and feminine alike; 1.2%), "the therapist" (1%), "chuckle" (1%), "coughing" (1%), etc.

An NLP researcher, a clinical psychologist and two interning therapists went over the labels and their frequencies together and characterized 11 categories of paralinguistic events that are meaningful in psychological treatment: *low tone*, *high tone*, *imitation tone*, *crying*, *smirk*, *tut-tut*, *sigh*, *body-related*, *humming*, *joy*, and *sarcasm*. Then, each of the labels was classified into these categories (classification was trivial with 100% agreement, see Figure 3).

An initial study we conducted found strong correlations between paralinguistic events to postive and negative emotion words within psychotherapy sessions, e.g., strong positive correlation (r=0.823, p <0.001) between *joy* paralinguistic events and positive emotion words within the therapist's text.

### 4.4 Depressive Characteristics

Depression is one of the most common mental disorders. In 2017, it was estimated that more than 300 million people worldwide (4.4% of the global population) were suffering of depression (WHO et al., 2017). Many studies have examined the relationship between depression and language (Trotzek et al., 2018; Yates et al., 2017; ODea et al., 2018; Ramirez-Esparza et al., 2008; Rude et al., 2004; Holtzman et al., 2017; Al-Mosaiwi and Johnstone, 2018; Ophir et al., 2020; Fineberg et al., 2016; Tackman et al., 2019; Guntuku et al., 2017; Morales et al., 2017; Tausczik and Pennebaker, 2010).

Referring to textual characteristics found in the above-mentioned literature, an NLP researcher and

| |
|---|
| **Self-reference**: first person singular, I words, changes belong to personal pronouns, possessive and pronouns based on POS tagging, Many third person pronouns, Unrelated personal pronouns ("it") |
| **Emotions**: Negative Emotions, Positive Emotions, Negative Content, Sadness, Anger, Anxiety, Negative attitude towards others compared to non-depressed with positive |
| **Absoluteness spectrum**: absolute, extreme, oath, hesitation, lack of fluency, tentative |
| **Time and space**: past, present, future, month of the writings, location |
| **Text length**: number of words, number of letters |
| **Direct expression related to depression**: "my depression", "my anxiety", "my therapist", "I was diagnosed with depression", Antidepressants e.g., "Zoloft", "Paxil" |
| **Data-driven top phrases**: "I went to", "my whole", "sometimes I", "I'm so sorry", "to scare you", "to have it", "my son was", "it wasn't" |
| **Lyrical and abstract writing** (life, time, values and religion) compared to non-depressed who are characterized by concrete prose writing (days, events, places, behaviors) and less reference to time |
| **Miscellaneous**: death related words, perceptual processes, article, contradiction (said, could have), attention to ingestion, curses, conditions ("if"), negation, interrupted and uncommitted, questions and question marks, necessity ("need") words compared to fewer words of desire ("love", "want"), swirls, not concrete (lots of words but little variety, short sentences, three points, fillers words as "like", unknown "don't know", shame, disappointment, repetitive, passive/active, numbers, helplessness, avoiding, repression, generalization (general talk and not about specific details), reputation, physical health, financial status, respect esteem, self-confidence |

Figure 4: Linguistic characteristics of depressive texts, grouped by characteristic categories. We created lexicons for 14 of these characteristics.

an interning therapist examined the sessions in the base dataset, and prepared a list of categories characterising depressive behavior, each category containing a list of characteristics. See Figure 4 for these characteristics.

Then, characteristic words were compiled in the following manner. A Random Forest classifier (Liaw and Wiener, 2002) was trained on all the clients' texts from the base data sessions, to predict the sadness-level label of a given text, as found in the POMS questionnaire of the corresponding session. A text was input to the classifier as a bag-of-words vector. Once the training completed, a few hundred of the most important features (words) were extracted from the trained classifier. These words were then categorized manually into 14 of the depressive characteristics, forming 14 new lexicons. One of these lexicons, for example, is called *tentativeness* (see under "Absoluteness spectrum" category in Figure 4), and consists of words such as כנראה (probably), אולי (maybe), and יתכן (perhaps). These word categorizations were then approved by two additional interning therapists.

## 5 Data-driven Word Lists

We next describe data-driven methods, applied on our base dataset, that extract lists of words for purposes of psychotherapetic analysis of session transcripts.

### 5.1 Well-Being

A potentially useful feature for automatically identifying outcome, i.e., improvement over psychotherapy treatment, is the client's well-being throughout

NON_CLINICAL_CONDITION = {punctuation} <PUNC>, {you} את, **{she} היא**, **{he} הוא**, {knows} יודע, {xxx} XXX, **{him} לו**, **{her} לה**, {really} באמת, {with} עם, {I said} אמרתי, {ah} אה, {and} ו, **{her} אותה**, {also} גם, **{his} שלו**, {on} על, **{and she} והיא**, {always} תמיד, **{she was} הייתה**

CLINICAL_CONDITION = {but} אבל, {know} יודע, {then} אז, **{I} אני**, {such} כזה, {as} כאילו, **{that I} שאני**, {something} משהו, {it} זה, {yes} כן, {this} הזה, {say} אגיד, {which} איזה, {number} <NUM>, **{to me} לי**, **{I was} הייתי**, {em} אמ, {you} אתה, {can} יכול, {already} כבר

Figure 5: Data-driven lists of words characterizing clients in *non-clinical* condition versus *clinical* condition.

the treatment. A collection of lexicons correlative to level of well-being (ranging from clinical, worst, to non-clinical condition, best) may assist in recognizing such patterns in treatment.

To extract data-driven lists of words that characterize client well-being, we followed the Marker Approach (Mergenthaler, 1996; Buchheim and Mergenthaler, 2000). First, the client texts from the base data sessions with the worst (0-8, clinical condition) and best (32-40, non-clinical condition) ORS questionnaire well-being scores were extracted. A total of 38 clinical and 139 non-clinical sessions were found in the data. Next, vocabularies were identified (Fertuck et al., 2012) for each of the two "worst" and "best" corpora in reference to each other. That is, words that are significantly more frequent in one text versus the other are marked. The top 20 words from each group was included in the final lexicons (see Figure 5). This set of lexicons did not go through an evaluation process yet.

Note that the emerging *clinical condition* lexicon includes words of first-person singular (FPS) form, which is consistent with the literature that finds an association between increased verbal use of the first-person and higher levels of distress (Tackman et al., 2019; Guntuku et al., 2017; Morales et al., 2017; Tausczik and Pennebaker, 2010). Moreover, this is in line with the theoretical literature that highlights the dominant role of self-focus and self-criticism in maintaining and intensifying individuals' negative affect, which in turn leads to increased symptoms of distress (Beck, 1967; Blatt, 1995; Pyszczynski and Greenberg, 1987; Shahar et al., 2020). Meanwhile, the *non-clinical condition* lexicon includes words of third-person singular (TPS), which might indicate a correlation to a healthier condition of well-being and speaking about others.

## 5.2 Conversation Topics in Psychotherapy

Therapists are driven to find methods for improving the quality of psychotherapy sessions, for example, by understanding whether the themes about which they converse with their clients influence the result-

| Topic 187 | Topic 58 | Topic 108 | Topic 30 | Topic 10 | Topic 94 | Topic 19 | Topic 177 |
|---|---|---|---|---|---|---|---|
| משפחה | עובד | בוקר | כסף | ללמוד | חרדה | מים | כלים |
| Family | Employee | Morning | Money | Learn | Anxiety | Water | Dishes |
| אמא | עבודה | לילה | לשלם | לימודים | שליטה | קפה | כביסה |
| Mother | Working | Night | Pay | Studies | Control | Coffee | Laundry |
| דודה | משרד | לישון | חשבון | תואר | פחד | כוס | מטבח |
| Aunt | Office | Sleep | Invoice | Degree | Fear | Glass | Kitchen |
| ילדים | אנשים | לקום | חודש | קורס | לשחרר | לשתות | מים |
| Children | People | Getting up | Month | Course | Release | Drink | Water |
| אחותי | מנהל | יום | בנק | אוניברסיטה | מובן | לקפוץ | מקלחת |
| Sister | Director | Day | Bank | University | Understandable | Jump | Shower |
| דודים | עסק | מיטה | מחיר | מבחן | זמן | יין | לשטוף |
| Uncles | Business | Bed | Price | Test | Time | Wine | Wash |
| אחים | בוס | שעה | דירה | תחום | עצבים | בקבוק | כיור |
| Brothers | Boss | Time | Apartment | Domain | Nerves | Bottle | Sink |
| סבתא | לקוחות | עייפה | עולה | מקצוע | גוף | בירה | מדיח |
| Grandmother | Customers | Tired | Costs | Profession | Body | Beer | Dishwasher |
| הורים | תחום | ללכת | סכום | שנה | התקף | שתייה | בגדים |
| Parents | Domain | Go | Amount | Year | Attack | Drink | Clothing |
| נכדים | שיווק | התעוררתי | משכורת | מתמטיקה | סטרס | קולה | מכונת כביסה |
| Grandchildren | Marketing | Woke | Salary | Math | Stress | Coca-Cola | Washing machine |

Figure 6: A sample of topics.

ing outcome of the treatment. Hence, we wish to explore the topics within the sessions, and examine what words are characteristic of those topics.

We applied Latent Dirichlet Allocation (LDA; Blei et al. (2003)) on the transcripts data to detect clusters of words, occurring similarly within the psychotherapy sessions. This resulted in a set of 200 topics and their probability of appearing in the data (signifying how much weight they have in the psychotherapy data), with each topic containing a list of 20 words. Figure 6 shows a few examples of topics and their words, as generated from the data.

We find, for example, that topics 72, 15, 152, and 171 describe "celebration", "leisure experience", "enjoyment", and "choice", which intuitively seem to be related to positive experiences and to high functioning. On the other hand, topics such as 81, 199, 166, and 61 seem to be about "loneliness", "suffering", "physical difficulties", and "anger", which intuitively seem related to negative experiences and to low functioning.

We explored which topics (clusters) best identified clients' well-being and alliance ruptures (see Appendices A.2.1, A.2.4) and whether changes in these topics were associated with changes in outcome. A sparse multinomial logistic regression model was run to predict which topics best identified clients' functioning levels, and the occurrence of alliance ruptures in the sessions. Additionally, multi-level growth models were used to explore the associations between changes in topics and changes in outcome. The model identified the ruptures and outcome labels above chance (65%-75% accuracy). Change trajectories in topics were associated with change trajectories in outcome. The first four topics best correlated to a negative outcome. The results suggest that topic models can exploit rich linguistic data within sessions to identify psychotherapy process and outcomes. For the detailed study see Atzil-Slonim et al. (2021).

It is important to note that the purpose of this section is to show a method for topic modeling, and not to produce topical-word lexicons for general use. The method should be reproduced on the data for which the analysis is required.

## 6 Lexicons Based on Expert Knowledge and Automatic Methods

This section describes lexicons that are *automatically* converted or expanded from existing *expert-based* lexicons.

### 6.1 Hebrew Translation for LIWC

Linguistic Inquiry and Word Count (LIWC) (Pennebaker et al., 2015) is the most famous lexicon collection in the field of psychological text analysis (tens of thousands of citations). LIWC contains 120 lexicons and has been incorporated in many research studies. A Hebrew translation of some of the LIWC lexicons, when possible, would contribute to aligned cross-lingual research. As LIWC is commercial, we cannot publicly release the translated lexicons described here, however the translation procedure we follow may be useful for other researchers seeking to translate certain lexicons.

Some of the categories are difficult or even impossible to translate into Hebrew. For example, the *articles* lexicon (e.g., "a", "an", "the", etc.) has no Hebrew equivalent,[11] nor does the *I words* lexicon (as explained in Section 2).

For lexicons that an equivalent can be produced (e.g. *family*, *work*, etc.), we suggest the translation process as follows: an LIWC lexicon contains a list of *prefixes* of words. In the first step, expand each prefix to all of its expanded forms using an English dictionary[12] (e.g., abandon* to: abandon, abandoned, abandoning, abandonment etc.). This provides a list of concrete words under each category (lexicon) instead of prefixes. In the second step, generate a list of optional translated words by translating each word via the word2word package[13] (Choe et al., 2019). This package provides 20 candidate translations for each word, hence each Hebrew-translated lexicon is 20 times the size of the respective English-LIWC lexicon. A total of about 150,000 words emerged for the translated lexicons. This number of words can be verified in about 1,000 hours by a three-judge verification process (estimating 500 words per judge per hour), which we are in the process of doing.

### 6.2 Expansions

As future work we plan to expand expert-knowledge-based lexicons, such as the *emotional variety* lexicon (Section 4.2), using automated methods. For example, we can automatically expand words on their inflection types, or find semantically similar words with, e.g., embedding-based expansions (for initial algorithm see Appendix A.3). Needless to say, the products of these methods will require expert validation procedures.

## 7 Limitations

The lexicons presented are based on a unique dataset of psychotherapy session transcripts. The language used by clients and therapists in these sessions do not necessarily reflect the language naturally occurring in other settings. Additionally, the statistical demographics of the participants in the utilized sessions are not fully balanced in terms of gender, age, education and relationship status (see Appendix A.1.1 for details). Again, this may influence the overall language observed, and in turn, the computations performed throughout our work in generating and verifying the lexicons.

## 8 Conclusion

We present a collection of novel Hebrew lexicons, based on psychological data and domain expert knowledge. We describe a variety of lexicon development methods: expert-knowledge-based, data-driven using labeled data and unsupervised learning. We address levels of reliability—agreement between three judges (expert knowledge) versus automatic methods that are vulnerable to noise. We describe the importance of the lexicons for psychology research, as well as initial uses cases with results.

The lexicons are released for the benefit of the community, contributing to psychological text-analysis research in Hebrew and cross-lingual research in general. Furthermore, we hope that the methods described will inspire the creation of additional lexicons in Hebrew and in other languages.

[11]The indefinite articles do not exist, while the definite article *the* is realized morphologically as a possibly ambiguous prefix which is attached to the token.

[12]E.g., the dictionary in SpaCy (Honnibal and Montani, 2017) or NLTK (Loper and Bird, 2002).

[13]Bilingual lexicons for 3,564 language pairs `https://github.com/kakaobrain/word2word`

## Acknowledgements

We thank the anonymous reviewers for their careful reading of our manuscript and their insightful comments and suggestions. This project has received funding from the Israel Science Foundation (grants 1348/15 and 1278/16); and from the European Research Council (ERC) under the European Union's Horizon 2020 research and innovation program, grant agreement No. 802774 (iEXTRACT).

## References

Mohammed Al-Mosaiwi and Tom Johnstone. 2018. In an absolute state: Elevated use of absolutist words is a marker specific to anxiety, depression, and suicidal ideation. *Clinical Psychological Science*, 6(4):529–542.

Aviad Albert, Brian MacWhinney, Bracha Nir, and Shuly Wintner. 2013. The hebrew childes corpus: transcription and morphological analysis. *Language resources and evaluation*, 47(4):973–1005.

Dana Atzil-Slonim, Daniel Juravski, Eran Bar-Kalifa, Eva Gilboa-Schechtman, Rivka Tuval-Mashiach, Natalie Shapira, and Yoav Goldberg. 2021. Using topic models to identify clients' functioning levels and alliance ruptures in psychotherapy. *Psychotherapy*.

Aaron T Beck. 1967. *Depression: Clinical, experimental, and theoretical aspects*. University of Pennsylvania Press.

Matthew D Blagys and Mark J Hilsenroth. 2000. Distinctive features of short-term psychodynamic-interpersonal psychotherapy: A review of the comparative psychotherapy process literature. *Clinical psychology: Science and practice*, 7(2):167–188.

Sidney J Blatt. 1995. The destructiveness of perfectionism: Implications for the treatment of depression. *American psychologist*, 50(12):1003.

David M Blei, Andrew Y Ng, and Michael I Jordan. 2003. Latent dirichlet allocation. *the Journal of machine Learning research*, 3:993–1022.

Anna Buchheim and Erhard Mergenthaler. 2000. The relationship among attachment representation, emotion-abstraction patterns, and narrative style: A computer-based text analysis of the adult attachment interview. *Psychotherapy Research*, 10(4):390–407.

Yo Joong Choe, Kyubyong Park, and Dongwoo Kim. 2019. word2word: A collection of bilingual lexicons for 3,564 language pairs. *arXiv preprint arXiv:1911.12019*.

Max Coltheart. 1981. The mrc psycholinguistic database. *The Quarterly Journal of Experimental Psychology Section A*, 33(4):497–505.

James A Cranford, Patrick E Shrout, Masumi Iida, Eshkol Rafaeli, Tiffany Yip, and Niall Bolger. 2006. A procedure for evaluating sensitivity to within-person change: Can mood measures in diary studies detect change reliably? *Personality and Social Psychology Bulletin*, 32(7):917–929.

Heiner Ellgring and Klaus R Scherer. 1996. Vocal indicators of mood change in depression. *Journal of Nonverbal Behavior*, 20(2):83–110.

Fredrik Falkenström, Robert L Hatcher, Tommy Skjulsvik, Mattias Holmqvist Larsson, and Rolf Holmqvist. 2015. Development and validation of a 6-item working alliance questionnaire for repeated administrations during psychotherapy. *Psychological Assessment*, 27(1):169.

Eric A Fertuck, Erhard Mergenthaler, Mary Target, Kenneth N Levy, and John F Clarkin. 2012. Development and criterion validity of a computerized text analysis measure of reflective functioning. *Psychotherapy Research*, 22(3):298–305.

SK Fineberg, J Leavitt, S Deutsch-Link, S Dealy, CD Landry, K Pirruccio, S Shea, S Trent, G Cecchi, and PR Corlett. 2016. Self-reference in psychosis and depression: a language marker of illness. *Psychological medicine*, 46(12):2605.

Joseph L Fleiss. 1971. Measuring nominal scale agreement among many raters. *Psychological bulletin*, 76(5):378.

Peter L Greenberg, Heinz Tuechler, Julie Schanz, Guillermo Sanz, Guillermo Garcia-Manero, Francesc Solé, John M Bennett, David Bowen, Pierre Fenaux, Francois Dreyfus, et al. 2012. Revised international prognostic scoring system for myelodysplastic syndromes. *Blood*, 120(12):2454–2465.

Edward Guadagnoli and Vincent Mor. 1989. Measuring cancer patients' affect: Revision and psychometric properties of the profile of mood states (poms). *Psychological Assessment: A Journal of Consulting and Clinical Psychology*, 1(2):150.

Sharath Chandra Guntuku, David B Yaden, Margaret L Kern, Lyle H Ungar, and Johannes C Eichstaedt. 2017. Detecting depression and mental illness on social media: an integrative review. *Current Opinion in Behavioral Sciences*, 18:43–49.

Richard M Harris and David Rubinstein. 1975. Paralanguage, communication, and cognition. *Organization of behavior in face-to-face interaction*, pages 251–276.

Nicholas S Holtzman et al. 2017. A meta-analysis of correlations between depression and first person singular pronoun use. *Journal of Research in Personality*, 68:63–68.

Matthew Honnibal and Ines Montani. 2017. spacy 2: Natural language understanding with bloom embeddings, convolutional neural networks and incremental parsing. *To appear*, 7(1):411–420.

Adam O Horvath and Leslie S Greenberg. 1989. Development and validation of the working alliance inventory. *Journal of counseling psychology*, 36(2):223.

Ralph B Hupka, Zbigniew Zaleski, Jurgen Otto, Lucy Reidl, and Nadia V Tarabrina. 1997. The colors of anger, envy, fear, and jealousy: A cross-cultural study. *Journal of cross-cultural psychology*, 28(2):156–171.

Clayton J Hutto and Eric Gilbert. 2014. Vader: A parsimonious rule-based model for sentiment analysis of social media text. In *Eighth international AAAI conference on weblogs and social media*.

Alon Itai and Shuly Wintner. 2008. Language resources for Hebrew. *Language Resources and Evaluation*, 42(1):75–98.

Daniel Juravski. 2020. Natural language processing methods for analysing textual psychotherapy data.

J Richard Landis and Gary G Koch. 1977. The measurement of observer agreement for categorical data. *biometrics*, pages 159–174.

Andy Liaw and Matthew Wiener. 2002. Classification and regression by randomforest. *R News*, 2(3):18–22.

Edward Loper and Steven Bird. 2002. Nltk: The natural language toolkit. *arXiv preprint cs/0205028*.

Guillem Massó, Patrik Lambert, Carlos Rodríguez Penagos, and Roser Saurí. 2013. Generating new liwc dictionaries by triangulation. In *Asia Information Retrieval Symposium*, pages 263–271. Springer.

Douglas M McNair. 1992. Profile of mood states. *Educational and Industrial Testing Service*.

Erhard Mergenthaler. 1996. Emotion–abstraction patterns in verbatim protocols: A new way of describing psychotherapeutic processes. *Journal of consulting and clinical psychology*, 64(6):1306.

Erhard Mergenthaler and Charles Stinson. 1992. Psychotherapy transcription standards. *Psychotherapy research*, 2(2):125–142.

Scott D Miller, BL Duncan, J Brown, JA Sparks, and DA Claud. 2003. The outcome rating scale: A preliminary study of the reliability, validity, and feasibility of a brief visual analog measure. *Journal of brief Therapy*, 2(2):91–100.

Saif M Mohammad and Peter D Turney. 2013. Crowdsourcing a word–emotion association lexicon. *Computational Intelligence*, 29(3):436–465.

Michelle Morales, Stefan Scherer, and Rivka Levitan. 2017. A cross-modal review of indicators for depression detection systems. In *Proceedings of the Fourth Workshop on Computational Linguistics and Clinical Psychology-From Linguistic Signal to Clinical Reality*, pages 1–12.

Amir More and Reut Tsarfaty. 2016. Data-driven morphological analysis and disambiguation for morphologically rich languages and universal dependencies. In *Proceedings of COLING 2016*.

J Christopher Muran, Jeremy D Safran, Bernard S Gorman, Lisa Wallner Samstag, Catherine Eubanks-Carter, and Arnold Winston. 2009. The relationship of early alliance ruptures and their resolution to process and outcome in three time-limited psychotherapies for personality disorders. *Psychotherapy: Theory, Research, Practice, Training*, 46(2):233.

JC Muran, JD Safran, LW Samstag, and A Winston. 2004. Patient and therapist postsession questionnaires, version 2004. *New York: Beth Israel Medical Center*.

Bridianne ODea, Tjeerd W Boonstra, Mark E Larsen, Thin Nguyen, Svetha Venkatesh, and Helen Christensen. 2018. The relationship between linguistic expression and symptoms of depression, anxiety, and suicidal thoughts: A longitudinal study of blog content. *arXiv preprint arXiv:1811.02750*.

Anthony D Ong, Lizbeth Benson, Alex J Zautra, and Nilam Ram. 2018. Emodiversity and biomarkers of inflammation. *Emotion*, 18(1):3.

Yaakov Ophir, Refael Tikochinski, Christa SC Asterhan, Itay Sisso, and Roi Reichart. 2020. Deep neural networks detect suicide risk from textual facebook posts. *Scientific reports*, 10(1):1–10.

James W Pennebaker, Ryan L Boyd, Kayla Jordan, and Kate Blackburn. 2015. The development and psychometric properties of liwc2015. Technical report.

Robert Plutchik. 2000. Emotions in the practice of psychotherapy-clinical implications of affect theories.

Tom Pyszczynski and Jeff Greenberg. 1987. Self-regulatory perseveration and the depressive self-focusing style: a self-awareness theory of reactive depression. *Psychological bulletin*, 102(1):122.

Nairan Ramirez-Esparza, Cindy K Chung, Ewa Kacewicz, and James W Pennebaker. 2008. The psychology of word use in depression forums in english and in spanish: Texting two text analytic approaches. In *ICWSM*.

Diego Rocco, Rachele Mariani, and Diego Zanelli. 2013. The role of non-verbal interaction in a short-term psychotherapy: Preliminary analysis and assessment of paralinguistic aspects. *Research in Psychotherapy: Psychopathology, Process and Outcome*, pages 54–64.

Stephanie Rude, Eva-Maria Gortner, and James Pennebaker. 2004. Language use of depressed and depression-vulnerable college students. *Cognition & Emotion*, 18(8):1121–1133.

H Andrew Schwartz, Salvatore Giorgi, Maarten Sap, Patrick Crutchley, Lyle Ungar, and Johannes Eichstaedt. 2017. Dlatk: Differential language analysis toolkit. In *Proceedings of the 2017 conference on empirical methods in natural language processing: System demonstrations*, pages 55–60.

Golan Shahar, Megan L Rogers, Hadar Shalev, and Thomas E Joiner. 2020. Self-criticism, interpersonal conditions, and biosystemic inflammation in suicidal thoughts and behaviors within mood disorders: A bio-cognitive-interpersonal hypothesis. *Journal of personality*, 88(1):133–145.

Natalie Shapira, Gal Lazarus, Yoav Goldberg, Eva Gilboa-Schechtman, Rivka Tuval-Mashiach, Daniel Juravski, and Dana Atzil-Slonim. 2020. Using computerized text analysis to examine associations between linguistic features and clients' distress during psychotherapy. *Journal of counseling psychology*.

Jonathan Shedler. 2010. The efficacy of psychodynamic psychotherapy. *American psychologist*, 65(2):98.

David V Sheehan, Yves Lecrubier, K Harnett Sheehan, Patricia Amorim, Juris Janavs, Emmanuelle Weiller, Thierry Hergueta, Roxy Baker, and Geoffrey C Dunbar. 1998. The mini-international neuropsychiatric interview (mini): the development and validation of a structured diagnostic psychiatric interview for dsm-iv and icd-10. *The Journal of clinical psychiatry*.

Wiesław Sikorski. 2012. Paralinguistic communication in the therapeutic relationship. *Arch Psychiatry Psychother*, 1:49–54.

Richard F Summers and Jacques P Barber. 2009. *Psychodynamic therapy: A guide to evidence-based practice*. Guilford Press.

Allison M Tackman, David A Sbarra, Angela L Carey, M Brent Donnellan, Andrea B Horn, Nicholas S Holtzman, To'Meisha S Edwards, James W Pennebaker, and Matthias R Mehl. 2019. Depression, negative emotionality, and self-referential language: A multi-lab, multi-measure, and multi-language-task research synthesis. *Journal of personality and social psychology*, 116(5):817.

Michael Tanana, Aaron Dembe, Christina S Soma, Zac Imel, David Atkins, and Vivek Srikumar. 2016. Is sentiment in movies the same as sentiment in psychotherapy? comparisons using a new psychotherapy sentiment database. In *Proceedings of the Third Workshop on Computational Linguistics and Clinical Psychology*, pages 33–41.

Yla R Tausczik and James W Pennebaker. 2010. The psychological meaning of words: Liwc and computerized text analysis methods. *Journal of language and social psychology*, 29(1):24–54.

George L Trager. 1961. The typology of paralanguage. *Anthropological Linguistics*, pages 17–21.

Marcel Trotzek, Sven Koitka, and Christoph M Friedrich. 2018. Utilizing neural networks and linguistic metadata for early detection of depression indications in text sequences. *IEEE Transactions on Knowledge and Data Engineering*.

Michel Valstar, Björn Schuller, Kirsty Smith, Florian Eyben, Bihan Jiang, Sanjay Bilakhia, Sebastian Schnieder, Roddy Cowie, and Maja Pantic. 2013. Avec 2013: the continuous audio/visual emotion and depression recognition challenge. In *Proceedings of the 3rd ACM international workshop on Audio/visual emotion challenge*, pages 3–10. ACM.

Leon Van Wissen and Peter Boot. 2017. An electronic translation of the liwc dictionary into dutch. In *Electronic lexicography in the 21st century: Proceedings of eLex 2017 conference*, pages 703–715. Lexical Computing.

World Health Organization WHO et al. 2017. Depression and other common mental disorders: global health estimates. Technical report, World Health Organization.

Anna Wierzbicka. 1985. Different cultures, different languages, different speech acts: Polish vs. english. *Journal of pragmatics*, 9(2-3):145–178.

Andrew Yates, Arman Cohan, and Nazli Goharian. 2017. Depression and self-harm risk assessment in online forums. *arXiv preprint arXiv:1709.01848*.

# A Appendices

## A.1 Base Dataset Description

### A.1.1 Clients

The dataset was drawn as a sample from a broader pool of clients who received individual psychotherapy at a university training outpatient clinic, located in a central city in Israel. Data were collected naturalistically between August 2014 and August 2016 as part of the clinic's regular practice of monitoring clients' progress. From an initial sample of 180 clients who provided their consent to participate in the study, 34 (18.88%) dropped out (deciding one-sidedly to end treatment before the planned termination date). Clients were selected from the larger sample to match two criteria: (1) treatment duration of at least 15 sessions, and (2) full data including audio recordings to be used for the transcriptions and session-by-session questionnaires available for each client. These criteria corresponded to our analytic strategy of detecting within-client associations between linguistic features and session processes and outcomes. Clients were also excluded, based on the M.I.N.I. 6.0 (Sheehan et al., 1998) if they were diagnosed as severely disturbed, either due to a current crisis, had severe trauma and accompanying post- traumatic stress disorder, a past or present psychotic or manic diagnosis, and/or current substance abuse. Based on these criteria we excluded 77 (42.7%) clients. Thus, of the total sample, the data for 68 (38.33%) clients who met the above-mentioned inclusion criteria were transcribed, for a total of 872 transcribed sessions.

The clients were all above the age of 18 ($M_{age}$=39.06, SD=13.67, range=20–77), majority of whom were women (58.9%). Of the clients, 53.5% had at least a bachelor's degree, 53.5% reported being single, 8.9% were in a committed relationship, 23.2% were married and 14.2% were divorced or widowed. Clients' diagnoses were established based on the Mini International Neuropsychiatric Diagnostic Interview for Axis I DSM-IV diagnoses (MINI 5.0; Sheehan et al., 1998). Of the entire sample, 22.9% of the clients had a single diagnosis, 20.0% had two diagnoses, and 25.7% had three or more diagnoses. The most common diagnoses were comorbid anxiety and affective disorders[14] (25.7%), followed by other comorbid disorders (17.1%), anxiety disorders (14.3%), and affective disorders (5.7%). A sizable group of clients (31.4%) reported experiencing relationship concerns, academic/occupational stress, or other problems but did not meet criteria for any Axis I diagnosis.

### A.1.2 Therapists and Therapy

Clients were treated by 59 therapists in various stages of their clinical training. Clients were assigned to therapists in an ecologically valid manner based on real-world issues, such as therapist availability and caseload. Most therapists treated one client each (47 therapists), but some (10) treated two clients and (2) more. Each therapist received one hour of individual supervision every two weeks and four hours of group supervision on a weekly basis. All therapy sessions were audiotaped for supervision. Supervisors were senior clinicians. Individual and group supervision focused heavily on reviewing audiotaped case material and technical interventions designed to facilitate the appropriate use of therapist interventions. Individual psychotherapy consisted of once- or twice-weekly sessions. The language of therapy was Modern Hebrew (MH). The dominant approach in the clinic includes a short-term psychodynamic psychotherapy treatment model (e.g.,Blagys and Hilsenroth,2000; Shedler, 2010; Summers and Barber, 2009). The key features of the model include: (a) a focus on affect and the experience and expression of emotions, (b) exploration of attempts to avoid distressing thoughts and feelings, (c) identification of recurring themes and patterns, (d) an emphasis on past experiences, (e) a focus on interpersonal experiences, (f) an emphasis on the therapeutic relationship, and (g) exploration of wishes, dreams, or fantasies (Shedler, 2010). On average, treatment length was 37 sessions (SD = 23.99, range = 18–157). Treatment was open- ended in length, but given that psychotherapy was provided by clinical trainees at a university-based outpatient community clinic, the treatment duration was often restricted to be 9 months.

### A.1.3 Transcriptions

To capture the treatment processes from session to session, and since the transcription process is highly expensive, transcriptions were conducted alternately (i.e., sessions 2, 4, 6, 8 and so on until

[14]The following DSM-IV diagnoses were assessed in the affective disorders cluster: major depressive disorder, dysthymia and bipolar disorder. The following DSM-IV diagnoses were assumed in the anxiety disorders cluster: panic disorder, agoraphobia, generalized anxiety disorder and social anxiety disorder.

one session before the last session). In cases where material was incomplete (such as the quality of the recordings, or the questionnaires for a specific session), the next session was transcribed instead. The transcriber team was composed of seven transcribers, all of whom were graduate students in the University's psychology department. The transcribers went through a one day training workshop and monthly meetings were held throughout the transcription process to supervise the quality of their work. The training included specific guidelines on how to handle confidential and sensitive information and the transcribers were instructed to replace names and places by pseudonyms and to substitute any other identifying information. The transcription protocol followed general guidelines, as described in (Mergenthaler and Stinson, 1992), and in Albert et al. (2013). The word forms, the form of commentaries, and the use of punctuation were kept as close as possible to the speech presentation. Everything was transcribed, including word fragments as well as syllables or fillers (such as "ums", "ahs", "uh huhs" and "you know"). The audiotape was transcribed in its entirety and provided a verbatim account of the session. The transcripts included elisions, mispronunciations, slang, grammatical errors, non-verbal sounds (e.g., laughs, cry, sighs), and background noises. The transcription rules were limited in number and simple (for example, each client and therapist utterances should be on a separate line ;each line begins with the specification of the speaker) and the format used several symbols to indicate comments (such as [...] to indicate the correct form when the actual utterance was mispronounced, or <number of minutes of silence >). The transcripts were proofread by the research coordinator. The final transcripts could be processed by human experts or automatically by computer.

There were 872 transcripts in total (the mean transcribed sessions per client was 12.56; SD=4.93) Each transcript incorporated metadata such as the client's code, which allowed the client data to be linked across sessions and for hierarchical analysis. The transcriptions totaled about four million words over 150,000 talk turns (i.e., switching between speakers). On average, there were 5800 words in a session, of which 4538 (78%; SD=1409.62; range 416-8176) were client utterances and 1266 (22%; SD=674.99; range 160-6048) were therapist utterances with a mean of 180.07 (SD=95.37; range 30-845) talk turns per session.

#### A.1.4 Procedure and Ethical Considerations

The procedures were part of the routine assessment and monitoring process in the clinic. All research materials were collected after securing the approval of the authors' university ethics committee. Only clients that gave their consent to participate were included in the study. Clients were told that they could choose to terminate their participation in the study at any time without jeopardizing treatment. The clients completed the ORS before each therapy session and the WAI after each session. The therapist completed the WAI after each therapy session. The sessions were audiotaped and transcribed according to a protocol described above. All data collected was anonymized (see Section A.1.3) and only then exposed to a very small number of researchers, as agreed upon by the participants. The data is stored encrypted.

### A.2 Outcome and Process Measurements

#### A.2.1 Outcome Rating Scale (ORS; (Miller et al., 2003))

The ORS is a 4-item visual analog scale developed as a brief alternative to the OQ-45. The scale is designed to assess change in three areas of client functioning that are widely considered to be valid indicators of progress in treatment: functioning, interpersonal relationships, and social role performance. Respondents complete the ORS by rating four statements on a visual analog scale anchored at one end by the word Low and at the other end by the word High. This scale yields four separate scores between 0 and 10 that sum to one score ranging from 0 to 40, with higher scores indicating better functioning. The ORS has strong reliability estimates ($\alpha$=0.87-0.96) and moderate correlations between the ORS items and the OQ-45 subscale and total scores (ORS total - OQ-45 total: r = 0.59).

#### A.2.2 Profile of Mood States (POMS; (McNair, 1992))

The POMS assesses mood variables and is widely used. For the purpose of this study, we used an abbreviated version of the measure, which was adapted for intensive repeated measurements (Cranford et al., 2006) and consists of 12 words that describe current emotional states. The negative affect scale includes depressed mood (2 items), anxious mood (2 items), and anger (2 items). The positive affect scale includes contentment (2 items), vigor

(2 items), and calmness (2 items). Examples of feelings on the POMS are 'anxious', 'sad', 'angry', 'happy', 'lively', and 'calm'. Clients were asked to evaluate how they felt during the session on a 5-point Likert scale ranging from 'Not at all' to 'Extremely'. The POMS has been tested on college students and was found to be both valid and reliable (Guadagnoli and Mor, 1989).

#### A.2.3 Working Alliance Inventory (WAI; (Horvath and Greenberg, 1989))

The WAI is a self report questionnaire (both for therapist and client). It is one of the most widely investigated common factors that was found positively correlated to treatment outcome in psychotherapy. It includes items ranging from 0 ("not at all") to 5 ("completely") to evaluate three components (1) agreement on treatment goals (2) agreement on therapeutic tasks and (3) a positive emotional bond between client and therapist (Falkenström et al., 2015)

#### A.2.4 Post-Session Questionnaire (PSQ; (Muran et al., 2004))

Alliance ruptures were assessed after each session with a single-item question from the therapist's perspective: "Did you experience any tension, any misunderstanding, conflict or disagreement in the relationship with your patient?" Both items are answered on a 5-point Likert scale ranging from 1 ("not at all") to 5 ("constantly"), reflecting the subjectively perceived intensity of a rupture. Following the recommendations provided by (Muran et al., 2009), a rupture was defined as any rating higher than 1 on the scale.

### A.3 Expansion of Complementing Word Sets

This section formally defines the problem of expanding complimentary lexicons and describes technique as a solution.

**Given:**

1. $positive\_seed$, $negative\_seed$ which are two complementing lexicon seeds. E.g., Enthusiastic=[mighty, wow, energetic, ...] and the compliment Not_Enthusiastic=[apathetic, oh, nothing, ...]
2. $confidence\_level$, float greater than 0
3. $expand\_rate$, integer greater than 0
4. $radius$, integer greater than 1

**Output:**

$positive\_expansion$, $negative\_expansion$, new lexicons, each containing the given respective lexicon and additional words that match the lexicon's semantic knowledge.

**Algorithm Intuition**

The expansion is performed in several rounds, where in each round the two seeds $positive\_seed$, $negative\_seed$ expand simultaneously on the basis of words semantically similar to words that already exist in the seed. The generation process of new semantically similar words candidates uses the *word-similarity* package[15] that is based on pretrained Hebrew Twitter word embeddings, and returns similar words for a given word, with similarity probabilities. The $expand\_rate$ parameter represents the number of similar words that the *word-similarity* returns (default configured as 30).

While expanding, care is taken not to deviate from the lexicon to its complementing lexicon (to get a feel for the importance of this step, see Figure 1 of positive and negative emotion words, showing how semantically close the words in the complementing lexicons can be). Each word in the seed list is used as a "witness" for similar words (weighted by similarity probability). In case there is more than one "witness" for a new candidate word, the similarity probabilities are summed. This "sieve" process is done by making sure that for each word that enters the expansion lexicon there are enough "witnesses", other close words already in the existing seed lexicon (i.e., their sum of probabilities for similarity to the candidate word is above threshold for filter criterion) and also does not appear in the complementary lexicon. The $confidence\_level$ parameter (default configured to 3) represents the threshold for filter criterion.

The result of the expansion is used as input for the next round. The $radius$ parameter represents the number of expansion rounds.

**Algorithm Steps**

1. For $radius$ times:
   (a) For each of $positive\_seed$ and $negative\_seed$ seeds, create new sets of candidate words $positive\_candidates$ and $negative\_candidates$, by expanding the words in the seeds with

[15] https://github.com/Ronshm/hebrew-word2vec

*word-similarity* with $expand\_rate$ parameter as number of similar words.

   (b) Each of $positive\_candidates$, $negative\_candidates$ passes a *candidates-sieve* process which creates $positive\_survivors$, $negative\_survivors$: filter out low-probability words (sum of probabilities less than $confidence\_level$) or words that appear in the complementary seed list (i.e., $negative\_candidates$ for the $positive\_candidates$ and vise versa) .

   (c) Update seed lists $positive\_seed$ and $negative\_seed$ with the corresponding lists $positive\_survivors$ and $negative\_candidates$.

2. return $positive\_seed$, $negative\_seed$

# Community-level Research on Suicidality Prediction in a Secure Environment: Overview of the CLPsych 2021 Shared Task

**Sean MacAvaney**
IR Lab, Georgetown University
sean@ir.cs.georgetown.edu
University of Glasgow

**Anjali Mittu**
University of Maryland
amittu@umd.edu

**Glen Coppersmith**
Qntfy
glen@qntfy.com

**Jeff Leintz**
NORC at the University of Chicago
Leintz-Jeff@norc.org

**Philip Resnik**
University of Maryland
resnik@umd.edu

## Abstract

Progress on NLP for mental health — indeed, for healthcare in general — is hampered by obstacles to shared, community-level access to relevant data. We report on what is, to our knowledge, the first attempt to address this problem in mental health by conducting a shared task using sensitive data in a secure data enclave. Participating teams received access to Twitter posts donated for research, including data from users with and without suicide attempts, and did all work with the dataset entirely within a secure computational environment. We discuss the task, team results, and lessons learned to set the stage for future tasks on sensitive or confidential data.

## 1 Introduction

In natural language processing, and in AI more generally, progress depends on data. The most significant progress on a problem takes place when an entire community is working on the same dataset at the same time; for example, the wide availability of speech recognition today is a result of decades of research using DARPA benchmark datasets and evaluations for speech-related tasks (Juang and Rabiner, 2005).

In healthcare, however, community-level activity is an enormous challenge. Laws and regulations related to data confidentiality create obstacles to access, including significant administrative overhead such as data use agreements and significant technical overhead involving arrangements for secure data distribution, storage, and management (Lane and Schur, 2010). In mental health and particularly crisis detection, missteps like Samaritans Radar raise highly public red flags despite well-intentioned goals (Horvitz and Mulligan, 2015; Resnik et al., 2021). All these legal, regulatory, operational, and public perception risks naturally make potential data providers skittish about data sharing. As a result, important research in healthcare is balkanized, with community efforts scattered among different datasets in *ad hoc* fashion as different teams work with the data they are able to gain access to. Or potentially it doesn't take place at all, as talented researchers go work on other problems where obtaining data is just easier.

Secure data enclaves are one solution to this problem (Lane and Schur, 2010). The key idea in a data enclave is to bring researchers to sensitive data, rather than disseminating data out to researchers. A data enclave provides secure remote access to data using carefully designed statistical, technical, legal and operational controls. Computation on an enclave is done using a copy of the data residing there without full networking access, meaning that nothing can be imported or exported without disclosure review. This does not replace necessary steps like IRB approvals, data use agreements, and record de-identification; for example, data enclave users can still look at private data within the enclave and need to agree not to attempt de-anonymization. However, it drastically simplifies community-level access. A single, comprehensive description of security provisions can be created for data providers and ethical review boards, and data providers need to enter into data use agreements only with the enclave, rather than with individual teams.

To our knowledge, the CLPsych 2021 shared task is a first-of-its-kind endeavor: as far as we know, it is the first time a community-level shared task with sensitive mental health data has been conducted on a data enclave, and more generally shared tasks on sensitive data are rare in the NLP and machine learning communities. In addition, although uses of data enclaves are often centered on the use of analytics tools, in this shared task the environment was designed to support the full arsenal of NLP and machine learning methods. We accomplished this by partnering with NORC at the



*Proceedings of the Seventh Workshop on Computational Linguistics and Clinical Psychology*, pages 70–80
June 11, 2021. 

University of Chicago. Since 2006, the NORC Data Enclave® has served U.S. state and federal agencies, research institutes, foundations, and universities by securely housing and providing remote access to confidential data. In a collaborative project with University of Maryland, NORC has developed the UMD/NORC Mental Health Data Enclave (henceforth the Enclave, for short), a subset of NORC Data Enclave infrastructure designed specifically with the requirements of mental health NLP and machine learning research in mind.

Data for this shared task were provided by Qntfy, which runs OurDataHelps.org, an online platform that permits donations of digital life data (including social media) for the purposes of advancing research in mental health and wellbeing. Individuals come from a range of lived experience with mental health, specifically related to this shared task: individuals who have survived suicide attempts, loved ones of people who have died by suicide, and people who just want to help. For this shared task, Qntfy established a data provider agreement with NORC, and NORC executed data use agreements with the participating teams. The University of Maryland, College Park IRB reviewed and approved a protocol for research with, and sharing of, the OurDataHelps data. The arrangement here therefore exemplifies the advantages of data enclaves discussed above. For the data provider, it was much easier to work out an agreement with just a single entity running an established secure infrastructure, which significantly lowered the bar for sharing data with multiple teams. In addition, NORC's platform and processes for team access, platform security, and import/export review created a far greater level of confidence in privacy controls than sending data out to a large number of far-flung teams with heterogeneous environments. For teams, this provided a rare opportunity to work with sensitive mental health data containing actual outcomes, not proxy data as is more common in social media mental health research and which can be problematic for a variety of reasons (Ernala et al., 2019).

The shared task itself involved assessment of suicide risk via prediction of suicide attempts, based on the natural language of users on Twitter. There were two subtasks: Subtask 1 involved assessing suicide risk given 30 days of tweets prior to the date of an attempt (or a corresponding date when no attempt was made), and Subtask 2 involved assessing suicide risk given the prior six months of tweets.

A set of 21 teams signed up and were onboarded on the Enclave. A total of five teams ultimately submitted systems by the deadline. All teams have been given several months of additional access and support on the Enclave, in order to permit continued experimentation. We are hopeful that results obtained during this extended time period will lead to publications beyond CLPsych.

In this overview paper, we provide not only a summary overview the shared task itself, in terms of the research problem and participating teams' findings about predicting suicide risk from Twitter data, but also a retrospective analysis of conducting a shared task in a secure enclave, including lessons learned and recommendations for future tasks of this kind.[1]

## 2 Background and Related Work

A number of recent articles discuss the use of NLP, machine learning, and social media in service of mental health. As important motivating background, a meta-analysis by Franklin et al. (2017) concludes that prediction of suicidal thoughts and behaviors has not improved in fifty years, encouraging a shift to algorithmic and machine learning approaches. Schafer et al. (2021) provide significant empirical support for this view via another meta-analysis looking specifically at traditional theory-driven versus machine learning approaches to prediction of suicide risk, demonstrating that the latter are significantly more effective at prediction.[2] Naslund et al. (2020) and Lee et al. (2021) provide overviews that include thoughtful, big-picture commentary on research and clinical applications for mental health taking advantage of NLP, machine learning, and social media. Resnik et al. (2021) offer an overview of issues more specifically focused on using naturally occurring language as a source of evidence in suicide prediction.

One running theme throughout discussions of

[1] We would be happy to discuss logistical issues, and share details and specific language from our IRB protocol, data provider, and data use agreements, in order to facilitate others who would like to organize shared tasks similar to this one. Interested readers should contact clpsych-2021-shared-task-organizers@googlegroups.com.

[2] In regard to the goals of prediction versus scientific explanation and understanding, it is worth noting the argument by Yarkoni and Westfall (2017) that psychology research as a whole, including research with explanatory goals, would benefit by taking a predictive approach.

this kind involves the availability of data to work with, and the interplay, or even tension, between the need for research and the need to respect privacy and other ethical considerations. Horvitz and Mulligan (2015) provide one short, useful discussion specifically focused on data and privacy, and Benton et al. (2017) and Chancellor et al. (2019) discuss ethical issues specifically with regard to social media and work on mental health. Lane and Schur (2010) provide a valuable entry point to the concept of data enclaves as a way to balance the need for data access in order to make progress in healthcare with respect for patient privacy — this concept ties in directly with the call by Schafer et al. (2021) for community-level mental health datasets to be easily available for research so that the predictive ability of models can be compared and research can be replicated. Those kinds of comparisons and replications are instrumental in modern data-driven research because without them it is impossible to gain insight into which approaches are most promising or to rule out the possibility that apparent differences are related to idiosyncratic differences in data.

Related, the most current paradigms in NLP and machine learning involve both general-purpose pre-training and task-specific fine-tuning. To some extent, pre-training data may capture generalizations about language that transfer well to problems in the mental health space. However, many off-the-shelf language resources that are commonly used, such as BERT (Devlin et al., 2019), are built from sources such as books and Wikipedia entries. These may translate poorly to systems dependent on social media posts from Twitter, Facebook, or an online discussion forum. It is well known that systems perform better when they are trained on materials similar to the materials the system will run on (Alsentzer et al., 2019; Beltagy et al., 2019). Therefore using task-specific data from immediately relevant sources as training data for social media based mental health tasks is a high priority that requires attention.

Another theme found in related literature involves the nature and quality of the variables being predicted. The sensitivity of mental health data has led to a proliferation of proxy variables taken from publicly available data rather than ground-truth clinical variables or real-world outcomes (e.g. De Choudhury and De, 2014; Coppersmith et al., 2014; Yates et al., 2017; Shing et al., 2018; Cohan et al., 2018; Thorstad and Wolff, 2019). As two particularly well known and influential examples, Coppersmith et al. (2014) infer mental health diagnoses of Twitter users by looking for publicly self-reported diagnoses, and De Choudhury et al. (2016) infer mental health progressions to suicidal ideation by examining when Reddit users shift from mental health subreddits to the SuicideWatch subreddit. Such data tend to have the advantages of being readily accessible and large in size. However, Ernala et al. (2019) note a variety of problems and limitations in using proxies rather than clinically grounded variables. Coppersmith et al. (2018) offer a rare exception in this kind of work, using an ethical process of data donation to obtain social media data *with outcomes* for research on prediction of suicide attempts; our shared task is based on a subset of their data.

## 3 Data

We briefly describe our data sources, and how we constructed the shared task datasets for binary classification tasks.

### 3.1 Data sources

We began with data donated to the OurDataHelps.org platform, discussed in greater detail by Coppersmith et al. (2018). Donations to the platform include data from people who have survived a suicide attempt, data from people who died by suicide that has been donated by loved ones, and data donated by people who have not attempted suicide but want to help. When donations take place, a questionnaire is filled out that collects basic demographic data and mental health history. This includes the number of past suicide attempts and dates associated with them, although dates are not provided in all cases.

Although the platform permits collection of a wide range of data, including, for example, social media, fitness, and wearable data, in this shared task we restricted our attention to Twitter data and a subset of basic information from the questionnaire. Only publicly available tweets are used, typically visible to friends and family, and these were de-identified before being provided to the Enclave.

On the Enclave, participants also had access to a copy of the UMD Reddit Suicidality Dataset (Shing et al., 2018; Zirikly et al., 2019). This dataset was used by one of the teams (NUSIDS) in their submission.

In addition, a non-sensitive practice dataset using the shared task data format was provided to participants so they could work on developing and debugging their systems outside of the Enclave. It was based on a modified version of the `depression-detection` dataset (Wang et al., 2019).[3]

### 3.2 Users with Suicide Attempts

In the version of the data we began with, there are 3,631 users, 1,613 of whom attempted (and possibly died by) suicide. From this version, we imposed several filters. We only considered users who had donated Twitter data and who had reported their gender and date of birth in the questionnaire, in order to match users with a suicide attempt to a control user. If a user had attempted suicide, we only included them if they had a date associated with the attempt, a necessary restriction in order to examine tweets in the time period leading up to the attempt. For users with multiple attempts, we only considered the most recent attempt having a date. Filtering in this way left 250 users with suicide attempts, associated dates, and data prior to the attempt. For Subtask 1, we restricted the set to users who had made posts in the 30 days prior to their suicide attempt, a total of 68. For Subtask 2, we restricted the set to users who had made Twitter posts during the six months prior to the attempt, which included a total of 97 users. Teams were provided with anonymized user IDs, the date of the most recent suicide attempt (if applicable), and a list of the user's de-identified tweets from the applicable time span.

### 3.3 Control Users

Similar to Coppersmith et al. (2018), we included a set of control users matched one-to-one with users who had attempted suicide, based on having the same gender, similar age (within 5 years), and similar number of tweets. These criteria resemble previous matching in the 2015 CLPsych shared task (Coppersmith et al., 2015) and in Coppersmith et al. (2018). Age and gender are common controls in the mental health space, and we chose to match using a similar number of tweets so that corresponding users in the dataset would be represented by similar quantities of social media evidence. For each user with a suicide attempt, we found a match by first finding all users matching age and gender, then selecting the user with the closest number of tweets. Tweets taken from the control user were from the same time frame as their match who had an attempt in order to minimize differences in context, such as tweets about world events.

| | Subtask 1 | Subtask 2 |
|---|---|---|
| Total # of Users | 114 / 22 | 164 / 30 |
| Users Under 30 | 104 / 15 | 138 / 23 |

Table 1: The total number of users in each subtask and the number of users under the age of 30. The numbers in the table are given as (training set) / (test set)

| | Subtask 1 | Subtask 2 |
|---|---|---|
| Female | 118 | 168 |
| Male | 12 | 20 |
| Non-Binary | 4 | 4 |
| Other | 2 | 2 |

Table 2: The distribution of gender across all users.

Table 1 shows the final number of users in each subtask and Table 2 shows the age distribution of users. In the shared task, we saved 15% of the users for the test set; these numbers are shown in the table. For both subtasks, most of the users were female between the ages of 18 to 24 and most of the users were under the age of 30. Within the time period, for Subtask 1, users had an average of 24 tweets per person and in Subtask 2, there were an average of 102 tweets per person.

## 4 Baseline

A baseline system was provided to shared task participants to use or build upon.[4] Baseline preprocessing includes several standard steps. First, we removed all URLs, user mentions, and emojis from the tweets. Whenever a user's tweet includes an image, GIF, or link, the links are removed. We tokenized the tweets using the Twitter-specific Twikenizer and removed stopwords from the tweets' text using the default SpaCy (Honnibal et al., 2020) stopword list.[5] Last, we split hashtags into the words they are made up of: first, we try to split by camel-case or by underscores; if that fails, we use a method from HashTagSplitter, attempting to split into the smallest subset of real words.[6]

[3] `https://github.com/seanmacavaney/clpsych2021-shared-task/tree/main/practice-dataset`

[4] https://github.com/anjmittu/clpsych2021-shared-task-baseline

[5] https://github.com/Guilherme-Routar/Twikenizer

[6] https://github.com/matchado/HashTagSplitter

The baseline classification model used logistic regression with the default parameters from SciKit Learn (Pedregosa et al., 2011), employing unigram and bigram count vectors.

## 5 The Enclave

As discussed in the introduction, data-driven research in mental health, and healthcare more generally, faces significant obstacles owing to important concerns about privacy and data confidentiality. Data enclaves offer a potential solution (Lane and Schur, 2010).

NORC at the University of Chicago, an independent, non-profit research institution, took on the operational aspects of running this shared task on their data enclave. Significant time was spent working with Qntfy, who were responsible for providing the OurDataHelps data, and the shared task organizers, to develop the data provider agreement, data use agreements, operational policies, supporting infrastructure, and technical and operational support for the organizers and shared task teams.

All aspects of the shared task on the Enclave were run using exactly the same procedures as for NORC's traditional Data Enclave clients, such as government agencies working with confidential databases. Teams that worked on the shared task executed a data use agreement with NORC and then were "onboarded" to the Enclave, being provided with account logins, passwords, documentation, procedures for uploading and export (both requiring human review of the material entering or leaving the Enclave), and contacts and procedures for technical support.

The Enclave environment includes two main parts. The first part is a secure virtual desktop (using Citrix), accessed via the Data Enclave login page through an internet browser. The second part of the Enclave is NORC's Mental Health Data Enclave (MHDE) Cluster on Amazon Web Services (AWS). From within the secure Citrix desktop, participants use PuTTY ssh to reach a gateway machine on this cluster. They can run code there or submit batch jobs using the Slurm cluster management and job scheduling system.[7] The AWS environment is configured to spin up a new instance for the duration of the job and then spin it down when completed, conserving compute resources to save cost.

Crucially, the Enclave is a closed environment. Neither the secure desktop nor the AWS cluster permit access to the Internet. It is not possible to `scp` or `sftp` data. It is not possible to open a socket in a program that connects externally. It is not possible to print, print screen, or even to copy/paste to or from the external environment.

The NORC Data Enclave's data security model integrates a portfolio approach with the Five Safes framework (Ritchie, 2017) to harden the security posture. This means that bringing materials in, such as code, data, or other resources, requires an import request process. Each request triggers a robust review process to provide safe passage of confidential micro-data and ensure imported material does not contain any virus or code aimed at disabling the capabilities or facilitating unauthorized access. In order to set up the Enclave environment and hopefully speed up this process for shared task participants, it was pre-loaded with major Python packages and tools (more than 4000 of them), the shared task baseline code, and shared task data; see further discussion in Section 8.

Similarly, as a data custodian for restricted data (e.g. confidential micro-data for federal, state and commercial clients), NORC must ensure that any data leaving the NORC Data Enclave is safe and free of inappropriate disclosures. This means that there is a request-based procedure for exporting any material from the Enclave, with formal review criteria that include both dataset-specific criteria and general guidelines applied globally across all requests.

## 6 Submissions

Each team was permitted up to three submissions for each subtask (30 days and 6 months). In each subtask, the numbered submissions for each team distinguish the "primary" submission (numbered 1) from additional contrastive runs (numbered 2 and 3). In total, we received 30 submissions, with five teams providing three runs each for both subtasks.

**NUSIDS** (Zagatti et al., 2021). For the shared task, NUSIDS designed SHTM, a Self-Harm Topic Model, which combines standard Latent Dirichlet Allocation (LDA) with a self-harm dictionary. This was tested using a combination of the shared task data, along with the practice dataset and the UMD Reddit Suicidality Dataset. In their submission to the task, the team used a combination of an LSTM and term feature vectors with SHTM-based fea-

[7] https://slurm.schedmd.com/

| Team (Sub.) | $F_1$ | $F_2$ | TPR | FAR | AUC |
|---|---|---|---|---|---|
| NUSIDS (1) | 0.583 | 0.648 | 0.700 | 0.636 | 0.645 |
| NUSIDS (2) | 0.615 | 0.714 | 0.800 | 0.727 | 0.664 |
| NUSIDS (3) | 0.300 | 0.300 | 0.300 | 0.636 | 0.373 |
| ScyLab (1) | 0.526 | 0.481 | 0.455 | **0.273** | 0.678 |
| ScyLab (2) | 0.526 | 0.481 | 0.455 | **0.273** | 0.678 |
| ScyLab (3) | 0.421 | 0.385 | 0.364 | 0.364 | 0.636 |
| sentimenT5 (1) | 0.455 | 0.455 | 0.455 | 0.545 | 0.438 |
| sentimenT5 (2) | 0.500 | 0.472 | 0.455 | 0.364 | 0.616 |
| sentimenT5 (3) | 0.571 | 0.656 | 0.727 | 0.818 | 0.413 |
| SoS (1) | 0.286 | 0.278 | 0.273 | 0.636 | 0.264 |
| SoS (2) | 0.400 | 0.377 | 0.364 | 0.455 | 0.529 |
| SoS (3) | 0.364 | 0.364 | 0.364 | 0.636 | 0.397 |
| UlyaLamia (1) | **0.692** | **0.763** | **0.818** | 0.545 | 0.702 |
| UlyaLamia (2) | 0.522 | 0.536 | 0.545 | 0.545 | 0.409 |
| UlyaLamia (3) | 0.636 | 0.636 | 0.636 | 0.364 | **0.740** |
| Our baseline | 0.636 | 0.636 | 0.636 | 0.364 | 0.661 |

Table 3: Results of participating systems and our baseline for Subtask 1 (30 days). The best result for each metric is listed in bold.

| Team (Sub.) | $F_1$ | $F_2$ | TPR | FAR | AUC |
|---|---|---|---|---|---|
| NUSIDS (1) | 0.684 | 0.812 | **0.929** | 0.786 | 0.663 |
| NUSIDS (2) | 0.703 | **0.823** | **0.929** | 0.714 | 0.648 |
| NUSIDS (3) | 0.649 | 0.759 | 0.857 | 0.786 | 0.480 |
| ScyLab (1) | 0.769 | 0.704 | 0.667 | **0.067** | 0.809 |
| ScyLab (2) | 0.769 | 0.704 | 0.667 | **0.067** | 0.791 |
| ScyLab (3) | **0.815** | 0.764 | 0.733 | **0.067** | **0.844** |
| sentimenT5 (1) | 0.467 | 0.467 | 0.467 | 0.533 | 0.618 |
| sentimenT5 (2) | 0.516 | 0.526 | 0.533 | 0.533 | 0.591 |
| sentimenT5 (3) | 0.727 | 0.769 | 0.800 | 0.400 | 0.720 |
| SoS (1) | 0.429 | 0.411 | 0.400 | 0.467 | 0.444 |
| SoS (2) | 0.533 | 0.533 | 0.533 | 0.467 | 0.640 |
| SoS (3) | 0.400 | 0.400 | 0.400 | 0.600 | 0.502 |
| UlyaLamia (1) | 0.595 | 0.671 | 0.733 | 0.733 | 0.582 |
| UlyaLamia (2) | 0.581 | 0.592 | 0.600 | 0.467 | 0.564 |
| UlyaLamia (3) | 0.645 | 0.658 | 0.667 | 0.400 | 0.569 |
| Our baseline | 0.710 | 0.724 | 0.733 | 0.333 | 0.764 |

Table 4: Results of participating systems and our baseline for Subtask 2 (6 months). The best result for each metric is listed in bold.

tures. Submissions varied in the hyper-parameters of the model (e.g., window size and number of topics), as well as the training data.

**ScyLab** (Gamoran et al., 2021). The ScyLab submission used Bayesian modeling over features grounded in domain knowledge. These features included behavioral information learned by Twitter activity, Linguistic Inquiry and Word Count (LIWC) (Pennebaker et al., 2015) based features using priors from Eichstaedt et al. (2018) and other dictionary-based approaches. The submissions varied the distributions for the priors and hyperparameters (type of regression) for the logistic-regression model.

**sentimenT5** (Morales et al., 2021). SentimenT5 took different approaches in their submissions to explore the performance of simple traditional models versus fine-tuned deep learning models. In both Subtasks 1 and 2, they submitted results from gradient-boosted classifiers. One used syntax features and the other character TF-IDF features. For Subtask 1, they also submitted results from a contextualized language model classifier, and, for Subtask 2, a voting ensemble method.

**SoS** (Wang et al., 2021). Team SoS introduced the C-Attention Network, which uses latent feature information implicitly in the embeddings. This was compared with submissions using KNN and SVM classifiers. Latent features included using Doc2vec embeddings (Lau and Baldwin, 2016). Hand-crafted features included emotion lexicons, part-of-speech tags, and a custom dictionary that models various stages of suicidal behavior.

**UlyaLamia** (Bayram and Benhiba, 2021). In the UlyaLamia submissions, the authors were motivated by real-life applicability of their model to use tweet-level classification. The team's submissions used a majority voting approach over individual tweets. In order to pick which machine learning method to use, the team experimented with multiple methods tuned on the training data using a leave-one-out strategy. Their final submissions were the top methods from the leave-one-out results.

## 7 Results

We evaluated each system in terms of $F_1$, $F_2$ (favoring recall), True Positive Rate (TPR), False Alarm (Positive) Rate (FAR), and Area Under the ROC Curve (AUC). We use $F_1$ score as the primary evaluation metric, though it is valuable to consider all metrics for a complete view of the system performance.

We present the results of the submissions in Tables 3 and 4. In Subtask 1, Team UlyaLamia ranked highest in $F_1$, $F_2$ and TPR; however, their FAR was higher than the baseline and in the middle of the other team's submissions. Team UlyaLamia was also the only team to exceed the baseline $F_1$ score, with NUSIDS being the next closest team. In Subtask 2, Team ScyLab ranked highest in $F_1$, FAR, and AUC. Their strongest submission beat or met

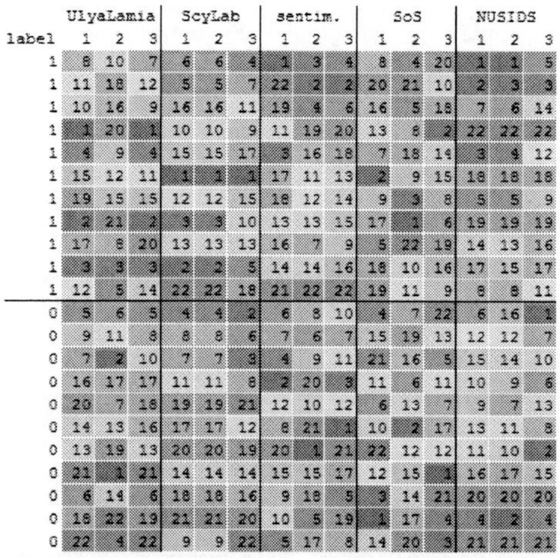


Figure 1: Rank comparison of the submissions for Subtask 1. A label of 1 indicates users with suicide attempts. Ranks closer to 1 indicate a higher score (more likely to have made a suicide attempt) given to the user. Rows are sorted by label, then median rank.

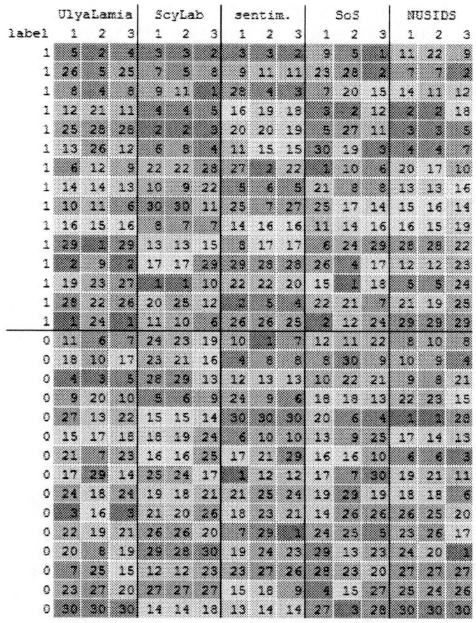


Figure 2: Rank comparison of the submissions for Subtask 2. A label of 1 indicates users with suicide attempts. Ranks closer to 1 indicate a higher score given to the user. Rows are sorted by label, then median rank.

the baseline in every metric and was notably low in their FAR. Five submissions came close or beat the baseline in $F_1$ score in Subtask 2.

The methods used by teams in the shared task had difficulties performing well in both subtasks. Given shorter-term information starting 30 days prior to an attempt, tweet-specific language (UlyaLamia) performed beste, but dictionary-based methods (e.g., ScyLab) worked best with the longer-term evidence (6 months prior to an attempt).

To gain a better understanding of the differences between the submissions, we plot the ranks of each test user for both subtasks in Figures 1 and 2. From these figures, we can see that some users easily classified by most systems, while others were notably difficult. For instance, in the last positive (label=1) row in Figure 2 (Subtask 2), the majority of systems were (incorrectly) very confident that the user did not make a suicide attempt. Nevertheless, three submissions gave this user the highest or second-highest likelihood. These results suggest that an ensemble method may be beneficial for this task.

This task is notably similar to Coppersmith et al. (2018), who performed experimentation including OurDataHelps.org data with similar restrictions, matching criteria, and the same binary outcomes. They found that a longer history of tweets led to slightly better predictions, but, unlike our shared task, they did not find a significant increase in performance between using tweets 90 to 0 days prior to an attempt and using tweets 180 to 90 days prior. In Coppersmith et al. (2018), the AUC score using tweets 30 days prior to an attempt is .89 and the AUC score using tweets six months prior to an attempt is .93.

At the same time, it is important to note that those results are not directly comparable to the present task, given differences in dataset size and composition. Coppersmith et al. (2018) used more OurDataHelps data, and this was augmented with a dataset of users who had made publicly self-stated suicide attempts, building on work in Coppersmith et al. (2016). In total, Coppersmith et al. (2018) performed their experimentation using a dataset containing 418 users with suicide attempts, compared to this task's 97 users.

## 8 Enclave Lessons Learned

We solicited feedback from all registered teams (both those who submitted results and those who did not) regarding the shared task experience. This discussion and our lessons learned for the future are informed by their comments.

**Onboarding.** Shared tasks are bursty by nature,

the first burst involving participants getting started. In contrast, the ongoing operations of a data enclave involve a more continuous scheduling process for new user account requests. This led to challenges in the onboarding process. As noted in Section 5, procedures for this shared task were identical to the procedures used when serving organizations like government agencies, with not one fewer *i* dotted, not one fewer *t* crossed. This meant that teams experienced longer than expected delays between completing their paperwork and actually being able to begin work on the Enclave. We would recommend more lead time in the future, leaving significant time for account requests and also having teams prioritize which members need access first.

**Importing code and dependencies.** Similarly, data enclaves require strict import policies and procedures; every import request is treated as though it could contain highly confidential data, a virus, or disabling code. Again, the bursty nature of shared task activity created challenges. Despite our attempts to anticipate and pre-load software and data resources that were likely to be needed (informed by an earlier survey of people engaged in CLPsych-related work), the burst of requests as teams got started created long delays as teams waited for their code and software dependencies to come online. Workarounds, such as recreating code manually, were complicated by the inability to copy/paste inside the environment.

**Time zones.** The CLPsych 2021 Shared Task received global interest, with teams participating on several continents. However, data enclaves rarely provide 24/7 support. While having a diverse set of teams work on the task is indispensable, having support concentrated in a single U.S. time zone disproportionately affected those working outside the U.S. We anticipate that these issues could be mitigated in part by greater lead time (again), and also by streamlining processes to require fewer round trips of communication.

**Slurm and Notebooks.** These days, many prefer to conduct NLP research in an interactive setting using Jupyter Notebooks. While these were supported on the head node of the cluster, they were not available when running jobs on compute nodes, including those with GPU resources. This is worth considering. While such an arrangement would run through one's compute budget faster (as compute nodes would remain running), the interactive benefits may be a tradeoff that teams are willing to make, and this would also avoid batch-job overhead for those who do not require the capabilities offered by a scheduler like Slurm.

**Connectivity and Enclave Maintenance.** Like any well supported infrastructure, the Enclave requires regular maintenance and has occasional downtime. Scheduled maintenance was easy to plan for, but unplanned downtime can be a real challenge in deadline-driven activities like a shared task.

Despite these challenges, which certainly gave rise to some frustration, a number of teams expressed gratitude for being able to work on data that would otherwise be unavailable, and others expressed that they were pleased with the overall responsiveness and speed of the Enclave. Some also expressed appreciation for having had ample of compute credits for conducting their experiments.[8]

If there is a unifying theme in our lessons learned, it is that the challenges we encountered are connected almost entirely with the gap between the typical flexibility of experimental computational work in NLP, particularly in the compressed time frame of a shared task, versus the more extended, carefully centralized, step-by-step, controlled processes that take place on a data enclave. But of course that's the whole point: those same careful, centralized processes are the things that guard against inappropriate use and disclosure of sensitive data.

As a particular note for the future, more advance planning and communication with participants would alleviate several of these challenges, especially onboarding and importing code and dependencies. For this shared task, we chose to prioritize allowing participants to start working on the task sooner, rather than requiring teams to commit long before they would begin work and start going through a more structured and scheduled process to prepare the Enclave with their specific team-level requests. We attempted to preload needed libraries and tools onto the Enclave even before teams began to register — but we could not predict all of the tools and resources participants would want, so even with our efforts there was still a gap. And although we tested the onboarding process and coding experience, any new, diverse group of people is going to discover unanticipated issues when using

[8]AWS credits supporting this activity were provided by Amazon.

a large production environment for a new purpose.

That said, it is worth noting that a time-bounded shared task is just one model for this type of collaborative work. In other domains, it is not uncommon for community shared activity to take place over the longer term, e.g use of the MIMIC dataset (Johnson et al., 2016) in research on electronic health records. A shorter-term, bursty event like a shared task may be the wrong model when navigating between the requirements of flexible research and the requirements of data privacy — many challenges would be mitigated if participants were not all attempting to meet the same deadline. Therefore, an alternative paradigm to consider would involve a more gradual intake of participants, reducing the backlogs and avoiding bottlenecks in account creation and handling of initial import requests. This would would also allow participants to more freely work in their own time zone, and factor in downtimes in their schedule.

## 9 Conclusion

In this effort, we introduced a mental health shared task using sensitive language data in a secure data enclave that offered broad NLP and machine learning capabilities. Participants conducted studies on the prediction of suicide risk based on tweets, using donated data containing actual outcomes rather than proxy data and matching individuals who attempted suicide with control users. Participants built systems that were able to achieve high predictive power (up to 0.823 $F_1$ score), while carefully balancing true positives and false alarms. Through the shared task, we learned more about the challenges of conducting such a task in an enclave environment, leading to observations that will help set the stage for future efforts of this kind.

## Acknowledgments

The shared task organizers would like to express deep gratitude to the individuals who donated data to OurDataHelps, without whom this research would not be possible. The organizers are also immensely grateful to all the participants for their efforts and patience; to the NORC partners and personnel (particularly co-author Jeff Leintz, Ron Jurek, Kyle Stufflebaum, Ramon Castillo, Rachel Miller, Sundeep Bhatia, Wesley Hale, Kim Le, John Nieszel, Jason Keller, and the Data Enclave Manager team) for their tremendous contributions and their willingness to step out onto the bleeding edge in making the Enclave and this shared task happen; to Tim Mulcahy, Scot Ausborn, and Christian Ilie for foundational discussions and effort getting the UMD/NORC Enclave collaboration off the ground; to co-author Glen Coppersmith, Tony Wood, Alex Yelskiy, and the rest of the Qntfy team for their leadership in suicide-related research and collecting and sharing OurDataHelps donated data; to Alexander Hoyle for technical assistance with AWS configuration; to Julia Lane for useful background on data enclaves; to NAACL for its support of CLPsych; and to the creators of the `depression-detection` github repository. This shared task received internal financial support at NORC and was also supported in part by Amazon through an AWS Machine Learning Research Award and by a University of Maryland AI + Medicine for High Impact (AIM-HI) Challenge Award.

## References

Emily Alsentzer, John R Murphy, Willie Boag, Wei-Hung Weng, Di Jin, Tristan Naumann, and Matthew McDermott. 2019. Publicly available clinical BERT embeddings. In *Proceedings of the 2nd Clinical Natural Language Processing Workshop (ClinicalNLP 2019)*.

Ulya Bayram and Lamia Benhiba. 2021. Determining a person's suicide risk by voting the short-term history of tweets for CLPsych 2021 shared task. In *Proceedings of the Seventh Workshop on Computational Linguistics and Clinical Psychology (CLPsych 2021)*.

Iz Beltagy, Kyle Lo, and Arman Cohan. 2019. SciBERT: A pretrained language model for scientific text. In *Proceedings of the 2019 Conference on Empirical Methods in Natural Language Processing and the 9th International Joint Conference on Natural Language Processing (EMNLP-IJCNLP 2019)*.

Adrian Benton, Glen Coppersmith, and Mark Dredze. 2017. Ethical research protocols for social media health research. In *Proceedings of the First ACL Workshop on Ethics in Natural Language Processing*, pages 94–102.

Stevie Chancellor, Michael L. Birnbaum, Eric D. Caine, Vincent M. B. Silenzio, and Munmun De Choudhury. 2019. A taxonomy of ethical tensions in inferring mental health states from social media. In *Proceedings of the conference on fairness, accountability, and transparency*, pages 79–88.

Arman Cohan, Bart Desmet, Andrew Yates, Luca Soldaini, Sean MacAvaney, and Nazli Goharian. 2018.

SMHD: A large-scale resource for exploring online language usage for multiple mental health conditions. In *Proceedings of the 27th International Conference on Computational Linguistics (COLING 2018)*.

Glen Coppersmith, Mark Dredze, and Craig Harman. 2014. Quantifying mental health signals in Twitter. In *Proceedings of the workshop on computational linguistics and clinical psychology: From linguistic signal to clinical reality (CLPsych 2014)*, pages 51–60.

Glen Coppersmith, Mark Dredze, Craig Harman, Kristy Hollingshead, and Margaret Mitchell. 2015. CLPsych 2015 shared task: Depression and PTSD on Twitter. In *Proceedings of the 2nd Workshop on Computational Linguistics and Clinical Psychology: From Linguistic Signal to Clinical Reality (CLPsych 2015)*, pages 31–39.

Glen Coppersmith, Ryan Leary, Patrick Crutchley, and Alex Fine. 2018. Natural language processing of social media as screening for suicide risk. *Biomedical informatics insights*, 10:1178222618792860.

Glen Coppersmith, Kim Ngo, Ryan Leary, and Tony Wood. 2016. Exploratory data analysis of social media prior to a suicide attempt. In *Proceedings of the Workshop on Computational Linguistics and Clinical Psychology: From Linguistic Signal to Clinical Reality (CLPsych 2016)*.

Munmun De Choudhury and Sushovan De. 2014. Mental health discourse on Reddit: Self-disclosure, social support, and anonymity. In *Proceedings of the International AAAI Conference on Web and Social Media*, volume 8.

Munmun De Choudhury, Emre Kiciman, Mark Dredze, Glen Coppersmith, and Mrinal Kumar. 2016. Discovering shifts to suicidal ideation from mental health content in social media. In *Proceedings of the 2016 CHI conference on human factors in computing systems*, pages 2098–2110.

Jacob Devlin, Ming-Wei Chang, Kenton Lee, and Kristina Toutanova. 2019. BERT: Pre-training of deep bidirectional transformers for language understanding. In *Proceedings of the 2019 Conference of the North American Chapter of the Association for Computational Linguistics: Human Language Technologies (NAACL-HLT 2019)*.

Johannes C Eichstaedt, Robert J Smith, Raina M Merchant, Lyle H Ungar, Patrick Crutchley, Daniel Preoţiuc-Pietro, David A Asch, and H Andrew Schwartz. 2018. Facebook language predicts depression in medical records. *Proceedings of the National Academy of Sciences*, 115(44):11203–11208.

Sindhu Kiranmai Ernala, Michael L Birnbaum, Kristin A Candan, Asra F Rizvi, William A Sterling, John M Kane, and Munmun De Choudhury. 2019. Methodological gaps in predicting mental health states from social media: triangulating diagnostic signals. In *Proceedings of the 2019 CHI conference on human factors in computing systems*, pages 1–16.

Joseph C. Franklin, Jessica D. Ribeiro, Kathryn R. Fox, Kate H. Bentley, Evan M. Kleiman, Xieying Huang, Katherine M. Musacchio, Adam C. Jaroszewski, Bernard P. Chang, and Matthew K. Nock. 2017. Risk factors for suicidal thoughts and behaviors: A meta-analysis of 50 years of research. *Psychological Bulletin*.

Avi Gamoran, Yonatan Kaplan, Almog Simchon, and Michael Gilead. 2021. Using psychologically-informed priors for suicide prediction in the CLPsych 2021 shared task. In *Proceedings of the Seventh Workshop on Computational Linguistics and Clinical Psychology (CLPsych 2021)*.

Matthew Honnibal, Ines Montani, Sofie Van Landeghem, and Adriane Boyd. 2020. spaCy: Industrial-strength Natural Language Processing in Python.

Eric Horvitz and Deirdre Mulligan. 2015. Data, privacy, and the greater good. *Science*, 349(6245):253–255.

Alistair EW Johnson, Tom J Pollard, Lu Shen, H Lehman Li-Wei, Mengling Feng, Mohammad Ghassemi, Benjamin Moody, Peter Szolovits, Leo Anthony Celi, and Roger G Mark. 2016. Mimic-iii, a freely accessible critical care database. *Scientific data*, 3(1):1–9.

Biing-Hwang Juang and Lawrence R Rabiner. 2005. Automatic speech recognition–a brief history of the technology development. *Georgia Institute of Technology. Atlanta Rutgers University and the University of California. Santa Barbara*, 1:67.

Julia Lane and Claudia Schur. 2010. Balancing access to health data and privacy: a review of the issues and approaches for the future. *Health services research*, 45(5p2):1456–1467.

Jey Han Lau and Timothy Baldwin. 2016. An empirical evaluation of doc2vec with practical insights into document embedding generation. In *Proceedings of the 1st Workshop on Representation Learning for NLP*.

Ellen E Lee, John Torous, Munmun De Choudhury, Colin A Depp, Sarah A Graham, Ho-Cheol Kim, Martin P Paulus, John H Krystal, and Dilip V Jeste. 2021. Artificial intelligence for mental healthcare: Clinical applications, barriers, facilitators, and artificial wisdom. *Biological Psychiatry: Cognitive Neuroscience and Neuroimaging*.

Michelle Morales, Prajjalita Dey, and Kriti Kohli. 2021. Team 9: A comparison of simple vs. complex models for suicide risk assessment. In *Proceedings of the Seventh Workshop on Computational Linguistics and Clinical Psychology (CLPsych 2021)*.

John A Naslund, Ameya Bondre, John Torous, and Kelly A Aschbrenner. 2020. Social media and mental health: Benefits, risks, and opportunities for research and practice. *Journal of technology in behavioral science*, 5(3):245–257.

F. Pedregosa, G. Varoquaux, A. Gramfort, V. Michel, B. Thirion, O. Grisel, M. Blondel, P. Prettenhofer, R. Weiss, V. Dubourg, J. Vanderplas, A. Passos, D. Cournapeau, M. Brucher, M. Perrot, and E. Duchesnay. 2011. Scikit-learn: Machine learning in Python. *Journal of Machine Learning Research*, 12:2825–2830.

James W Pennebaker, Ryan L Boyd, Kayla Jordan, and Kate Blackburn. 2015. The development and psychometric properties of LIWC2015. Technical report.

Philip Resnik, April Foreman, Michelle Kuchuk, Katherine Musacchio Schafer, and Beau Pinkham. 2021. Naturally occurring language as a source of evidence in suicide prevention. *Suicide and Life-Threatening Behavior*, 51(1):88–96.

Felix Ritchie. 2017. The 'five safes': a framework for planning, designing and evaluating data access solutions. In *Data for Policy 2017: Government by Algorithm? (Data for Policy)*. Zenodo.

Katherine M Schafer, Grace Kennedy, Austin Gallyer, and Philip Resnik. 2021. A direct comparison of theory-driven and machine learning prediction of suicide: A meta-analysis. *PLoS one*, 16(4):e0249833.

Han-Chin Shing, Suraj Nair, Ayah Zirikly, Meir Friedenberg, Hal Daumé III, and Philip Resnik. 2018. Expert, crowdsourced, and machine assessment of suicide risk via online postings. In *Proceedings of the Fifth Workshop on Computational Linguistics and Clinical Psychology: From Keyboard to Clinic (CLPsych 2018)*, pages 25–36.

Robert Thorstad and Phillip Wolff. 2019. Predicting future mental illness from social media: A big-data approach. *Behavior research methods*, 51(4):1586–1600.

Ning Wang, Fan Luo, Yuvraj Shivtare, Varsha Badal, K.P. Subbalakshmi, R. Chandramouli, and Ellen Lee. 2021. Learning models for suicide prediction from social media posts. In *Proceedings of the Seventh Workshop on Computational Linguistics and Clinical Psychology (CLPsych 2021)*.

Susan Wang, Labiba Kanij Rupty, Mahfuza Hu mayra Mohona, Aarthi Alagammai, Munira Omar, and Marwa Qabee. 2019. Depression detection using Twitter data - group project for udacity private and secure ai project showcase. `https://github.com/swcwang/depression-detection`.

Tal Yarkoni and Jacob Westfall. 2017. Choosing Prediction Over Explanation in Psychology: Lessons From Machine Learning. *Perspectives on Psychological Science*.

Andrew Yates, Arman Cohan, and Nazli Goharian. 2017. Depression and self-harm risk assessment in online forums. In *Proceedings of the 2017 Conference on Empirical Methods in Natural Language Processing (EMNLP 2017)*, pages 2968–2978, Copenhagen, Denmark. Association for Computational Linguistics.

Guilherme Augusto Zagatti, Sujatha Das Gollapalli, and See-Kiong Ng. 2021. Suicide risk prediction by tracking self-harm aspects in tweets. In *Proceedings of the Seventh Workshop on Computational Linguistics and Clinical Psychology (CLPsych 2021)*.

Ayah Zirikly, Philip Resnik, Özlem Uzuner, and Kristy Hollingshead. 2019. CLPsych 2019 shared task: Predicting the degree of suicide risk in Reddit posts. In *Proceedings of the Sixth Workshop on Computational Linguistics and Clinical Psychology (CLPsych 2019)*.

# Determining a Person's Suicide Risk by Voting on the Short-Term History of Tweets for the CLPsych 2021 Shared Task

**Ulya Bayram**
Department of Electronics Engineering
Çanakkale Onsekiz Mart University
Çanakkale, Turkey
ulya.bayram@comu.edu.tr

**Lamia Benhiba**
IAD Department, ENSIAS,
Mohammed V University in Rabat
Rabat, Morocco
lamia.benhiba@um5.ac.ma

## Abstract

In this shared task, we accept the challenge of constructing models to identify Twitter users who attempted suicide based on their tweets 30 and 182 days before the adverse event's occurrence. We explore multiple machine learning and deep learning methods to identify a person's suicide risk based on the short-term history of their tweets. Taking the real-life applicability of the model into account, we make the design choice of classifying on the tweet level. By voting the tweet-level suicide risk scores through an ensemble of classifiers, we predict the suicidal users 30-days before the event with an 81.8% true-positives rate. Meanwhile, the tweet-level voting falls short on the six-month-long data as the number of tweets with weak suicidal ideation levels weakens the overall suicidal signals in the long term.

## 1 Introduction

Suicide is amongst the most pressing public health issues facing today's society, stressing the need for rapid and effective detection tools. As people are increasingly self-expressing their distress on social media, an unprecedented volume of data is currently available to detect a person's suicide risk (Roy et al., 2020; Tadesse et al., 2020; Luo et al., 2020). In this shared task, we aim to construct tools to identify suicidal Twitter users (who attempted suicide) based on their tweets collected from spans of 30-days (subtask 1) and six months (subtask 2) before the adverse event's occurrence date (Macavaney et al., 2021). The small number of users in the labeled collections of subtask 1 (57 suicidal/57 control) and subtask 2 (82 suicidal/82 control) and the scarcity of tweets for some users pose these tasks as small-dataset classification challenges. On that note, Coppersmith et al. (2018) reported high performance with deep learning (DL) methods on these collections after enriching them with additional data (418 suicidal/418 control).

When formulating the strategy to attack the challenge, we were motivated by the real-life applicability of the methods. Some social media domains already started implementing auto-detection tools to prevent suicide (Ji et al., 2020). These tools continuously monitor the presence of suicide risk in new posts. Therefore, we chose to train the models at the tweet level. Next, we develop a majority voting scheme over the classified tweets to report an overall suicide risk score for a user. We employ simple machine learning (ML) methods and create an ensemble. We also experiment with DL methods to assess whether complexity would improve the results. Since successful ML applications thrive on feature engineering (Domingos, 2012), we conduct feature selection to evaluate and determine the best feature sets for the models.

Our experiments suggest that majority voting (MV) over tweet-level classification scores is a viable approach for the short-term prediction of suicide risk. We observe that DL methods require plentiful resources despite the small size of the datasets. Simple ML methods with feature selection return satisfactory results, and the performance further improves by the ensemble classifier. We also observe that the MV approach falls short on the six-month-long data regardless of the applied model. Yet this limitation provides the invaluable insight that suicidal ideation signals are more significant when the date of the suicidal event is closer, which stresses the need for more complex, noise immune models for longer time-spanning data. In this context, we consider a noise-immune model as a suicidal ideation detection model that is not affected by tweets lacking suicidal ideation.

## 2 Methods

**Pre-processing:** We clean the tweets by removing user mentions, URLs, punctuation, and non-ASCII characters, then normalize hashtags into words using a probabilistic splitting tool based on English



*Proceedings of the Seventh Workshop on Computational Linguistics and Clinical Psychology*, pages 81–86
June 11, 2021. ©2021 Association for Computational Linguistics

Wikipedia unigram frequencies (Anderson, 2019). We maintain stopwords and emojis, as they might provide clues regarding the suicidal ideation of the users.

**Experimentation Framework:** Before designing the experiments, we face a critical choice: Should we merge all tweets per user, or should we perform the assessment per tweet and then aggregate the scores? To answer this, we consider a real-life risk assessment system. The system should provide a score every time someone posts a tweet. Some social media domains already implement these systems (Ji et al., 2020). Hence, we select to train the models to classify tweets, then apply majority voting (MV) per user to compute a risk score based on the tweet scores. Our framework is described in Figure 1.

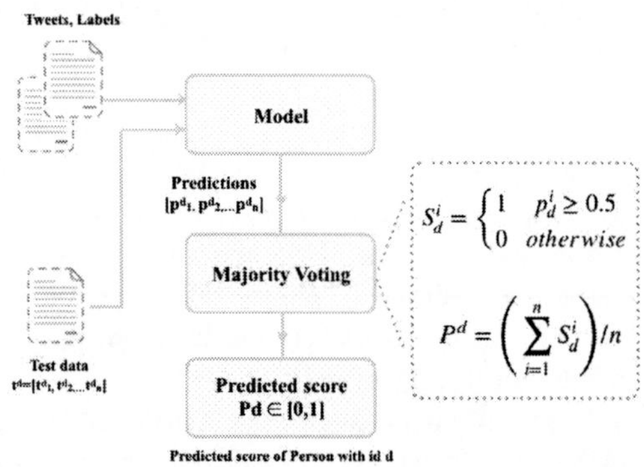


Figure 1: Classification framework used to compute person-level risk scores from the tweet-level scores.

**Experiments with Standard ML methods:** Before ML experiments, we initially explore a simple approach that constructs graphs from training sets and computes how well the given texts match the graphs (Bayram et al., 2018). However, tweets proved to be unfit for the method due to low word counts.

As most ML methods depend on learning from features, we select n-gram features where $n \leq 2$ for their popularity in suicide studies (O'Dea et al., 2015; De Choudhury et al., 2016; Pestian et al., 2020). For bigrams ($n = 2$), we apply a sliding window over concurrent words using the NLTK library (Bird et al., 2009). Next, we eliminate infrequent n-grams from the training set to reduce uninformative features (occurring in $\leq$3 tweets in 30-days, $\leq$10 tweets in 182-days training sets). Subsequently, we scale the features by row-normalizing them with the root of the sum of the square (i.e. variation) of the feature values.

Among the popular ML methods in suicide literature is logistic regression (LR) (Walsh et al., 2017; De Choudhury et al., 2016; O'Dea et al., 2015). We select the "liblinear" solver with default settings for being recommended for small datasets (Buitinck et al., 2013). To cover diverse mathematical frameworks and assumptions, we also include two naive Bayes methods (Gaussian (GNB) and Multinomial (MNB) with default settings) (Buitinck et al., 2013). We also experiment with K-Nearest Neighbors with different distance (uniform, weighted) and neighborhood ($k \in \{3, 5, 8\}$) settings, but we eliminate it for low within-dataset results. Similarly, ensemble-learning methods (Adaboost, XGBoost, Random Forest) also return underwhelming performance despite the parameter tuning, and thus, were eliminated. Additionally, we evaluate support vector machines (SVM) for their popularity in suicide research (Zhu et al., 2020; Pestian et al., 2020; O'Dea et al., 2015). SVM with rbf kernel proves to be successful but requires costly parameter tuning, while linear SVM (lSVM) shows success on within-dataset evaluations with less cost. Consequently, we select lSVM of sklearn (default settings) for the shared task (Buitinck et al., 2013), which returns only binary classification results. To convert them to probabilities, we apply probability calibration with logistic regression (CalibratedClassifierCV).

**Feature selection:** Following the ML method selections, we evaluate the effect of feature selection on ML performance. To compute feature importance scores, we also use the LR. For each selected number of features, we gather top suicidal and control features. Next, we train and evaluate the ML methods in a leave-one-out (LOO) framework using those features. The feature selection results of the selected ML methods for two subtasks are in Figure 2. We select the best ML models from these plots.

**Experiments with Ensemble:** Ensemble classifiers previously showed success in ML challenges (Niculescu-Mizil et al., 2009). Since every classifier renders predicted probabilities for every data point, we build an ensemble classifier to optimize the results of four selected ML methods (LR, GNB, MNB, lSVM). We adopt a weighting ensemble method where the weight of each classifier is set proportional to its performance (Rokach, 2010). We call this method weighted Ensemble (wEns).

**Experiments with DL:** To measure whether re-

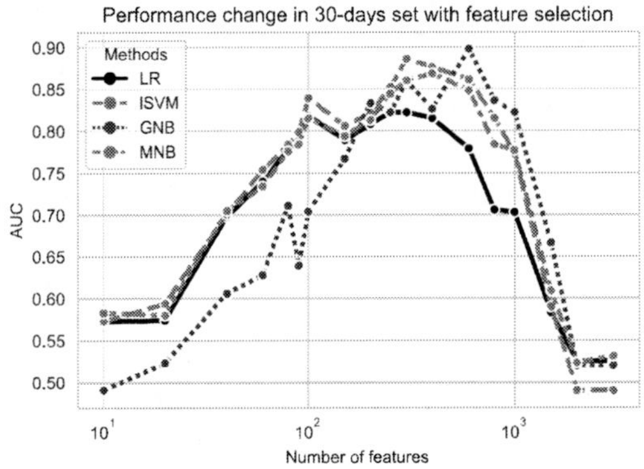


(a) Subtask 1

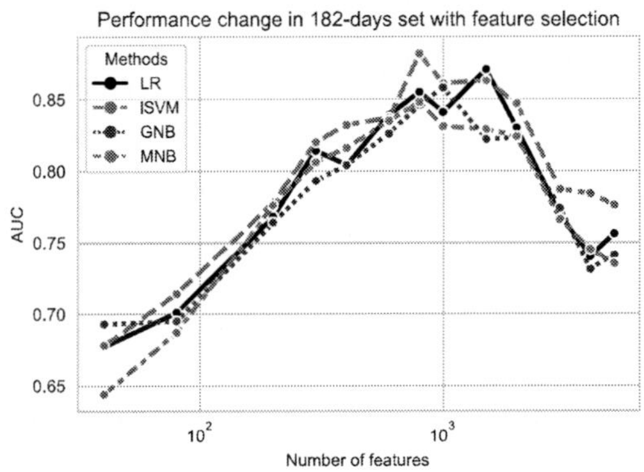


(b) Subtask 2

Figure 2: Feature selection evaluations on the labeled datasets of two subtasks.

sults would improve with complexity, we also evaluate shallow DL methods. We use the pre-trained transformer model Bert-base-uncased (Devlin et al., 2018) to catch the linguistics features of the tweets. The embeddings are then fed to a DL Recurrent Units-based architecture to learn text sequence orders. We experiment with two types of recurrent neural networks (RNNs): Long Short Term Memory (LSTM) (Gers et al., 1999), and Gated Recurrent Unit (GRU) known for overcoming vanishing and exploding gradient problems faced by vanilla RNNs during training (Cho et al., 2014). After assessing various configurations of both architectures, we settle on a multi-layer bi-directional GRU with the following characteristics: embedding dimension=256, number of layers=2, batch size=32. We call this model GRU-Bert. We include a drop-out to regularise learning and a fully connected layer with a Sigmoid activation to produce the classification for each tweet. Finally, we include the same majority voting framework to infer the classification on the user level. We use Pytorch (Paszke et al., 2019) and scikit-learn (Buitinck et al., 2013) libraries for implementation.

## 3 Results

Before training each classifier, we employ the best performing top features from the Figure 2, where every classifier has its most fitting top features for each subtask. Next, we construct a LOO cross-validation framework for within-dataset evaluations.[1] It is important to note that, in each step of the LOO, we choose new user ids for evaluation and completely exclude all of their tweets from the training sets to evade ML methods potentially learning the way a person drafts tweets. That means the within-dataset LOO results of a subtask are reported for all users of the labeled set. Moreover, the labeled datasets have more users than the unlabeled test sets per subtask (e.g. 57 vs. 11 suicidal users in subtask1). Ergo, we expect a high magnitudinal difference between the within-dataset and the test results.

Table 1: Within-dataset evaluation results.

| | F1 | F2 | TPR | FPR | AUC |
|---|---|---|---|---|---|
| **Subtask 1: (30 days)** | | | | | |
| LR | 78.0 | 81.6 | 84.2 | 31.6 | 80.8 |
| GNB | 81.2 | 88.8 | **94.7** | 38.6 | 89.3 |
| MNB | 83.1 | 84.8 | 86.0 | **21.0** | 86.8 |
| lSVM | 81.9 | 87.2 | 91.2 | 31.6 | 88.6 |
| wEns | **85.0** | **90.6** | **94.7** | 28.1 | **93.2** |
| GRU-Bert | 81.2 | 82.2 | 83.1 | **21.7** | 84.0 |
| **Subtask 2: (6 months)** | | | | | |
| LR | **81.9** | 83.9 | 85.4 | **23.2** | 85.5 |
| GNB | 69.6 | 83.0 | **95.1** | 78.0 | 81.5 |
| MNB | 75.7 | 77.1 | 78.0 | 28.0 | 82.8 |
| lSVM | 78.6 | 87.1 | 93.9 | 45.1 | 84.6 |
| wEns | **81.7** | **88.0** | 92.7 | 34.1 | **88.5** |
| GRU-Bert | 74.5 | 75.4 | 76.0 | 28.6 | 77.5 |

The within-dataset evaluation results of the selected methods are in Table 1. For subtask 1, we obtain the best LOO cross-validation score from the wEns method that combines the results of four ML methods (LR, MNB, GNB, lSVM) in a way that improves the results obtained from each of them. Meanwhile, GRU-Bert and MNB return the lowest false positive rates (FPR) for this subtask,

[1] Within-dataset evaluation results of the selected ML and weighted ensemble methods are obtained from LOO cross-validation. While for GRU-Bert, collections were split into training-validation-test sets in 70:10:20 ratios.

which might be a critical rate to consider in real-life applications in social media domains. LOO results of subtask 2 in Table 1 show that wEns returns the best scores for the longer-spanning dataset as well, where LR returns the best FPR, and GBN returns the highest true positives rate (TPR).

Table 2: Test results over unlabeled data and the results from the baseline method of CLPsych2021.

| | F1 | F2 | TPR | FPR | AUC |
|---|---|---|---|---|---|
| **Subtask 1: (30 days)** | | | | | |
| Baseline | 63.6 | 63.6 | 63.6 | **36.4** | 66.1 |
| LR | 63.6 | 63.6 | 63.6 | **36.4** | **74.0** |
| wEns | **69.2** | **76.3** | **81.8** | 54.5 | 70.2 |
| **Subtask 2: (6 months)** | | | | | |
| Baseline | **71.0** | **72.4** | **73.3** | **33.3** | **76.4** |
| LR | 64.5 | 65.8 | 66.7 | 40.0 | 56.9 |
| wEns | 59.5 | 67.1 | **73.3** | 73.3 | 58.2 |

Based on the LOO results, we select three different methods we were allowed to submit for the evaluation of the test set: LR, wEns, and GRU-Bert. We choose LR and wEns for their high performance on LOO experiments, while we select GRU-Bert for measuring how a DL method would generalize over the test sets. The baseline classifier provided by the organizers is also a logistic regression. However, it performs the classification over merged tweets of users - therefore is different from our implementation of LR. In Table 2, wEns appears to provide the best F1, F2, and TPR scores over the test set of subtask 1, while our LR outperforms the AUC of the baseline method. While these methods show the success of generalizability on the 30-days test set, the results are not that successful for subtask 2. The wEns method performs the same as the baseline in terms of TPR, but the rest of the scores are lower than the baseline results.

## 4 Discussion

In subtask 1, the test set results show that feature selection can considerably enhance the performance of ML models compared to the baseline. We also find that the ensemble classifier is comparably better than the baseline in this subtask. Meanwhile, though the baseline of CLPsych2021 is the same as our LR, our additional MV and feature selection together enable LR to substantially outperform the baseline. These successes of simple ML methods indicate that a collection of tweets from within the 30-days of a suicidal event is good enough to capture the existence of suicidal ideation, which is an important finding for future real-life suicide prevention applications.

In contrast to the observations from subtask 1, our test results on subtask 2 are unsatisfactory. Yet, they provide the valuable insight that suicidal signals are more significant in the short-term, and older tweets lacking suicidal ideation generate noise. This insight suggests the need to account for a time-domain aspect. To investigate the viability of this claim, we experiment with a simple time-decay coefficient in the MV framework and evaluate it through LR on the test set. We multiply each vote by the coefficient $2^{\frac{-timeDiff}{halfLife}}$ where $timeDiff$ is the number of days between the current and last tweets, and $halfLife$ (=7 days) is a hyperparameter that reflects the weight of a vote in the final suicide risk score of a user. Initial experiments show that even this simple time-decay coefficient improves the test results significantly. This observation suggests that tweet dates are critical features for this subtask and should be included in future work.

Notwithstanding, on both subtasks, the shallow DL methods we experimented with perform poorly. These results could be attributed to overfitting on the small dataset and noise sensitivity for the larger time-spanning dataset. Additionally, regardless of the dataset size, these methods proved to be computationally expensive. As within-dataset experiments using simple ML methods outperformed these expensive shallow DL methods, we excluded the latter from the test set evaluation. Future work on DL will include deeper, more complex, and noise immune methods that could integrate Convolutional neural networks (CNN), deeper LSTM or GRU layers, and experiments with various word embedding models.

If we compare our findings with those in Coppersmith et al. (2018), we observe different results in terms of short-term versus long-term dataset classifications. We attribute these different outcomes to the fact that the original study optimizes the design for detecting trait-level (relevant to risk for any point in time) suicide risk when we endeavor to identify suicidal ideation at the state level (immediate risk presence). This design choice, along with tweet-level classification, enabled our model to recognize suicidal nuances in short-term tweets. Meanwhile, we were unable to detect any suicidal ideation through manual inspection (reading and interpreting the tweets) over most of these tweets due to their noisy and ambiguous nature.

## 5 Conclusion

In this shared task, we investigate various models for identifying suicide risk based on user's tweets. Inspired by real-life applications, we focus on assessing suicide risk on the tweet level. Experimental results reveal that the ensemble classifier can identify suicidal users from 30-days tweets with a high performance rate, demonstrating the power of majority voting over tweet-level classifications for short-term suicide risk detection. Meanwhile, we construe from the underwhelming results on the six-month dataset that these models were more sensitive to the signals relevant to short term risk than those relevant to long term risk. In future work, we will incorporate a temporal aspect to improve the noise immunity of our models, and we will continue experimenting with more complex models.

## Ethics Statement

Secure access to the shared task dataset was provided with IRB approval under University of Maryland, College Park protocol 1642625.

## Acknowledgements

The organizers are particularly grateful to the users who donated data to the OurDataHelps project without whom this work would not be possible, to Qntfy for supporting the OurDataHelps project and making the data available, to NORC for creating and administering the secure infrastructure, and to Amazon for supporting this research with computational resources on AWS. The authors are thankful to the anonymous reviewers for their constructive comments and valuable suggestions.

## References

Derek Anderson. 2019. wordninja Python library. `https://github.com/keredson/wordninja`. [Online; accessed 11-March-2021].

Ulya Bayram, Ali A Minai, and John Pestian. 2018. A lexical network approach for identifying suicidal ideation in clinical interview transcripts. In *International Conference on Complex Systems*, pages 165–172. Springer.

Steven Bird, Ewan Klein, and Edward Loper. 2009. *Natural language processing with Python: analyzing text with the natural language toolkit.* " O'Reilly Media, Inc.".

Lars Buitinck, Gilles Louppe, Mathieu Blondel, Fabian Pedregosa, Andreas Mueller, Olivier Grisel, Vlad Niculae, Peter Prettenhofer, Alexandre Gramfort, Jaques Grobler, Robert Layton, Jake VanderPlas, Arnaud Joly, Brian Holt, and Gaël Varoquaux. 2013. API design for machine learning software: experiences from the scikit-learn project. In *ECML PKDD Workshop: Languages for Data Mining and Machine Learning*, pages 108–122.

Kyunghyun Cho, Bart Van Merriënboer, Dzmitry Bahdanau, and Yoshua Bengio. 2014. On the properties of neural machine translation: Encoder-decoder approaches. *arXiv preprint arXiv:1409.1259.*

Glen Coppersmith, Ryan Leary, Patrick Crutchley, and Alex Fine. 2018. Natural language processing of social media as screening for suicide risk. *Biomedical informatics insights*, 10:1178222618792860.

Munmun De Choudhury, Emre Kiciman, Mark Dredze, Glen Coppersmith, and Mrinal Kumar. 2016. Discovering shifts to suicidal ideation from mental health content in social media. In *Proceedings of the 2016 CHI Conference on Human Factors in Computing Systems*, pages 2098–2110. ACM.

Jacob Devlin, Ming-Wei Chang, Kenton Lee, and Kristina Toutanova. 2018. Bert: Pre-training of deep bidirectional transformers for language understanding. *arXiv preprint arXiv:1810.04805.*

Pedro Domingos. 2012. A few useful things to know about machine learning. *Communications of the ACM*, 55(10):78–87.

Felix A Gers, Jürgen Schmidhuber, and Fred Cummins. 1999. Learning to forget: Continual prediction with lstm. In *9th International Conference on Artificial Neural Networks: ICANN '99*, pages 850–855. IET.

Shaoxiong Ji, Shirui Pan, Xue Li, Erik Cambria, Guodong Long, and Zi Huang. 2020. Suicidal ideation detection: A review of machine learning methods and applications. *IEEE Transactions on Computational Social Systems.*

Jianhong Luo, Jingcheng Du, Cui Tao, Hua Xu, and Yaoyun Zhang. 2020. Exploring temporal suicidal behavior patterns on social media: Insight from twitter analytics. *Health informatics journal*, 26(2):738–752.

Sean Macavaney, Anjali Mittu, Glen Coppersmith, Jeff Leintz, and Philip Resnik. 2021. Community-level research on suicidality prediction in a secure environment: Overview of the CLPsych 2021 shared task. In *Proceedings of the Seventh Workshop on Computational Linguistics and Clinical Psychology (CLPsych 2021)*. Association for Computational Linguistics.

Alexandru Niculescu-Mizil, Claudia Perlich, Grzegorz Swirszcz, Vikas Sindhwani, Yan Liu, Prem Melville, Dong Wang, Jing Xiao, Jianying Hu, Moninder Singh, et al. 2009. Winning the kdd cup orange challenge with ensemble selection. In *KDD-Cup 2009 Competition*, pages 23–34. PMLR.

Bridianne O'Dea, Stephen Wan, Philip J Batterham, Alison L Calear, Cecile Paris, and Helen Christensen. 2015. Detecting suicidality on twitter. *Internet Interventions*, 2(2):183–188.

Adam Paszke, Sam Gross, Francisco Massa, Adam Lerer, James Bradbury, Gregory Chanan, Trevor Killeen, Zeming Lin, Natalia Gimelshein, Luca Antiga, Alban Desmaison, Andreas Kopf, Edward Yang, Zachary DeVito, Martin Raison, Alykhan Tejani, Sasank Chilamkurthy, Benoit Steiner, Lu Fang, Junjie Bai, and Soumith Chintala. 2019. Pytorch: An imperative style, high-performance deep learning library. In H. Wallach, H. Larochelle, A. Beygelzimer, F. d'Alché-Buc, E. Fox, and R. Garnett, editors, *Advances in Neural Information Processing Systems 32*, pages 8024–8035. Curran Associates, Inc.

John Pestian, Daniel Santel, Michael Sorter, Ulya Bayram, Brian Connolly, Tracy Glauser, Melissa DelBello, Suzanne Tamang, and Kevin Cohen. 2020. A machine learning approach to identifying changes in suicidal language. *Suicide and Life-Threatening Behavior*, 50(5):939–947.

Lior Rokach. 2010. Ensemble-based classifiers. *Artificial intelligence review*, 33(1):1–39.

Arunima Roy, Katerina Nikolitch, Rachel McGinn, Safiya Jinah, William Klement, and Zachary A Kaminsky. 2020. A machine learning approach predicts future risk to suicidal ideation from social media data. *NPJ digital medicine*, 3(1):1–12.

Michael Mesfin Tadesse, Hongfei Lin, Bo Xu, and Liang Yang. 2020. Detection of suicide ideation in social media forums using deep learning. *Algorithms*, 13(1):7.

Colin G Walsh, Jessica D Ribeiro, and Joseph C Franklin. 2017. Predicting risk of suicide attempts over time through machine learning. *Clinical Psychological Science*, 5(3):457–469.

H Zhu, X Xia, J Yao, H Fan, Q Wang, and Q Gao. 2020. Comparisons of different classification algorithms while using text mining to screen psychiatric inpatients with suicidal behaviors. *Journal of psychiatric research*, 124:123–130.

# Learning Models for Suicide Prediction from Social Media Posts

**Ning Wang**[1] **Fan Luo**[1] **Yuvraj Shivtare**[1*] **Varsha Badal**[2]
**K.P. Subbalakshmi**[1] **R. Chandramouli**[1] **Ellen Lee**[2]
Stevens Institute of Technology[1]
Hoboken, NJ 07030
{nwang7, fluo4, yshivtar, ksubbala, mouli}@stevens.edu
University of California San Diego[2]
San Diego, CA 92161
{vbadal, eel013}@health.ucsd.edu

## Abstract

We propose a deep learning architecture and test three other machine learning models to automatically detect individuals that will attempt suicide within (1) 30 days and (2) six months, using their social media post data provided in (Macavaney et al., 2021) via the CLPsych 2021 shared task. Additionally, we create and extract three sets of handcrafted features for suicide risk detection based on the three-stage theory of suicide and prior work on emotions and the use of pronouns among persons exhibiting suicidal ideations. Extensive experimentations show that some of the traditional machine learning methods outperform the baseline with an F1 score of 0.741 and F2 score of 0.833 on subtask 1 (prediction of a suicide attempt 30 days prior). However, the proposed deep learning method outperforms the baseline with F1 score of 0.737 and F2 score of 0.843 on subtask 2 (prediction of suicide 6 months prior).

## 1 Introduction

According to World Health Organization (WHO) [1], close to 800,000 people die due to suicide every year, which is one person every 40 seconds. The US Centers for Disease Control and Prevention (CDC) [2] claimed that suicide was the tenth leading cause of death overall in the United States. Recently, there has been a trend in using natural language processing (NLP) techniques on unstructured physician notes from electronic health record (EHR) data to detect high-risk patients (Fernandes et al., 2018).

With the proliferation of social media where there is free sharing of information, mining data from these platforms has become a natural way to extend the above body of work in more natural settings. Consequently, researchers have started to apply machine learning and NLP based techniques to detect suicide ideation on social media platforms (Ramírez-Cifuentes et al., 2020; Roy et al., 2020). Some of them focused on handcrafted features, including TF-IDF (Zhang et al., 2011), LIWC (Tausczik and W, 2010), N-gram, Part-of-Speech (PoS) and emotions (Shah et al., 2020; Zirikly et al., 2019; Zhang et al., 2015; Ji et al., 2020), while others explored language embeddings (Cao et al., 2019; Jones et al., 2019; Sawhney et al., 2018; Coppersmith et al., 2018).

In this paper, we present several approaches to detect suicide ideation from Twitter posts (1) 30 days before the attempt and (2) six months before the attempt. We use the dataset provided by the CLPsych 2021 Shared Tasks Macavaney et al. (2021) towards this goal.

The main contributions of our work are:

- Explored and generated multiple handcrafted feature sets motivated by prior work in this area
- Proposed a new deep learning architecture that uses latent features from tweets to detect suicide attempts
- Tested several machine learning algorithms using only handcrafted features and only latent features
- Achieved better performance than baseline in terms of F1, F2 and True Positive Rate (TPR) on both subtasks

**Summary of Findings:** The main takeaways from this work are:

- Extensive testing on the dataset shows that latent feature (Doc2Vec (Lau and Baldwin, 2016)), is better at detecting suicide attempts from the tweets than handcrafted features
- Most of our models performed better on detecting individuals who have attempted suicide or were a victim of suicide than on detecting control individuals who have not

*Equal contribution to 2nd author

[1]https://www.who.int/

[2]https://www.cdc.gov/



*Proceedings of the Seventh Workshop on Computational Linguistics and Clinical Psychology*, pages 87–92
June 11, 2021. 

- The KNN and SVM with latent features perform best on subtask 1, with respect to F1, F2 and TPR; while our proposed C-Attention (C-Att) network performs best on subtask 2, with respect to F1, F2 and TPR

## 2 Method

Before we describe the methods in detail we provide a summary of the features used in our work. We use two classes of features: latent features and handcrafted features. These are described in the sections below.

### 2.1 Latent Features

Latent features are typically obtained as language embeddings. In our case, we used the Doc2vec (Lau and Baldwin, 2016) to generate both word embeddings and document embeddings on each post. Doc2Vec creates a vectorized representation of a group of words (or a single word, when used in that mode) taken collectively as a single unit. For every document in the corpus, Doc2Vec computes a feature vector. There are two models for implementing Doc2vec: Distributed Memory version of Paragraph Vector (PV-DM) and Distributed Bag of Words version of Paragraph Vector (PV-DBOW). For our experimentation, we used Distributed Memory (DM) version. DM randomly samples consecutive words from a sentence and predicts a center word using these randomly sampled set of context words and the feature vector.

### 2.2 Handcrafted Features

#### 2.2.1 Emotions

Emotions can be good indicators of depression and suicide ideation (Desmet and Hoste, 2013; Coppersmith et al., 2016; Cao et al., 2020; Ghosh et al., 2020), so we include emotions as one of the handcrafted features. We used the method proposed in (Shao et al., 2019) to generate 12 emotion tags, including contentment, pride, fear, anxiety, sadness, disgust, relief, shame, anger, interest, agreeableness and joy. Apart from that we also generated emotion intensity scores using NRC lexicon (Mohammad, 2018), for the emotions like anger, anticipation, disgust, fear, joy, sadness, surprise and trust. After removing duplicates, we selected 17 emotion tags.

#### 2.2.2 Parts of Speech

We use NLTK (Bird et al., 2009) to generate Part-of-Speech tags. PoS tags can detect the syntactic structure difference between users that attempt suicide and the control group (Ji et al., 2020). It has been shown (Roubidoux, 2012) that persons attempting suicide use more first person pronouns. Therefore, we also calculate the number of occurrences of first person pronouns like "I", "me", "mine" and "myself" and include this count as another PoS related handcrafted feature. In total, we generated 34 PoS tags per post for the "30 days prior prediction" subtask and 37 PoS tags for the "6 months prior prediction" subtask.

#### 2.2.3 Three-step theory of suicide and suicide dictionary

We then generate a dictionary of words based on the three-step theory of suicide (3ST) (Klonsky and May, 2015) beginning with the ideation, followed by unmitigated strengthening of the idea due to insufficient social support and precipitated by an attempt. These stages are underpinned by feelings of hopelessness (Dixon et al., 1991), thwarted belongingness and burdensomeness (Chu et al., 2018; Forkmann and Teismann, 2017). Violence usually differentiates attempters and non-attempters (Stack, 2014). Surviving an attempt is expected to be accompanied by feelings of shame (Wiklander et al., 2012; Wolk-Wasserman, 1985). We expect these feelings to be out of phase with each other creating a leading, inline and lagging indicator of suicide attempt. We used Word2vec (Mikolov et al., 2013b,a,c) software to construct these dictionaries using the accompanying utility (also available in online versions) by evaluating closest neighbors of words (gloom and burden, violence, hurt and shame), each containing about 100 words with some manual cleanup and editing. The manual cleanup involved removing stop-words, words with hyphens, special characters, some vernacular tokens, and words that differed in capitalization alone. We generated this feature set by counting each keyword in each post. In addition, we manually created a dictionary of suicide keywords based on suicide-related words published in (Low et al., 2020; Yao et al., 2020), and counted how many suicide-related keywords occurred in each post. [3]

### 2.3 Models

In this work, we proposed a deep learning model and used a few other machine learning models for

[3]Available at: `https://sites.google.com/stevens.edu/infinitylab/suicide-risk-detection`

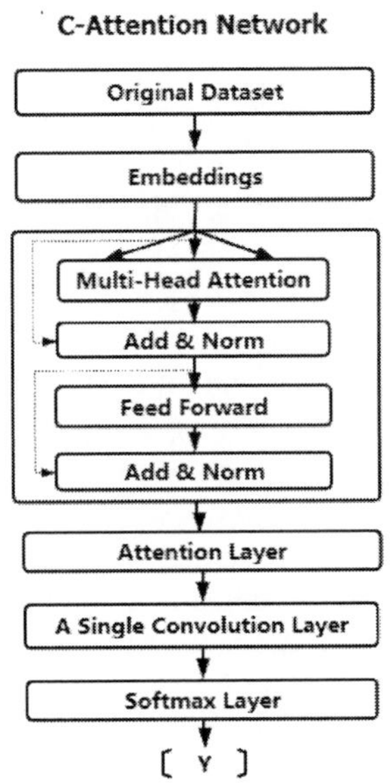


Figure 1: The proposed architecture of C-Attention Network

each subtask. The proposed deep learning model, which we refer to as the C-Attention Network, is our primary model.

#### 2.3.1 C-Attention Network

Figure 1 depicts our C-Attention network which uses latent features to detect suicide attempts. This network is similar to our prior C-Attention Embedding model (Wang et al., 2020) with the following differences:

- In this work we consider each post as a small document, and use Doc2Vec to generate a 100-dimension embedding representation for each post; whereas the work in (Wang et al., 2020) generated a sentence embedding for each sentence in a speech.
- We removed the positional encoding layer since there is no positional dependency among posts.

In summary, the architecture first calculates the embeddings of the dataset, then processes it via a multi head self-attention (MHA) module that captures the intra-feature relation-ships; an attention layer followed by a single convolution layer and a softmax layer. The MHA module is the same as that proposed in (Vaswani et al., 2017) for the popular transformer architecture.

#### 2.3.2 Latent Features with Other Machine Learning Models

In this approach we combined all the posts for each user. Stop words were removed from the posts and lemmatized. The average length of posts was found to be 140 words. Long posts were chunked into 150 words segments to retain meaningful information in each post. A single 200-dimension embedding vector is generated for each segment using the Doc2Vec as described in Section 2.1.

We applied linear discriminant analysis (LDA) (McLachlan, 2004) for dimensionality reduction before classification. The output of LDA was fed to machine learning models. $K$-Nearest Neighbors (KNN) (Jiang et al., 2012) with $K$=3, Support Vector Machine (SVM) (Ríssola et al., 2019) with linear kernel (referred to as SVM(EB) in the rest of the paper) and Decision Tree (D-Tree) (Song and Ying, 2015) classifier models were considered.

#### 2.3.3 Handcrafted Features with Other Machine Learning Models

We used three other machine learning models on the handcrafted features described in Sec 2.2 to address both challenges. The three machine learning models were: Random Forest Classifier (RF) (Breiman, 2001), Logistic Regression (LR) (Aladağ et al., 2018) and Support Vector Machine (SVM) (Ríssola et al., 2019) (referred to as SVM(HF) for the rest of the paper). We used the entire handcrafted features since we found that leaving out any of those handcrafted feature sets would introduce a performance drop. We fine-tuned the parameters of each ML model, for example, we set the kernel as rbf (radial basis function) on SVM(HF) model; set the solver as liblinear (limited to one-versus-rest schemes) on LR model; and set the max depth to 4 on RF model to get the best predictions.

## 3 Results

Table 1 and Table 2 show the performance results. The results reported in Table 1 were obtained by running the KNN, SVM(EB) and SVM(HF) models which were trained on the entire training set. The performance of the models are measured in terms of F1 and F2 scores, True Positive Rates (TPR), False Positive Rates (FPR) and Area Under the ROC Curve (AUC).

| | F1 | F2 | TPR | FPR | AUC |
|---|---|---|---|---|---|
| **Subtask 1 (30 days)** | | | | | |
| Baseline | 0.636 | 0.636 | 0.636 | 0.364 | 0.661 |
| KNN | 0.286 | 0.278 | 0.273 | 0.636 | 0.264 |
| SVM(EB) | 0.400 | 0.377 | 0.364 | 0.455 | 0.529 |
| SVM(HF) | 0.364 | 0.364 | 0.364 | 0.636 | 0.397 |
| **Subtask 2 (6 months)** | | | | | |
| Baseline | 0.710 | 0.724 | 0.733 | 0.333 | 0.764 |
| KNN | 0.429 | 0.411 | 0.400 | 0.467 | 0.444 |
| SVM(EB) | 0.533 | 0.533 | 0.533 | 0.467 | 0.640 |
| SVM(HF) | 0.400 | 0.400 | 0.400 | 0.600 | 0.502 |

Table 1: Results obtained by running the KNN, SVM(EB) and SVM(HF) models trained on the entire training set.

| | F1 | F2 | TPR | FPR | AUC |
|---|---|---|---|---|---|
| **Subtask 1 (30 days)** | | | | | |
| Baseline | 0.636 | 0.636 | 0.636 | 0.364 | 0.661 |
| C-Att | 0.690 | 0.806 | **0.909** | 0.727 | 0.504 |
| SVM(HF) | 0.621 | 0.726 | 0.818 | 0.818 | 0.570 |
| LR | 0.571 | 0.556 | 0.545 | 0.364 | 0.434 |
| RF | 0.444 | 0.392 | 0.364 | **0.273** | 0.603 |
| KNN | **0.741** | **0.833** | **0.909** | 0.545 | **0.694** |
| D-Tree | 0.667 | 0.750 | 0.818 | 0.636 | 0.591 |
| SVM(EB) | **0.741** | **0.833** | **0.909** | 0.545 | 0.653 |
| **Subtask 2 (6 months)** | | | | | |
| Baseline | 0.710 | 0.724 | 0.733 | 0.333 | **0.764** |
| C-Att | **0.737** | **0.843** | **0.933** | 0.600 | 0.76 |
| SVM(HF) | 0.600 | 0.706 | 0.800 | 0.867 | 0.518 |
| LR | 0.563 | 0.584 | 0.600 | 0.533 | 0.542 |
| RF | 0.417 | 0.362 | 0.333 | **0.267** | 0.558 |
| KNN | 0.500 | 0.479 | 0.467 | 0.400 | 0.536 |
| D-Tree | 0.500 | 0.479 | 0.467 | 0.400 | 0.533 |
| SVM(EB) | 0.444 | 0.417 | 0.400 | 0.400 | 0.489 |

Table 2: Results obtained when the training dataset was split into training and validation set as described. HF represents handcrafted features. EB represents word embeddings.

## 4 Analysis/Discussion

The results reported in Table 1 were generated by the KNN, SVM(EB) and SVM(HF) models, which performed best on the training set. From Table 1, we can see that the baseline provided by the CLPsych 2021 shared task outperformed all of these methods. After a thorough investigation of the results, we observed that those models that did not perform best on the training set, performed better on the test set. It probably indicates that we over-trained our models on the training set.

As a result, in the following experiments, we randomly split the training set into 80% for training and 20% for validation, and use the models that performed best on the validation set to predict suicide in the test set. The new performance results on the test set are shown in Table 2.

We noted that in subtask 1, KNN and SVM(EB) performed best in terms of F1, F2 and TRP. The best AUC was achieved by KNN only, and the best FPR was achieved by RF. In subtask 2, C-Att performed best in terms of F1, F2 and TRP; the best FPR was achieved by RF; and the best AUC was achieved by Baseline.

Our experiment results would indicate that:

- In general, latent features perform better than handcrafted features in this shared task
- C-Att model performs better on longer range suicide predictions and KNN and SVM(EB) work better on shorter range suicide predictions
- Besides RF, our other models perform better on detecting suicide individuals than control individuals

## 5 Conclusion

In this work, we introduce C-Attention model and test other machine learning models to automatically detect suicidal individuals based on the latent feature (Doc2Vec) and handcrafted features including emotions, PoS, and three-step theory of suicide and suicide dictionary. Our results show that both KNN and SVM(EB) achieved the best F1 score of 0.741 and F2 score of 0.833 on subtask 1 (prediction of a suicide attempt 30 days prior), and C-Att reached the best F1 score of 0.737 and F2 score of 0.843 on subtask 2 (prediction of suicide 6 months prior).

Ultimately, this work supports the use of social media as an avenue to better predict and understand the experience of suicidal thoughts. However more work is needed to better decipher why certain features and models best predict suicidality in large, diverse, representative samples.

## Ethics Statement

Secure access to the shared task dataset was provided with IRB approval under University of Maryland, College Park protocol 1642625.

## Acknowledgements

We appreciate the efforts of the organizers of this challenge to make the data and computational resources available to us.

The organizers are particularly grateful to the users who donated data to the OurDataHelps project without whom this work would not be possible, to Qntfy for supporting the OurDataHelps project and making the data available, to NORC for creating and administering the secure infrastructure, and to Amazon for supporting this research with computational resources on AWS.

## References

Ahmet Emre Aladağ, Serra Muderrisoglu, Naz Berfu Akbas, Oguzhan Zahmacioglu, and Haluk O Bingol. 2018. Detecting suicidal ideation on forums: Proof-of-concept study. *JMIR*.

Steven Bird, Ewan Klein, and Edward Loper. 2009. *Natural language processing with Python: analyzing text with the natural language toolkit.* " O'Reilly Media, Inc.".

Leo Breiman. 2001. Random forests. *Springer*.

Lei Cao, Huijun Zhang, and Ling Feng. 2020. Building and using personal knowledge graph to improve suicidal ideation detection on social media. *IEEE*.

Lei Cao, Huijun Zhang, Ling Feng, Zihan Wei, Xin Wang, Ningyun Li, and He Xiaohao. 2019. Latent suicide risk detection on microblog via suicide-oriented word embeddings and layered attention. *ArXiv*.

Carol Chu, Megan L Rogers, Anna R Gai, and Thomas E Joiner. 2018. Role of thwarted belongingness and perceived burdensomeness in the relationship between violent daydreaming and suicidal ideation in two adult samples. *Journal of aggression, conflict and peace research.*

Glen Coppersmith, Ryan Leary, Patrick Crutchley, and Alex Fine. 2018. Natural language processing of social media as screening for suicide risk. *Biomedical informatics insights*, 10:1178222618792860.

Glen Coppersmith, Kim Ngo, Ryan Leary, and Anthony Wood. 2016. Exploratory analysis of social media prior to a suicide attempt. In *Proceedings of the third workshop on computational linguistics and clinical psychology*, pages 106–117.

Bart Desmet and Véronique Hoste. 2013. Emotion detection in suicide notes. *ScienceDirect*.

Wayne A Dixon, P Paul Heppner, and Wayne P Anderson. 1991. Problem-solving appraisal, stress, hopelessness, and suicide ideation in a college population. *Journal of Counseling Psychology*, 38(1):51.

Andrea C Fernandes, Rina Dutta, Sumithra Velupillai, Jyoti Sanyal, Robert Stewart, and David Chandran. 2018. Identifying suicide ideation and suicidal attempts in a psychiatric clinical research database using natural language processing. *Scientific reports*, 8(1):1–10.

Thomas Forkmann and Tobias Teismann. 2017. Entrapment, perceived burdensomeness and thwarted belongingness as predictors of suicide ideation. *Psychiatry research*, 257:84–86.

Soumitra Ghosh, Asif Ekbal, and Pushpak Bhattacharyya. 2020. A multitask framework to detect depression, sentiment and multi-label emotion from suicide notes. *Springer*.

Shaoxiong Ji, Shirui Pan, Xue Li, Erik Cambria, Guodong Long, and Zi Huang. 2020. Suicidal ideation detection: A review of machine learning methods and applications. *IEEE*.

Shengyi Jiang, Guansong Pang, Meiling Wu, and Limin Kuang. 2012. An improved k-nearest-neighbor algorithm for text categorization. *Expert Systems with Applications*, 39(1):1503–1509.

Noah Jones, Natasha Jaques, Pat Pataranutaporn, Asma Ghandeharioun, and Picard Rosalind. 2019. Analysis of online suicide risk with document embeddings and latent dirichlet allocation. *IEEE*.

E David Klonsky and Alexis M May. 2015. The three-step theory (3st): A new theory of suicide rooted in the "ideation-to-action" framework. *International Journal of Cognitive Therapy*, 8(2):114–129.

Jey Han Lau and Timothy Baldwin. 2016. An empirical evaluation of doc2vec with practical insights into document embedding generation. *arXiv preprint arXiv:1607.05368*.

Daniel M Low, Laurie Rumker, Tanya Talkar, John Torous, Guillermo Cecchi, and Satrajit S Ghosh. 2020. Natural language processing reveals vulnerable mental health support groups and heightened health anxiety on reddit during covid-19: Observational study. *Journal of medical Internet research*, 22(10):e22635.

Sean Macavaney, Anjali Mittu, Glen Coppersmith, Jeff Leintz, and Philip Resnik. 2021. Community-level research on suicidality prediction in a secure environment: Overview of the CLPsych 2021 shared task. In *Proceedings of the Seventh Workshop on Computational Linguistics and Clinical Psychology (CLPsych 2021)*. Association for Computational Linguistics.

Geoffrey J McLachlan. 2004. *Discriminant analysis and statistical pattern recognition*, volume 544. John Wiley & Sons.

Tomas Mikolov, Kai Chen, Greg Corrado, and Jeffrey Dean. 2013a. Efficient estimation of word representations in vector space. *arXiv preprint arXiv:1301.3781*.

Tomas Mikolov, Ilya Sutskever, Kai Chen, Greg Corrado, and Jeffrey Dean. 2013b. Distributed representations of words and phrases and their compositionality. *arXiv preprint arXiv:1310.4546*.

Tomas Mikolov, Ilya Sutskever, Kai Chen, Greg Corrado, and Jeffrey Dean. 2013c. Distributed representations of words and phrases and their compositionality. *arXiv preprint arXiv:1310.4546*.

Saif M. Mohammad. 2018. Word affect intensities. In *Proceedings of the 11th Edition of the Language Resources and Evaluation Conference (LREC-2018)*, Miyazaki, Japan.

Diana Ramírez-Cifuentes, Freire Ana, Ricardo Baeza-Yates, Joaquim Puntí, Pilar Medina-Bravo, Diego Alejandro Velazquez, Josep Maria Gonfaus, and Jordi Gonzàlez. 2020. Detection of suicidal ideation on social media: multimodal, relational, and behavioral analysis. 22(7):e17758.

Susan M. Roubidoux. 2012. Linguistic manifestations of power in suicide notes: An investigation of personal pronouns.

Arunima Roy, Katerina Nikolitch, Rachel McGinn, Safiya Jinah, William Klement, and Zachary A Kaminsky. 2020. A machine learning approach predicts future risk to suicidal ideation from social media data. *NPJ digital medicine*, 3(1):1–12.

EstebanRíssola, Diana Ramírez-Cifuentes, Ana Freire, and Fabio Crestani. 2019. Suicide risk assessment on social media: Usi-upf at the clpsych 2019 shared task. *ACL*.

Ramit Sawhney, Prachi Manchanda, Puneet Mathur, Raj Singh, and Shah Rajiv Ratn. 2018. Exploring and learning suicidal ideation connotations on social media with deep learning. *ACL*.

Faisal Muhammad Shah, Farsheed Haque, Ragib Un Nur, Shaeekh Al Jahan, and Zarar Mamud. 2020. A hybridized feature extraction approach to suicidal ideation detection from social media post. *IEEE*.

Zongru Shao, Rajarathnam Chandramouli, KP Subbalakshmi, and Constantine T Boyadjiev. 2019. An analytical system for user emotion extraction, mental state modeling, and rating. *Expert Systems with Applications*, 124:82–96.

Yan-Yan Song and LU Ying. 2015. Decision tree methods: applications for classification and prediction. *Shanghai archives of psychiatry*, 27(2):130.

Steven Stack. 2014. Differentiating suicide ideators from attempters: Violence—a research note. *Suicide and Life-Threatening Behavior*, 44(1):46–57.

Yla R Tausczik and Pennebaker James W. 2010. The psychological meaning of words: Liwc and computerized text analysis methods. *SAGE*.

Ashish Vaswani, Noam Shazeer, Niki Parmar, Jakob Uszkoreit, Llion Jones, Aidan N Gomez, Lukasz Kaiser, and Illia Polosukhin. 2017. Attention is all you need. *arXiv preprint arXiv:1706.03762*.

Ning Wang, Mingxuan Chen, and Koduvayur P Subbalakshmi. 2020. Explainable CNN-attention networks (c-attention network) for automated detection of alzheimer's disease. ACM SIGKDD.

Maria Wiklander, Mats Samuelsson, Jussi Jokinen, Åsa Nilsonne, Alexander Wilczek, Gunnar Rylander, and Marie Åsberg. 2012. Shame-proneness in attempted suicide patients. *BMC psychiatry*, 12(1):1–9.

Danuta Wolk-Wasserman. 1985. The intensive care unit and the suicide attempt patient. *Acta Psychiatrica Scandinavica*, 71(6):581–595.

Hannah Yao, Sina Rashidian, Xinyu Dong, Hongyi Duanmu, Richard N Rosenthal, and Fusheng Wang. 2020. Detection of suicidality among opioid users on reddit: Machine learning–based approach. *Journal of medical internet research*, 22(11):e15293.

Lei Zhang, Xiaolei Huang, Tianli Liu, Ang Li, Zhenxiang Chen, and Zhu Tingshao. 2015. Using linguistic features to estimate suicide probability of chinese microblog users. *Springer*.

Wen Zhang, Taketoshi Yoshida, and Xijin Tang. 2011. A comparative study of tf*idf, lsi and multi-words for text classification. *ScienceDirect*.

Ayah Zirikly, Philip Resnik, Ozlem Uzuner, and Kristy Hollingshead. 2019. Clpsych 2019 shared task: Predicting the degree of suicide risk in reddit posts. *ACL*.

# Suicide Risk Prediction by Tracking Self-Harm Aspects in Tweets: NUS-IDS at the CLPsych 2021 Shared Task

**Sujatha Das Gollapalli, Guilherme Augusto Zagatti, See-Kiong Ng**
idssdg@nus.edu.sg, gzagatti@u.nus.edu, seekiong@nus.edu.sg
Institute of Data Science, National University of Singapore, Singapore

## Abstract

We describe our system for identifying users at-risk for suicide based on their tweets developed for the CLPsych 2021 Shared Task. Based on research in mental health studies linking self-harm tendencies with suicide, in our system, we attempt to characterize self-harm aspects expressed in user tweets over a period of time. To this end, we design $SHTM$, a Self-Harm Topic Model that combines Latent Dirichlet Allocation with a self-harm dictionary for modeling daily tweets of users. Next, differences in moods and topics over time are captured as features to train a deep learning model for suicide prediction.

## 1 Introduction

Social media portals provide outlets for people to express their thoughts and emotions, and researchers have noted that user writings on social media contain signs and symptoms of various mental disorders (Coppersmith et al., 2014). Due to this reason, automated methods for identifying individuals "at risk" for various conditions such as depression, suicide, and addiction based on their online activity is an upcoming, recent research topic (Niederhoffer et al., 2019; Losada et al., 2020a).

In this paper, we focus on *suicide*, a leading cause of mortality among younger population (Patton et al., 2009) and address the problem of identifying individuals at-risk for suicide as part of the CLPsych 2021 Shared Task. In particular, we make use of the well-established link between self-harm tendencies and suicide (Kidger et al., 2012; Losada et al., 2020b) and study the expression of self-harm moods in user tweets. Our contributions are as follows:

- We propose $SHTM$, a topic model for capturing the self-harm aspects expressed in user writings. $SHTM$ uses self-harm dictionaries in a novel way within the Latent Dirichlet Allocation model to represent the topical as well as self-harm content expressed in a given text. $SHTM$ extracts self-harm word groups that may be indicative of various mental health issues seen in at-risk persons.
- Next, we characterize mood changes captured in the writings using $SHTM$ and show that the topic and mood profiles of the "control" and "at risk" individuals over time are different. We use this information to design features for our deep learning based classification model and test them on the tweet datasets from the CLPsych 2021 Shared Task.

## 2 Methods

### 2.1 $SHTM$: Our Topic Model

Probabilistic topic models are widely-used in text mining and NLP research for their ability to extract latent topics from a given document collection in an unsupervised manner (Koltcov et al., 2014; Lin and He, 2009; Wei and Croft, 2006). In particular, topic models based on Latent Dirichlet Allocation (Blei et al., 2003) were effectively used to characterize temporal topical trends and topical evolution (Bolelli et al., 2009; Lau et al., 2012; He et al., 2009). We describe our extension to the well-known LDA model for handling self-harm content changes through *SHTM* our Topic Model for Self-Harm content.

The document generative process in standard LDA is based on the assumption that a given document can be viewed as a mixture of latent topics. To model self-harm aspects expressed in text, we make use of a dictionary comprising of expert-compiled words commonly-used by individuals engaging in self-harm activities ($\mathcal{D}_{\mathcal{SH}}$) and "split" the document text based on whether a word is found in $\mathcal{D}_{\mathcal{SH}}$ or $\mathcal{V}$ (the rest of the vocabulary). That is, we assume that the presence of a word from $\mathcal{D}_{\mathcal{SH}}$ indicates a **S**elf-**H**arm **M**ood (SHM) expressed by the user whereas other words express "regular" topics.



*Proceedings of the Seventh Workshop on Computational Linguistics and Clinical Psychology*, pages 93–98
June 11, 2021. ©2021 Association for Computational Linguistics

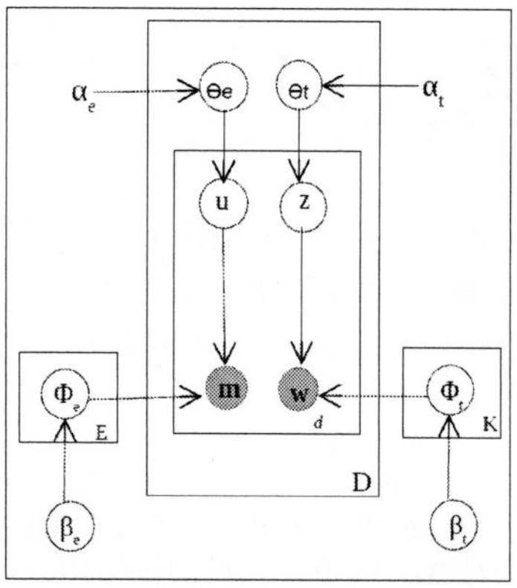


Figure 1: Plate diagram illustrating the graphical model for $SHTM$. $D$ is the number of tweets. $K$ and $E$ refer to the number of topics and self-harm aspects, respectively, while $z$ and $u$ refer to their corresponding latent variables for a particular tweet, respectively. The words sampled from the latent SHM and topics distributions are represented by $m$ and $w$ respectively. $\alpha_t, \alpha_e, \beta_t, \beta_e$ are Dirichlet hyperparameters. (Heinrich, 2005)

Based on the above premise, each word in the text generation process of $SHTM$ is either conditioned on a *latent* topic $t$, or a *latent* self-harm mood $e$, and a given document is a mixture of topics $\theta_t$ (as in regular LDA) as well as a mixture of SHMs $\theta_e$ (which includes "NoSH or no self-harm" mood). The plate diagram for $SHTM$ is shown in Figure 1. We refer the interested reader to Heinrich (2005) for the derivations for the sampling equations due to space constraints.

In $SHTM$, the topic assignment process (operating on all words in $\mathcal{V}$) is exactly the same as in standard LDA, whereas the self-harm mood assignments though similar, work only on words from $\mathcal{D}_{\mathcal{SH}}$. Furthermore, input texts with no words from $\mathcal{D}_{\mathcal{SH}}$ are directly assigned the "NoSH" mood. We posit that via this distinction of words based on their presence in $\mathcal{D}_{\mathcal{SH}}$, we can capture both the topical content and self-harm moods of a text directly via $SHTM$'s topical and mood dimensions. That is, similar to how a given document can be represented using its topic proportion vector (in a reduced dimension) in standard LDA, using $SHTM$, each user-generated text can be represented using a topic proportion vector as well as an SHM proportion vector and these vectors can be used to track changes along time when temporal information is available.

That is, let $\ldots w_{t-1}, w_t, w_{t+1} \ldots$ represent a sequence of writings for a given user. To track the change in mood for the user at timepoint $t$, given a context window w, we use the averaged SHM vectors for $w_{t-\text{w}} \ldots w_{t-1}$ and compute the difference between this average vector and the SHM vector for $w_t$ using measures such as cosine distance or KL divergence (Hall et al., 2008; Gollapalli and Li, 2015).

## 2.2 Our LSTM Classification Model

We used a deep learning model based on Long Short-Term Memory (LSTM) shown in Figure 2. Since both LSTMs and term feature vectors are effective for text classification problems (Aggarwal and Zhai, 2012; Pouyanfar et al., 2018), our model aims to combine the benefits of both via a two-part setup in which the output from the LSTM which captures the sequence information present in textual content is combined with aggregate features such as normalized term frequencies and $SHTM$-based features.

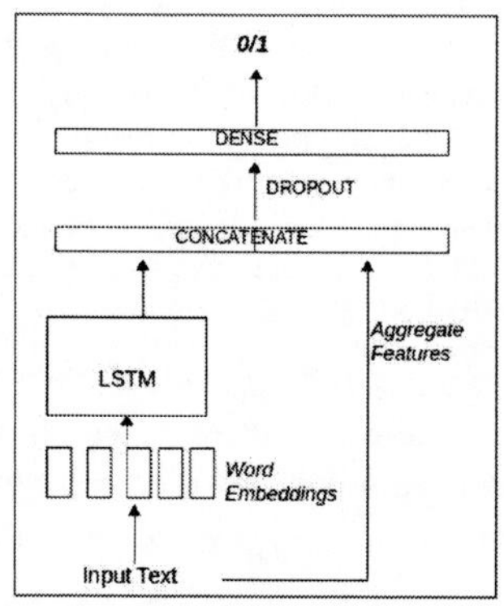


Figure 2: Schematic diagram of our model

# 3 Experiments and Results

**Data**: The dataset for the CLPsych 2021 Shared Task contains Twitter posts of users who attempted or committed suicide, and control individuals collected from OurDataHelps (ODH).[1] The competition involves two subtasks: "Prediction of a suicide attempt 30 days prior" (ODH30) and "Prediction of suicide attempt 6 months prior" (ODH182). We refer the reader to the overview paper of the CLPsych 2021 Shared Task (Macavaney et al., 2021) for further details on the data.

Briefly, the datasets for both tasks are fairly balanced containing roughly equal number of positive and control users as well as tweets. For the ODH182 and ODH30 subtasks, the training

[1] https://ourdatahelps.org

datasets comprise 162 and 109 users and 13K and 2K tweets, respectively. The test datasets comprise about 20 percent of the number of users available for training. The Shared Task also provides access to two other datasets: (1) a Practice Dataset (PD) comprising of tweets of users with '#depression' or similar hashtag[2] and (2) the University of Maryland (UMD) Suicidality Dataset based on Reddit posts (Zirikly et al., 2019; Shing et al., 2018).

As part of the task setup, all data was only accessible within a secure computing environment known as the UMD/National Opinion Research Center (NORC) Mental Health Data Enclave and all experiments were to be performed in this space. We refer the reader to MacAvaney, et al (2021) for details of the Enclave and the challenges involved in performing experiments in such environments.

**Implementation Details**: $SHTM$ was implemented in Java by extending the topic model code provided in the Mallet toolkit (McCallum, 2002). Default settings in Mallet were used for hyperparameter initialization and probability sampling. We tested three options including (a) All ODH data including the data provided for ODH30 and ODH182 tasks (ODH-only), (b) All ODH data and UMD data (ODH+UMD), and (c) All ODH and tweets from the Practice Dataset (ODH+PD). We used only data from relevant subreddits (picked manually based on term filters 'suicide', 'self-harm' and 'depression') for the UMD collection. Based on the word clusters extracted by $SHTM$for each SHM on a few choices of number of topics and SHM, we set the values of the number of topics and SHMs, respectively to $(20, 5)$ for ODH-only, $(15, 5)$ for ODH+UMD and $(50, 10)$ for ODH+PD. $SHTM$assignments from these runs were used for computing features for classification.

We employed standard text mining normalization steps to process the tweets. That is, all stopwords, punctuation and tokens starting with "@", referring to URLs, and non-alphanumeric ones were removed and all content was lowercased. After employing a term frequency threshold of 3, the vocabulary size ($\mathcal{V}$) is approximately 13K. For our self-harm word dictionary ($\mathcal{D}_{\mathcal{SH}}$), we curated words from the sources for Pyscholinguistic features used by Trifan et al (2020) to assemble a small list of 50 phrases corresponding to self-harm activities. Words in $\mathcal{D}_{\mathcal{SH}}$ include "self-image" "bruises", "numbing", and "trauma".[3]

**Incorporating Context and Sampling**: In our tasks, while predictions need to be made at user-level, we are given a sequence of time-stamped tweets with each user. Rather than create a single training instance clubbing all tweets available for a user, or creating a separate instance per tweet, we choose a middle ground based on the notion that from a practical standpoint, a classifier should be able to handle partial data availability rather than the entire 30 or 182 day periods. We enable this by creating multiple instances per user based on a context window parameter (w).

Let $T_t$ represents the set of all tweets posted on date $t$. For each user, we select all tweets generated from $T_{t-w+1}$ to $T_t$ inclusive to create a training instance. Starting from the last tweet posted by the user, we slide the window n times to obtain a maximum of n overlapping instances for each user. In this way, we can sample user tweets along different timepoints for training our models.[4]

**Classifier Settings**: We experimented with emotion-enriched word embeddings (Agrawal et al., 2018) and GloVE (Pennington et al., 2014) word embeddings for representing text within LSTMs. The number of LSTM units were set to 50 with the sequence length set to 1000. The output from LSTMs and aggregate features were concatenated and input to a subsequent dense layer of size 100. The dropout rate was set to 0.2 and we used the Adam optimizer for training all models with cross-entropy loss.[5]

### 3.1 Results and Discussion

We briefly summarize our results in this section. Note that we have several tunable parameters: number of topics/SHM, clusters for $SHTM$ model, learning model parameters such as LSTM and layer dimensions, as well as the $n$ and w parameters that affect number of training instances added per user and the context window for aggregating tweets. We tune these parameters using validation experiments. That is, the training data is randomly split into 80/20% train/validation portions of the data using three different random seeds. All parameter

[2] `https://github.com/swcwang/depression-detection`

[3] `https://github.com/NUS-IDS/clpsych21-sharedtask`

[4] All available sliding windows are considered during prediction and we predict a user as "positive" if any instance associated with the user is classified as positive.

[5] Classification models were implemented using Python 3.9.1 and associated Torch libraries provided on the Enclave.

| Setting/Model | F1 | F2 | TP | FP | AUC |
|---|---|---|---|---|---|
| **ODH-30** | | | *Averaged Validation Performance* | | |
| Competition Baseline | 0.228±0.108 | 0.259±0.135 | 0.285±0.159 | 0.729±0.115 | 0.335±0.169 |
| Best Validation: w=3, n=3 | 0.706±0.181 | 0.749±0.196 | 0.783±0.214 | 0.270±0.115 | 0.800±0.192 |
| | | | *Test Performance* | | |
| Competition Baseline | 0.636 | 0.636 | 0.636 | 0.364 | 0.661 |
| Our Top-2 submitted runs: w=3, n=3 | 0.615 | **0.714** | **0.8** | 0.727 | 0.664 |
| w=5, n=2 | 0.583 | **0.648** | **0.7** | 0.636 | 0.645 |
| **ODH-182** | | | *Averaged Validation Performance* | | |
| Competition Baseline | 0.547±0.034 | 0.597±0.049 | 0.643±0.105 | 0.483±0.178 | 0.654±0.033 |
| Best Validation, w=10, n=7 | 0.623±0.044 | 0.783±0.012 | 0.950±0.042 | 0.780±0.088 | 0.587±0.076 |
| | | | *Test Performance* | | |
| Competition Baseline | 0.71 | 0.724 | 0.733 | 0.333 | 0.764 |
| Our Top-2 submitted runs: w=10, n=7 | 0.684 | **0.812** | **0.929** | 0.786 | 0.663 |
| w=10, n=7 * | 0.703 | **0.823** | **0.929** | 0.714 | 0.648 |

Table 1: Performance of our classification is compared against the baseline model for the two subtasks of CLPsych 2021. $SHTM$ was trained on ODH-only with 20 topics and 5 SHMs for all our selected models, except for * which was trained on ODH + PD with 50 topics and 10 SHMs.

choices are based on the averaged F1 scores from these three runs.

The best models did not use large values for the context or sliding window. Rather, when instances for a user are extracted in reverse chronological order, values of w and n in the range 3-10 closest to the last available date for a user perform the best for classification on both the subtasks. This observation indicates that the content generated closest to the attempt date is highly informative in identifying a user's suicidality risk.

Word embeddings from EWE performed better than GloVE, and topic/SHM assignments from ODH-only corpus performed the best among our the three choices. The word clusters extracted from this corpus for the self-harm aspects are shown below:

| SHMID | Top-words |
|---|---|
| 1 | death shame bipolar relationships disgust bruises emotional obesity |
| 2 | cut emotional panic doubt disorder hopeless |
| 3 | suicide stress sadness relationships bleak helpless |
| 4 | anxiety worry depression accident friendships scratch guilt |

**Mood and Topic Profiles**: To analyze the differences in mood and topic profiles among the two groups of users ('positive' and 'control'), we examined the mean and variance of the KL-divergence between the SHM vector representing tweets on date $t$ and the average SHM vector of tweets from the past w−1 dates available for a user. We proceeded similarly for the corresponding topic vectors. For the positive class, we observe higher mean and variance for the KL-divergence of SHM vectors. In contrast, we observe a lower mean and variance for the KL-divergence in topics. Taken together, these trends suggest that there is expressive variation in SHM within the positive class which might explain the high false positive rate and warrants further investigation in future work.

**Classification Performance**: Table 1 illustrates the validation and test performances using our best configurations compared against the competition provided baseline model based on Logistic Regression. For the competition, the suggested measures include F1 (the standard measure combining precision and recall), F2 (which values recall twice as much as precision), true and false positive rates (TP and FP) as well as AUC which measures how the predictions are ranked.

Our model does significantly well in the validation runs on all measures for the ODH30 dataset but has significantly higher false positive rate and significantly lower AUC score for ODH182. For test performance, our model obtains a significantly higher F2 and true positive rates over the baseline model but is unable to beat the baseline on the F1 and AUC measures. We observe a significantly high number of false positives in all test runs with our model. The baseline performs surprisingly well on the test set as compared to training, while our model shows a higher degree of consistency.

Due to criticality of this prediction task, we would like to err on the side of caution. However, a high false positive rate is not useful in a practical prediction system. In future work, we aim to fully investigate this dataset specifically for reducing the FP rate, improving the overall prediction performance using other deep learning models and augmenting with related datasets (Losada et al., 2020a). We would also like to further investigate the capacity of SHM to act as a discriminant in other learning models (SVMs were not as successful as LSTMs in our experiments).

## 4 Conclusions and Future Work

We presented $SHTM$, our topic model for representing self-harm aspects expressed in social media texts. We used features based on self-harm mood changes and topic changes in tweets over time within a deep learning model to predict suicidal users. To the best of our knowledge, we are the first to employ topic models for studying mood characterization in context of suicide risk.

Several topic models were proposed in previous works for incorporating label information and improving prediction tasks (Blei and McAuliffe, 2007; Ramage et al., 2009; Nguyen et al., 2013; Ren et al., 2020). In future, we aim to incorporate emotion lexicons (Mohammad and Turney, 2010) into these models and suitably extend them to characterize temporal mood trends (Bolelli et al., 2009) of users with mental health issues such as depression, PTSD, and suicide (Chen et al., 2018).

## Ethics Statement

Secure access to the shared task dataset was provided with IRB approval under University of Maryland, College Park protocol 1642625.

## Acknowledgments

The authors are particularly grateful to the users who donated data to the OurDataHelps project without whom this work would not be possible, to Qntfy for supporting the OurDataHelps project and making the data available, to NORC for creating and administering the secure infrastructure, and to Amazon for supporting this research with computational resources on AWS.

This research/project was supported by the National Research Foundation, Singapore under its AI Singapore Programme (AISG Award No: AISG-GC-2019-001). Any opinions, findings and conclusions or recommendations expressed in this material are those of the author(s) and do not reflect the views of National Research Foundation, Singapore.

## References

Charu C. Aggarwal and ChengXiang Zhai. 2012. *A Survey of Text Classification Algorithms*.

Ameeta Agrawal, Aijun An, and Manos Papagelis. 2018. Learning emotion-enriched word representations. In *Proceedings of the 27th International Conference on Computational Linguistics*, pages 950–961.

David M. Blei and Jon D. McAuliffe. 2007. Supervised topic models. In *NIPS*, page 121–128.

David M. Blei, Andrew Y. Ng, and Michael I. Jordan. 2003. Latent dirichlet allocation. *J. Mach. Learn. Res.*, 3:993–1022.

Levent Bolelli, Şeyda Ertekin, and C Lee Giles. 2009. Topic and trend detection in text collections using latent dirichlet allocation. In *European conference on information retrieval*, pages 776–780. Springer.

Xuetong Chen, Martin D. Sykora, Thomas W. Jackson, and Suzanne Elayan. 2018. What about mood swings: Identifying depression on twitter with temporal measures of emotions. In *WWW*.

Glen Coppersmith, Mark Dredze, and Craig Harman. 2014. Quantifying mental health signals in Twitter. In *Proceedings of the Workshop on Computational Linguistics and Clinical Psychology*, pages 51–60.

Sujatha Das Gollapalli and Xiaoli Li. 2015. EMNLP versus ACL: Analyzing NLP research over time. In *EMNLP*, pages 2002–2006.

David Hall, Daniel Jurafsky, and Christopher D. Manning. 2008. Studying the history of ideas using topic models. In *EMNLP*.

Qi He, Bi Chen, Jian Pei, Baojun Qiu, Prasenjit Mitra, and Lee Giles. 2009. Detecting topic evolution in scientific literature: How can citations help? In *CIKM*, page 957–966.

G. Heinrich. 2005. Parameter estimation for text analysis. *Web: http://www. arbylon. net/publications/text-est. pdf*.

Judi Kidger, Jon Heron, Glyn Lewis, Jonathan Evans, and David Gunnell. 2012. Adolescent self-harm and suicidal thoughts in the alspac cohort: a self-report survey in england. In *BMC Psychiatry 12, 69*.

Sergei Koltcov, Olessia Koltsova, and Sergey Nikolenko. 2014. Latent dirichlet allocation: Stability and applications to studies of user-generated content. In *Proceedings of the 2014 ACM Conference on Web Science*, page 161–165.

Jey Han Lau, Nigel Collier, and Timothy Baldwin. 2012. On-line trend analysis with topic models:# twitter trends detection topic model online. In *Proceedings of COLING 2012*, pages 1519–1534.

Chenghua Lin and Yulan He. 2009. Joint sentiment/topic model for sentiment analysis. In *CIKM*, page 375–384.

David E. Losada, Fabio Crestani, and Javier Parapar. 2020a. erisk 2020: Self-harm and depression challenges. In *Advances in Information Retrieval*.

David E. Losada, Fabio Crestani, and Javier Parapar. 2020b. Overview of erisk 2020: Early risk prediction on the internet. In *Experimental IR Meets Multilinguality, Multimodality, and Interaction*, Lecture Notes in Computer Science.

Sean Macavaney, Anjali Mittu, Glen Coppersmith, Jeff Leintz, and Philip Resnik. 2021. Community-level research on suicidality prediction in a secure environment: Overview of the CLPsych 2021 shared task. In *CLPsych*.

Andrew Kachites McCallum. 2002. Mallet: A machine learning for language toolkit. Http://mallet.cs.umass.edu.

Saif Mohammad and Peter Turney. 2010. Emotions evoked by common words and phrases: Using Mechanical Turk to create an emotion lexicon. In *Proceedings of the NAACL HLT 2010 Workshop on Computational Approaches to Analysis and Generation of Emotion in Text*.

Viet-An Nguyen, Jordan L Ying, and Philip Resnik. 2013. Lexical and hierarchical topic regression. In *Advances in Neural Information Processing Systems*, volume 26.

Kate Niederhoffer, Kristy Hollingshead, Philip Resnik, Rebecca Resnik, and Kate Loveys, editors. 2019. *Proceedings of the Sixth Workshop on Computational Linguistics and Clinical Psychology*.

G.C. Patton, C. Coffey, S.M. Sawyer, Viner R.M., Haller D.M., Bose K., Vos T., Ferguson J., and Mathers C.D. 2009. Global patterns of mortality in young people: a systematic analysis of population health data. In *Lancet*.

Jeffrey Pennington, Richard Socher, and Christopher Manning. 2014. GloVe: Global vectors for word representation. In *EMNLP*.

Samira Pouyanfar, Saad Sadiq, Yilin Yan, Haiman Tian, Yudong Tao, Maria Presa Reyes, Mei-Ling Shyu, Shu-Ching Chen, and S. S. Iyengar. 2018. A survey on deep learning: Algorithms, techniques, and applications. *ACM Comput. Surv.*

Daniel Ramage, David Hall, Ramesh Nallapati, and Christopher D. Manning. 2009. Labeled LDA: A supervised topic model for credit attribution in multi-labeled corpora. In *EMNLP*.

Jason Ren, Russell Kunes, and Finale Doshi-Velez. 2020. Prediction focused topic models via feature selection. In *AISTATS*.

Han-Chin Shing, Suraj Nair, Ayah Zirikly, Meir Friedenberg, Hal Daumé III, and Philip Resnik. 2018. Expert, crowdsourced, and machine assessment of suicide risk via online postings. In *Proceedings of the Fifth Workshop on Computational Linguistics and Clinical Psychology: From Keyboard to Clinic*, pages 25–36.

Alina Trifan, Pedro Salgado, and José Luís Oliveira. 2020. Bioinfo@uavr at erisk 2020: on the use of psycholinguistics features and machine learning for the classification and quantification of mental diseases. In *Working Notes of CLEF*.

Xing Wei and W. Bruce Croft. 2006. Lda-based document models for ad-hoc retrieval. In *SIGIR*, page 178–185.

Ayah Zirikly, Philip Resnik, Özlem Uzuner, and Kristy Hollingshead. 2019. CLPsych 2019 shared task: Predicting the degree of suicide risk in Reddit posts. In *Proceedings of the Sixth Workshop on Computational Linguistics and Clinical Psychology*.

# A Comparison of Simple vs. Complex Models for Suicide Risk Assessment

**Michelle Morales**
Global Business Services
IBM

**Prajjalita Dey**
Global Business Services
IBM

**Kriti Kohli**
Corporate Technical Strategy
IBM

{michelle.morales, prajjalita.dey, kkohli}@ibm.com

## Abstract

This work presents the systems explored as part of the CLPsych 2021 Shared Task. More specifically, this work explores the relative performance of models trained on social media data for suicide risk assessment. For this task, we aim to investigate whether or not simple traditional models can outperform more complex fine-tuned deep learning models. Specifically, we build and compare a range of models including simple baseline models, feature-engineered machine learning models, and lastly, fine-tuned deep learning models. We find that simple more traditional machine learning models are more suited for this task and highlight the challenges faced when trying to leverage more sophisticated deep learning models.

## 1 Introduction

Globally 800,000 people die from suicide each year, which makes it one of the leading causes of death (Hannah Ritchie and Ortiz-Ospina, 2015). Despite decades of substantial efforts to analyze risk factors for suicidal thoughts and behaviors (Franklin et al., 2017), models have produced predictions only slightly better than random chance (AUCs=0.56-0.58) (Ophir et al., 2020). Recent progress in Natural Language Processing (NLP) and Machine Learning systems to predict suicide risk have been shown to have higher AUC 0.9 (Coppersmith et al., 2018), however it is still a complicated task particularly due to the sensitivity and difficulty in obtaining high quality labeled datasets.

This work is part of the 2021 CLPsych Shared Task (Macavaney et al., 2021), which provides secure and ethical access to sensitive data in order to work on the problem of predicting suicide risk from social media data. The shared task has two main objectives: prediction of a suicide attempt 30 days prior, and prediction of a suicide attempt 6 months prior. In this paper, we present our team's results from the Shared Task using a variety of methods to improve performance. We focus on exploring various machine learning ensemble models, feature engineering approaches and compare to deep learning architectures and Transfer Learning methods in NLP. We find that baseline models such as Term Frequency, used in combination with simple machine learning models outperform fine-tuned deep learning Transformer-based models.

## 2 Methods

Our goal for this task was to compare the results of models across different levels of complexity, and see how they perform in the context of a small dataset in the mental health space. All Tweets were aggregated at the user level, and each of the classification methods were implemented and compared at that level.

### 2.1 Dataset

This work leverages the data provided by the 2021 CLPsych Workshop organizers (Macavaney et al., 2021). Data was provided for a series of Twitter users and all their Tweets for a certain timeframe of history: in Subtask 1 that timeframe was 30 days, while in Subtask 2 the timeframe was 182 days. The dataset also provided true binary labels about past suicide attempts as well as the date of attempt if applicable - a first for this type of shared task, only possible because of the secure computing environment that was provided. Real world binary outcomes have been used in other types of work (Coppersmith et al., 2018).

### 2.2 Baseline Model

The baseline model provided by the organizers involved a Term Frequency model in conjunction with a Logistic Regression classifier. This method involved simple preprocessing: cleaning hashtags, removing stopwords, and tokenizing Tweets. In



*Proceedings of the Seventh Workshop on Computational Linguistics and Clinical Psychology*, pages 99–102
June 11, 2021. 

addition, all of the models described in Section 2.3 leveraged the same preprocessing approach.

## 2.3 Machine Learning Models

### 2.3.1 Gradient Boosting - Syntax Features

This model used a gradient boosting classifier with an emphasis on manually created grammatical features. Prior research in this space has shown that grammatical and syntactic patterns are a consistent differentiator between individuals characterized with suicide risk and those who are not (O'dea et al., 2017). The features created were intended to measure this, and focused on length and syntax patterns prominent within the user's Tweets. The length features comprised of both average word and sentence count. The syntax related features quantified pronoun usage, differentiating between first, second, and third-person pronouns as well as singular and plural pronouns.

### 2.3.2 Gradient Boosting - Character TF-IDF

This model used the same gradient boosting model as above, but used a different feature set. Also, this model stemmed the data as an additional preliminary preprocessing step. Instead of manually creating features from the text, this model utilized a character TF-IDF vector. Both gradient boosting models were applied to both Subtasks.

### 2.3.3 Ensemble Voting Classifier

Our third model used a voting method to create an ensemble machine learning model. Features were created using an n-gram Term Frequency with unigrams and bigrams, across the entire training set, with 5,000 maximum features. We then trained three machine learning models: a Logistic Regression classifier, a Multinomial Naive Bayes classifier, and a Random Forest classifer. We used a soft voting classifier - where the predicted class probabilities for each classifier are collected and averaged - and weighted each classifier equally. The final class label is then derived from the class label with the highest average probability between the three models. We picked conceptually different machine learning classifiers in order to balance out individual weaknesses in the average predicted probabilities.

## 2.4 Deep Learning Models

Lastly, we explored the effect of using NLP transfer learning methods and fine-tuning deep learning models. For this system, we used BERTweet (Nguyen et al., 2020) - a language model pre-trained on an 80GB corpus of 850M English Tweets - and fine-tuned it on the Shared Task dataset. BERTweet uses the same architecture as BERTbase (Devlin et al., 2018), with a pre-training procedure based on RoBERTa (Liu et al., 2019); it has generally proven to do better than its competitors on Tweet NLP tasks, including text classification. We only applied this deep learning system to Subtask 1, due to the limit on maximum sequence length at 512 and 128 for BERT and BERTweet respectively. Since Subtask 2 comprised of 6 months worth of Tweets its sequence length was above the maximum requirements of BERT and BERTweet, and therefore not included in this part of our investigation.

### 2.4.1 BERTweet Preprocessing

Before applying BERTweet to the classification task, we normalized the Tweets by following the same preprocessing steps applied to the BERT pre-training corpus. This included tokenizing the Tweets using TweetTokenizer from the NLTK toolkit and using the emoji package to translate emotion icons into text strings. In addition, raw Tweets were normalized by converting user mentions and web/url links into special tokens as provided through the normalization argument in the BERTweet Transformers package (Wolf et al., 2019).

### 2.4.2 Fine-tuned Model

We explored two fine-tuning methods. In Method 1, we created a BERTweet model instance with a randomly initialized sequence classification head on top of the encoder, of output size 2. In Method 2, we froze the entire architecture and attached a dense neural network layer, updating only the weights of the attached layers.

Both fine-tuning approaches used a maximum sequence length of 128 tokens, and models were optimized using AdamW (Loshchilov and Hutter, 2017), which implements gradient bias correction as well as weight decay. We followed the recommended hyperparameters for fine-tuning as described in Appendix A3 of (Devlin et al., 2018): batch size 16, fixed learning rate of 2e-5, 4 epochs for fine-tuning Method 1 and 10 epochs for Method 2.

In our fine-tuning Method 2, we kept all the weights of the pre-trained BERTweet model frozen and appended a dense linear layer, a dropout layer

| | F1 | F2 | TPR | FPR | AUC |
|---|---|---|---|---|---|
| **Subtask 1 (30 days)** | | | | | |
| Task Baseline | 0.636 | 0.636 | 0.636 | 0.364 | 0.661 |
| Run 1: Char. TF-IDF GB | 0.455 | 0.455 | 0.455 | 0.545 | 0.438 |
| Run 2: Syntax GB | 0.500 | 0.472 | 0.455 | 0.364 | 0.616 |
| Run 3: BERTweet | 0.571 | **0.656** | 0.727 | 0.818 | 0.413 |
| **Subtask 2 (6 months)** | | | | | |
| Task Baseline | 0.710 | 0.724 | 0.733 | 0.333 | 0.764 |
| Run 1: Syntax GB | 0.467 | 0.467 | 0.467 | 0.533 | 0.618 |
| Run 2: Char. TF-IDF GB | 0.516 | 0.526 | 0.533 | 0.533 | 0.591 |
| Run 3: Voting Classifier | 0.727 | **0.769** | 0.800 | 0.400 | 0.720 |

Table 1: Model results on CLPsych test set as compared to the task baseline system.

to reduce overfitting, and a softmax layer. The model was trained using a cross-entropy loss function. We computed the task performance after each training epoch on a validation set and selected the best model checkpoint to compute the performance on the test set.

## 3 Results

In Subtask 1, our models are as follows: Run 1 refers to the character TF-IDF gradient boosting model, Run 2 refers to the syntax gradient boosting model and Run 3 refers to the BERTweet model using fine-tuned Method 1. In the validation experiments, we found BERTweet fine-tuned Method 1 to outperform Method 2. In Subtask 2, Run 1 refers to the syntax gradient boosting model, Run 2 the character TF-IDF model, and Run 3 the voting classifier.

We see that in the case where the BERTweet model could be applied, it outperformed more simple machine learning models. However, although the BERTweet model had a high F1, F2, and TPR, it has a high FPR and a low AUC score - this implies that the model is overfitting, and has a tendency to predict 1s.

In the case where BERTweet could not be applied (Subtask 2), having an ensemble model fared better than the single gradient boosting models. The voting classifier outperformed the baseline in most metrics (F1, F2, TPR) but also had a nominally higher FPR and lower AUC score than the baseline. The increased FPR corresponds to misclassifying one negative sample as a positive sample. For assessing suicide risk though, we feel that it is better to overpredict suicide risk than underpredict, since the consequences of underpredicting are much more severe.

F2 score gives less weight to precision and more weight to recall therefore prioritizing the proportion of actual positives that were correctly identified. Both BERTweet (Subtask 1) and the voting classifier (Subtask 2) have higher F2 score than the baseline, however F2-score alone is an unsuitable metric as a classifier that predicts all 1s would have a recall of 1. The AUC is widely used to as a measure for predictive modeling accuracy, however, AUC is not recommended for small sample sizes (Hanczar et al., 2010). Overall, looking at all the metrics in Table 1 holistically is recommended.

## 4 Discussion

For Subtask 1, in the Transfer Learning methods, we tried two fine-tuning techniques. In the first approach, i.e. Method 1, we instantiate a BERTweet model with an added single linear layer on top for classification. In this approach, the entire pre-trained BERTweet model and the additional untrained classification layer is trained on our specific task. The average accuracy with the validation set was 0.51 and 0.45 for the test set, suggesting overfitting of the model. For the second approach, i.e. Method 2, we freeze all the layers of BERTweet and only update the weights of the attached layers. While the training loss decreased for the first 4 epochs, it did not decrease further, suggesting that the model was trained for too long and is also overfitting on the training data. While both approaches suggested that such a small dataset caused overfitting, a simple fine-tuning approach through adding one fully-connected layer to BERTweet and training the whole model end-to-end for a few epochs (Method 1) showed better results than appending a custom architecture to the frozen BERTweet model (Method 2). As all Tweets were aggregated into one large Tweet at the user level and the sequence length was limited to 128, effectively this approach reduced the dataset from Tweets of the last 30 days to the last 1-3 days depending on the Tweet length. This causes loss of potentially valuable data and features that may be missed as these particular models cannot learn from the older Tweets. As the machine learning models do not have these limiting properties, they are more suitable for this task. A recommendation for future work is to transform the dataset in an alternate manner, for example, creating a classification task at the Tweet level instead of the aggregated User-Tweet level.

## 5 Conclusion

The main question we sought to explore in this paper was the following, *would a classical machine learning model approach outperform a more sophisticated deep learning model for the suicide risk assessment task?* Given past research in this space that struggled with this task as well as the small nature of the datasets, it was our hypothesis that keeping it simple would lead to better performance. Our findings support this hypothesis. We found that BERTweet struggled with overfitting and demonstrated limitations, such as sequence length, that made it difficult to leverage for this task. In our evaluations, we found that a simple baseline model, or an ensemble of machine learning models can outperform the more sophisticated models. In addition, the short time period inherent in building a model for a Shared Task made it difficult to investigate alternate data transformations that are more appropriate for a complex model like fine-tuned BERT/BERTweet. However, we do find some promise in the test performance of BERTweet for Subtask 1 and believe with more time and exploration a variation of Transfer Learning models can be built and leveraged in a task of this nature.

## Ethics Statement

Secure access to the shared task dataset was provided with IRB approval under University of Maryland, College Park protocol 1642625.

## References

Glen Coppersmith, Ryan Leary, Patrick Crutchley, and Alex Fine. 2018. Natural language processing of social media as screening for suicide risk. *Biomedical Informatics Insights*, 10:1–11.

Jacob Devlin, Ming-Wei Chang, Kenton Lee, and Kristina Toutanova. 2018. BERT: pre-training of deep bidirectional transformers for language understanding. *CoRR*, abs/1810.04805.

Joseph Franklin, Jessica Ribeiro, KR Fox, KH Bentley, EM Kleiman, X Huang, KM Musaccchio, AC Jaroszewski, BP Chang, and MK Nock. 2017. Risk factors for suicidal thoughts and behaviors: A meta-analysis of 50 years of research. In *Psychol Bull.*, pages 143(2):187–232.

Blaise Hanczar, Jianping Hua, Chao Sima, John Weinstein, Michael Bittner, and Edward R. Dougherty. 2010. Small-sample precision of ROC-related estimates. *Bioinformatics*, 26(6):822–830.

Max Roser Hannah Ritchie and Esteban Ortiz-Ospina. 2015. Suicide. Https://ourworldindata.org/suicide.

Yinhan Liu, Myle Ott, Naman Goyal, Jingfei Du, Mandar Joshi, Danqi Chen, Omer Levy, Mike Lewis, Luke Zettlemoyer, and Veselin Stoyanov. 2019. Roberta: A robustly optimized BERT pretraining approach. *CoRR*, abs/1907.11692.

Ilya Loshchilov and Frank Hutter. 2017. Fixing weight decay regularization in adam. *CoRR*, abs/1711.05101.

Sean Macavaney, Anjali Mittu, Glen Coppersmith, Jeff Leintz, and Philip Resnik. 2021. Community-level research on suicidality prediction in a secure environment: Overview of the CLPsych 2021 shared task. In *Proceedings of the Seventh Workshop on Computational Linguistics and Clinical Psychology (CLPsych 2021)*. Association for Computational Linguistics.

Dat Quoc Nguyen, Thanh Vu, and Anh Tuan Nguyen. 2020. BERTweet: A pre-trained language model for English Tweets. In *Proceedings of the 2020 Conference on Empirical Methods in Natural Language Processing: System Demonstrations*, pages 9–14.

Bridianne O'dea, Mark E Larsen, Philip J Batterham, Alison L Calear, and Helen Christensen. 2017. A linguistic analysis of suicide-related twitter posts. *Crisis: The Journal of Crisis Intervention and Suicide Prevention*, 38(5):319.

Yaakov Ophir, Refael Tikochinski, Christa S. C. Asterhan, Itay Sisso, and Roi Reichart. 2020. Deep neural networks detect suicide risk from textual facebook posts. In *Scientific Reports. 10*, page 16685.

Thomas Wolf, Lysandre Debut, Victor Sanh, Julien Chaumond, Clement Delangue, Anthony Moi, Pierric Cistac, Tim Rault, Rémi Louf, Morgan Funtowicz, and Jamie Brew. 2019. Huggingface's transformers: State-of-the-art natural language processing. *CoRR*, abs/1910.03771.

# Using Psychologically-Informed Priors for Suicide Prediction in the CLPsych 2021 Shared Task

**Avi Gamoran** * and **Yonatan Kaplan** *
**Ram Isaac Orr** and **Almog Simchon** † and **Michael Gilead** †
Ben-Gurion University of the Negev, Israel
{avigam, kaplay, ramor, almogsi}@post.bgu.ac.il
mgilead@bgu.ac.il

## Abstract

This paper describes our approach to the CLPsych 2021 Shared Task, in which we aimed to predict suicide attempts based on Twitter feed data. We addressed this challenge by emphasizing reliance on prior domain knowledge. We engineered novel theory-driven features, and integrated prior knowledge with empirical evidence in a principled manner using Bayesian modeling. While this theory-guided approach increases bias and lowers accuracy on the training set, it was successful in preventing over-fitting. The models provided reasonable classification accuracy on unseen test data ($0.68 \leq AUC \leq 0.84$). Our approach may be particularly useful in prediction tasks trained on a relatively small data set.

## 1 Introduction

Suicide is a troubling public health issue (Haney et al., 2012), with an estimated prevalence of over 800,000 cases per year worldwide (Arensman et al., 2020). Suicide rates have been climbing steadily over the past two decades (Curtin et al., 2016; Naghavi, 2019; Glenn et al., 2020), especially in high-income countries (Arensman et al., 2020; Haney et al., 2012). Research has identified many risk factors linked to suicide (Franklin et al., 2017; Ribeiro et al., 2018), and suicide attempts (Yates et al., 2019; Miranda-Mendizabal et al., 2019). Despite these advances, directing these insights into real-life risk identification and suicide prevention remains challenging (Large et al., 2017b,a). Early identification is crucial, as direct, brief, and acute interventions are helpful in preventing suicide attempts (Doupnik et al., 2020).

For the sake of early detection, there are increasing attempts to try and find warning signs in publicly-available social media data. As part of this effort, the 2021 Computational Linguistics and Clinical Psychology Workshop (CLPysch), have provided access to de-identified Twitter feeds of individuals who have made suicide attempts (as well as others who have not), with the task of predicting suicide attempts based on tweets up to 30 days (*Subtask 1*) or 182 days (*Subtask 2*) before such attempts.

Machine-learning algorithms and natural language processing ("NLP") methods have proven highly useful on many prediction problems. Current approaches typically rely on inductive algorithms that learn regularities in the data. When data are noisy (as is the case in human behavior), the ability to generalize predictions often depends on the size of the training set. Given the sensitive nature of suicide-related data, labeled data on this matter are scarce. This relative scarcity of training examples (e.g., 114/164 individuals in the current task) presents a difficult prediction problem, and increased risk of model over-fitting.

In light of the unique properties of this problem, we reasoned that an emphasis on domain knowledge (rather than on algorithmic solution) is warranted, and may help reduce over-fitting. Therefore, we adopted the following principles for the prediction task: 1. We used logistic regression rather than potentially more complex models that are often more prone to over-fitting (e.g., DNN, SVM, RF). 2. We engineered and evaluated many theory-driven features, based on our domain expertise in psychology (e.g., Simchon and Gilead, 2018). 3. We integrated prior knowledge and the empirical evidence in a principled manner. Using Bayesian modeling, we incorporated empirical priors from past findings in psychology literature. When we lacked specific priors for a feature of interest, we regularized our parameters using general, domain-level empirical priors (van Zwet and Gelman, 2020), derived from a meta-analysis of replication studies in psychology (Open Science Collaboration et al., 2015).

* These authors contributed equally.
† These authors contributed equally.



*Proceedings of the Seventh Workshop on Computational Linguistics and Clinical Psychology*, pages 103–109
June 11, 2021. 

## 2 Methodology

Participants in the Shared Task were given a training set which consisted of 2485 tweets from 114 individuals, 57 having attempted suicide and 57 controls, in the 30-day set, and 15928 tweets from 164 individuals, 82 in each group, in the 182-day set.

### 2.1 Features

| Feature with Informed Priors | Effect-Size ($r$) |
|---|---|
| Adverbs-SD | 0.113 |
| Anger-M | 0.068 |
| Anger-SD | 0.068 |
| Body-SD | 0.07 |
| Female-M | 0.105 |
| Female-SD | 0.105 |
| Focus-On-Present-SD | 0.095 |
| Informal-SD | 0.041 |
| Ingest-SD | 0.021 |
| I-Pronouns-M | 0.046 |
| Negative-Emotion-M | 0.141 |
| Negative-Emotion-SD | 0.141 |
| Pronouns-M | 0.137 |
| Personal-Pronouns-M | 0.015 |
| Sexual-M | 0.073 |
| Sexual-SD | 0.073 |
| Swear-Words-M | 0.055 |
| Swear-Words-SD | 0.055 |
| Verbs-M | 0.101 |
| Work-M | -0.099 |
| They-M | 0.025 |

Table 1: LIWC Features with Informed Priors (Effect sizes from Eichstaedt et al., 2018). Effect sizes entered the model on the log odds scale. Shown here in Pearson's $r$ for convenience.

**Twitter behavioral aspects:** We counted the number of replies to others, and the number of unique fellow users mentioned in replies. The intuition behind these metrics being that they reflect on the social engagement of users. Loneliness and social isolation are robust risk factors for suicide (Leigh-Hunt et al., 2017; Franklin et al., 2017). The proportion of tweets written late at night (23:00 – 5:00) was measured, as sleep disorders are related to depression and suicidal ideation (Liu et al., 2020).

**LIWC:** The Linguistic Inquiry and Word Count (Pennebaker et al., 2015), is a widely used dictionary-based program for automatic text analysis. LIWC scales tap into psychological and linguistic features, and provide a good overview into an individual's psychological makeup (Chung and Pennebaker, 2018). LIWC has been used in analyzing social media prior to suicide attempts (Coppersmith et al., 2016), as well as in analysis of suicide notes (Pestian et al., 2012) and poems of poets who later committed suicide (Stirman and Pennebaker, 2001). A central finding from LIWC analyses on suicidal populations is an increase in words pertaining to the self, and a decrease in words regarding others. We therefore measured the ratio of self words ('I') to group-words ('We'). Most of the LIWC-derived features were given priors based on previous gold-standard findings in depression prediction, see Table 1 (Eichstaedt et al., 2018).

**The Mind-Perception Dictionary:** a dictionary tailored for mind perception which includes a category of agent-related emotions (Schweitzer and Waytz, 2020). The guiding idea was that individuals at risk of committing suicide may differ in their sense of agency from non-suicidal individuals. This feature was given a weakly-informed prior with center = 0.

**Custom Dictionaries:** We constructed custom dictionaries based on themes assumed to be linked with mental vulnerability, depression and suicide. The themes included were Social Longing, Fatigue, Self-destructive Behavior, and Unmet Desires and Needs. These features were given weakly-informed priors with center = 0.

### 2.2 Bayesian Modeling

Due to the large amount of potential predictive features, as a first step, we manually excluded variables which did not differ between suicidal individuals and controls in a univariate statistical analysis. A total of 30 significant variables were retained for the modeling stage (Table 1).

Using the 'rstanarm' package, an R wrapper for Stan (Carpenter et al., 2017; Goodrich et al., 2020), we deployed logistic-regression models with Bayesian MCMC estimation. The Bayesian infrastructure was chosen in order to formally determine custom priors for the various predictive features, based on existing psychological literature, and to regularize parameters based on the distribution of effect sizes in the field.

In order to assess the validity of this approach and its performance relative to inductive "bottom-up" methods, we chose to submit one psychologically informed model, one "default" weakly-informed Bayesian model, and one regularized regression model.

Our models were: *a)* Informed priors with centers of distributions according to effect sizes found in previous studies (Table 1). In *Subtask 1* the priors were from Cauchy distributions, with centers according to existing effect sizes, and scales set to 2.5 (the 'rstanarm' defaults): $\sim \text{Cauchy}(\mu, 2.5)$. In *Subtask 2* the priors were from Laplace distributions with centers according to effect sizes, and scales of 1.687 as an approximation of a mixture prior, recommended for use in a database of 86 psychological replication studies (van Zwet and Gelman, 2020): $\sim \mathcal{L}(\mu, 1.687)$. For an example of the Bayesian approach see Figure 1. *b)* Weakly-informed priors based on the 'rstanarm' defaults without any formal customizing. *c)* A regularized regression algorithm, using the 'glmnet' (Friedman et al., 2010) and 'caret' (Kuhn, 2020) R packages. In *Subtask 1* the model with optimal accuracy included $\alpha = 0$, ("Ridge" regression), and in *Subtask 2* it included $\alpha = 1$ ("Lasso" regression).

## 3 Results

### 3.1 Subtask 1

In *Subtask 1* the goal was to predict which Individuals were likely to attempt suicide based on tweets up to 30 days prior. Model performances on the training set are displayed in Table 2. The first model (M1) was a Bayesian logistic-regression model using psychologically informed priors. We compared 2 types of distributions for the priors (around the custom centers). The first, a Cauchy distribution with scales set at 2.5. The second, a Laplace distribution with scales of 1.687 (see "Bayesian Modeling" above). In the *Subtask 1* training set, the Informed-Priors Cauchy distribution slightly outperformed the Informed-Priors Laplace distribution in a 5-fold cross-validation.

The second model (M2) was a weakly-informed Bayesian logistic-regression model with priors drawn from a Cauchy Distribution with center = 0 and scale = 2.5.

The third model (M3) was logistic-regression model with regularization. We conducted 5-fold cross validation, with 3 repeats for hyper-parameter tuning of the penalty type ($\alpha$), and the regularization parameter ($\lambda$). In the *Subtask 1* training set, the optimal prediction accuracy included the hyper-parameters $\alpha = 0$ ("Ridge"), and $\lambda = 10$.

| | F1 | F2 | TPR | FPR | AUC |
|---|---|---|---|---|---|
| **Subtask 1 (30 days)** | | | | | |
| M1 | 0.466 | 0.452 | 0.447 | 0.423 | 0.543 |
| M2 | 0.480 | 0.474 | 0.476 | 0.436 | 0.546 |
| M3 | 0.589 | 0.580 | 0.573 | 0.374 | 0.599 |
| **Subtask 2 (6 months)** | | | | | |
| M1 | 0.586 | 0.529 | 0.499 | 0.187 | 0.739 |
| M2 | 0.668 | 0.626 | 0.602 | 0.184 | 0.745 |
| M3 | 0.710 | 0.670 | 0.646 | 0.175 | 0.735 |

Table 2: 5-fold CV Results. M1: Informed priors; M2: Weakly-informed priors; M3: Ridge/Lasso regression.

| | F1 | F2 | TPR | FPR | AUC |
|---|---|---|---|---|---|
| **Subtask 1 (30 days)** | | | | | |
| BL | 0.636 | 0.636 | 0.636 | 0.364 | 0.661 |
| M1 | 0.526 | 0.481 | 0.455 | 0.273 | 0.678 |
| M2 | 0.526 | 0.481 | 0.455 | 0.273 | 0.678 |
| M3 | 0.421 | 0.385 | 0.364 | 0.364 | 0.636 |
| **Subtask 2 (6 months)** | | | | | |
| BL | 0.710 | 0.724 | 0.733 | 0.333 | 0.764 |
| M1 | 0.769 | 0.704 | 0.667 | 0.067 | 0.809 |
| M2 | 0.769 | 0.704 | 0.667 | 0.067 | 0.791 |
| M3 | 0.815 | 0.764 | 0.733 | 0.067 | 0.844 |

Table 3: Official Test Results.BL: Task Baseline; M1: Informed priors; M2: Weakly-informed priors; M3: Ridge/Lasso regression.

### 3.2 Subtask 2

In *Subtask 2* the goal was to predict which Individuals were likely to attempt suicide from tweets up to 6 months (182 days) prior. M1 was a Bayesian logistic-regression model using psychologically informed priors. Like in *Subtask 1*, We compared 2 types of distributions for the priors: Cauchy and Laplace. In the *Subtask 2* training set, the Informed-Priors Laplace distribution outperformed the Informed-Priors Cauchy.

M2 again included a weakly-informed Bayesian logistic-regression model.

M3 was once more a regularized logistic-regression model. In the *Subtask 2* training set, the optimal prediction accuracy included $\alpha = 1$ ("Lasso"), and $\lambda = 0.1$.

Results on the test set are displayed in Table 3. In both tasks models yielded above-chance predictions, and performed better on the test set than the

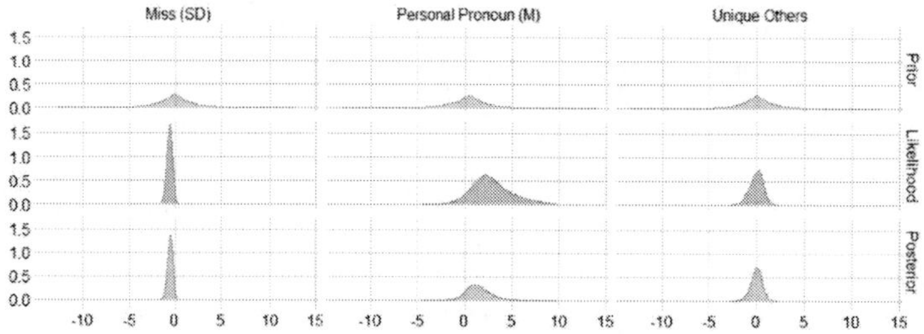


Figure 1: Example of the Bayesian approach using informed (Personal Pronouns) and weakly-informed (Miss, Unique Others) priors and likelihood of the evidence to estimate posterior distributions of three example parameters.

training set. In *Subtask 1*, the models only slightly outperformed the task's baseline model, but in *Subtask 2*, the models yielded high AUC scores.

## 4 Discussion

We trained simple classification models, based on psychological features, to determine which individuals may attempt suicide. We used Psychologically-informed and weakly-informed Bayesian models as well as regularized regression models. Our models yielded moderately successful predictions on *Subtask 1*, and considerably better predictions on *Subtask 2* ($0.791 \leq AUC \leq 0.844$, comparable to Cohen's d of $1.145 - 1.430$). In this task, the informed Bayesian model (M1) was more successful than the weakly-informed (M2). The data-driven regularized regression models (M3) were slightly less accurate in *Subtask 1* than the informed model (M1), and slightly more accurate in *Subtask 2*, perhaps due to the fact that *Subtask 2* included more data than *Subtask 1*.

In addition, in both tasks the Bayesian models (M1, M2) were particularly successful in avoiding False Positive prediction outcomes. Admittedly, in the case of suicide detection, it may be prudent to "err on the side of caution", to avoid missing patients in need of care. However, language-based screening on social media tends to be targeted more for broad risk-detection (Cook et al., 2016). In the case of early risk detection it may also be valid to avoid false alarms in order to reduce unwarranted alarm, especially given the potential for suicidal suggestibility.

Our theory-driven features, as well as the informed Bayesian models, were reliant on domain knowledge to help overcome the problem posed by working with small data sets. Indeed, incorporating knowledge gained from previous research seemed to have aided in forming a generalized model that did not exhibit over-fitting. Another benefit of this approach lies in model interpretability and in its conduciveness to cumulative scientific discovery. We relied on prior empirical findings, and produced updated empirical priors—in light of the task data—which are simple to interpret and share with others (refer to table 4 for feature importance analysis).

The majority of previous work in suicide prediction was done by using proxies to suicidal behavior such as clinical risk assessment and suicidal ideation, (see Fodeh et al., 2019; Ophir et al., 2020; Coppersmith et al., 2018). Thanks to the CLPsych workshop, and the access to valuable data directly indicative of suicidal behavior, we were able to present similar prediction accuracies on actual suicide attempts. The findings derived from this data show great promise for the use of NLP in suicide prevention.

## 5 Conclusion

Our current work provides a synthesis between classic scientific and novel data-driven paradigms. Future research is needed to further explore how psychological knowledge and data science methods can be combined to aid in the gradual accumulation of scientific knowledge, and produce actionable predictions that may help save lives.

## Ethics Statement

Secure access to the shared task dataset was provided with IRB approval under University of Maryland, College Park protocol 1642625.

## Acknowledgements

The authors wish to thank Yhonatan Shemesh, Inon Raz, Mattan S. Ben-Shachar, the CLPsych 2021

| Features | Effect-Size ($log - odds$) |
|---|---|
| **Subtask 1 (30 days)** | |
| **M1** | |
| Negative-Emotion-SD | 2.36 [0.83,4.59] |
| Negative-Emotion-M | -1.68 [-4.05,-0.05] |
| Swear-Words-M | 1.67 [-1.13,6.84] |
| Female-M | 1.06 [0.08,2.64] |
| Want-M | 1.04 [0.29,1.86] |
| **M2** | |
| Negative-Emotion-SD | 2.39 [0.88,4.19] |
| Negative-Emotion-M | -1.72 [-3.69,-0.13] |
| Swear-Words-M | 1.53 [-1.24,4.63] |
| Female-M | 1.15 [0.07,2.62] |
| Want-M | 1.04 [0.29,1.88] |
| **M3** | |
| They-M | 0.009 |
| I-Pronouns-M | 0.009 |
| Personal-Pronouns-M | 0.009 |
| Want-M | 0.009 |
| Negative-Emotion-SD | 0.008 |
| **Subtask 2 (6 months)** | |
| **M1** | |
| Informal-SD | 2.02 [0.32,4.17] |
| I-Pronouns-M | -1.5 [-2.85,-0.27] |
| Female-M | 1.45 [-0.10,0.4.84] |
| Personal-Pronouns-M | 1.345 [-0.50,3.87] |
| Sexual-M | -1.26 [-2.66,0.09] |
| **M2** | |
| Informal-SD | 2.99 [01.13,4.93] |
| Female-M | 2.59 [0.25,5.61] |
| Negative-Emotion-SD | 1.98 [-0.17,4.19] |
| I-Pronouns-M | -1.89 [-3.46,-0.31] |
| Personal-Pronouns-M | 1.87 [-0.80,4.51] |
| **M3** | |
| Personal-Pronouns-M | 0.51 |
| Negative-Emotion-SD | 0.11 |

Table 4: Most Important Features based on model coefficient values. Model coefficients are on the log-odds scale. Values in brackets denote 95% posterior uncertainty intervals.

Shared Task organizers, and the anonymous reviewers for their help and insight. The organizers are particularly grateful to the users who donated data to the OurDataHelps project without whom this work would not be possible, to Qntfy for supporting the OurDataHelps project and making the data available, to NORC for creating and administering the secure infrastructure, and to Amazon for supporting this research with computational resources on AWS.

## References

Ella Arensman, Vanda Scott, Diego De Leo, and Jane Pirkis. 2020. Suicide and suicide prevention from a global perspective. *Crisis*.

Bob Carpenter, Andrew Gelman, Matthew D Hoffman, Daniel Lee, Ben Goodrich, Michael Betancourt, Marcus A Brubaker, Jiqiang Guo, Peter Li, and Allen Riddell. 2017. Stan: a probabilistic programming language. *Grantee Submission*, 76(1):1–32.

Cindy K Chung and James W Pennebaker. 2018. What do we know when we liwc a person? text analysis as an assessment tool for traits, personal concerns and life stories. *The Sage handbook of personality and individual differences*, pages 341–360.

Benjamin L Cook, Ana M Progovac, Pei Chen, Brian Mullin, Sherry Hou, and Enrique Baca-Garcia. 2016. Novel use of natural language processing (nlp) to predict suicidal ideation and psychiatric symptoms in a text-based mental health intervention in madrid. *Computational and mathematical methods in medicine*, 2016.

Glen Coppersmith, Ryan Leary, Patrick Crutchley, and Alex Fine. 2018. Natural language processing of social media as screening for suicide risk. *Biomedical informatics insights*, 10:1178222618792860.

Glen Coppersmith, Kim Ngo, Ryan Leary, and Anthony Wood. 2016. Exploratory analysis of social media prior to a suicide attempt. In *Proceedings of the third workshop on computational linguistics and clinical psychology*, pages 106–117.

Sally C Curtin, Margaret Warner, and Holly Hedegaard. 2016. *Increase in suicide in the United States, 1999-2014*. 2016. US Department of Health and Human Services, Centers for Disease Control and ....

Stephanie K Doupnik, Brittany Rudd, Timothy Schmutte, Diana Worsley, Cadence F Bowden, Erin McCarthy, Elliott Eggan, Jeffrey A Bridge, and Steven C Marcus. 2020. Association of suicide prevention interventions with subsequent suicide attempts, linkage to follow-up care, and depression symptoms for acute care settings: a systematic review and meta-analysis. *JAMA psychiatry*, 77(10):1021–1030.

Johannes C Eichstaedt, Robert J Smith, Raina M Merchant, Lyle H Ungar, Patrick Crutchley, Daniel Preoţiuc-Pietro, David A Asch, and H Andrew Schwartz. 2018. Facebook language predicts depression in medical records. *Proceedings of the National Academy of Sciences*, 115(44):11203–11208.

Samah Fodeh, Taihua Li, Kevin Menczynski, Tedd Burgette, Andrew Harris, Georgeta Ilita, Satyan Rao, Jonathan Gemmell, and Daniela Raicu. 2019. Using machine learning algorithms to detect suicide risk factors on twitter. In *2019 International Conference on Data Mining Workshops (ICDMW)*, pages 941–948. IEEE.

Joseph C Franklin, Jessica D Ribeiro, Kathryn R Fox, Kate H Bentley, Evan M Kleiman, Xieyining Huang, Katherine M Musacchio, Adam C Jaroszewski, Bernard P Chang, and Matthew K Nock. 2017. Risk factors for suicidal thoughts and behaviors: a meta-analysis of 50 years of research. *Psychological bulletin*, 143(2):187.

Jerome Friedman, Trevor Hastie, and Robert Tibshirani. 2010. Regularization paths for generalized linear models via coordinate descent. *Journal of Statistical Software*, 33(1):1–22.

Catherine R Glenn, Evan M Kleiman, John Kellerman, Olivia Pollak, Christine B Cha, Erika C Esposito, Andrew C Porter, Peter A Wyman, and Anne E Boatman. 2020. Annual research review: a meta-analytic review of worldwide suicide rates in adolescents. *Journal of child psychology and psychiatry*, 61(3):294–308.

Ben Goodrich, Jonah Gabry, Imad Ali, and Sam Brilleman. 2020. rstanarm: Bayesian applied regression modeling via Stan. R package version 2.21.1.

Elizabeth M Haney, Maya E O'Neil, Susan Carson, A Low, K Peterson, LM Denneson, C Oleksiewicz, and D Kansagara. 2012. Suicide risk factors and risk assessment tools: A systematic review.

Max Kuhn. 2020. *caret: Classification and Regression Training*. R package version 6.0-86.

Matthew Large, Cherrie Galletly, Nicholas Myles, Christopher James Ryan, and Hannah Myles. 2017a. Known unknowns and unknown unknowns in suicide risk assessment: evidence from meta-analyses of aleatory and epistemic uncertainty. *BJPsych bulletin*, 41(3):160–163.

Matthew Michael Large, Christopher James Ryan, Gregory Carter, and Nav Kapur. 2017b. Can we usefully stratify patients according to suicide risk? *Bmj*, 359.

Nicholas Leigh-Hunt, David Bagguley, Kristin Bash, Victoria Turner, Stephen Turnbull, N Valtorta, and Woody Caan. 2017. An overview of systematic reviews on the public health consequences of social isolation and loneliness. *Public health*, 152:157–171.

Richard T Liu, Stephanie J Steele, Jessica L Hamilton, Quyen BP Do, Kayla Furbish, Taylor A Burke, Ashley P Martinez, and Nimesha Gerlus. 2020. Sleep and suicide: A systematic review and meta-analysis of longitudinal studies. *Clinical psychology review*, page 101895.

Andrea Miranda-Mendizabal, Pere Castellví, Oleguer Parés-Badell, Itxaso Alayo, José Almenara, Iciar Alonso, Maria Jesús Blasco, Annabel Cebria, Andrea Gabilondo, Margalida Gili, et al. 2019. Gender differences in suicidal behavior in adolescents and young adults: systematic review and meta-analysis of longitudinal studies. *International journal of public health*, 64(2):265–283.

Mohsen Naghavi. 2019. Global, regional, and national burden of suicide mortality 1990 to 2016: systematic analysis for the global burden of disease study 2016. *bmj*, 364.

OSF Open Science Collaboration et al. 2015. Estimating the reproducibility of psychological science. *Science*, 349(6251).

Yaakov Ophir, Refael Tikochinski, Christa SC Asterhan, Itay Sisso, and Roi Reichart. 2020. Deep neural networks detect suicide risk from textual facebook posts. *Scientific reports*, 10(1):1–10.

James W Pennebaker, Ryan L Boyd, Kayla Jordan, and Kate Blackburn. 2015. The development and psychometric properties of liwc2015. Technical report.

John P Pestian, Pawel Matykiewicz, Michelle Linn-Gust, Brett South, Ozlem Uzuner, Jan Wiebe, K Bretonnel Cohen, John Hurdle, and Christopher Brew. 2012. Sentiment analysis of suicide notes: A shared task. *Biomedical informatics insights*, 5:BII–S9042.

Jessica D Ribeiro, Xieyining Huang, Kathryn R Fox, and Joseph C Franklin. 2018. Depression and hopelessness as risk factors for suicide ideation, attempts and death: meta-analysis of longitudinal studies. *The British Journal of Psychiatry*, 212(5):279–286.

Shane Schweitzer and Adam Waytz. 2020. Language as a window into mind perception: How mental state language differentiates body and mind, human and nonhuman, and the self from others. *Journal of Experimental Psychology: General*.

Almog Simchon and Michael Gilead. 2018. A psychologically informed approach to CLPsych shared task 2018. In *Proceedings of the Fifth Workshop on Computational Linguistics and Clinical Psychology: From Keyboard to Clinic*, pages 113–118, New Orleans, LA. Association for Computational Linguistics.

Shannon Wiltsey Stirman and James W Pennebaker. 2001. Word use in the poetry of suicidal and non-suicidal poets. *Psychosomatic medicine*, 63(4):517–522.

Erik van Zwet and Andrew Gelman. 2020. A proposal for informative default priors scaled by the standard error of estimates. *arXiv preprint arXiv:2011.15037*.

Kathryn Yates, Ulla Lång, Martin Cederlöf, Fiona Boland, Peter Taylor, Mary Cannon, Fiona McNicholas, Jordan DeVylder, and Ian Kelleher. 2019.

Association of psychotic experiences with subsequent risk of suicidal ideation, suicide attempts, and suicide deaths: a systematic review and meta-analysis of longitudinal population studies. *JAMA psychiatry*, 76(2):180–189.

(?)

# Analysis of Behavior Classification in Motivational Interviewing

**Leili Tavabi[1], Trang Tran[1], Kalin Stefanov[2],**
**Brian Borsari[3], Joshua D Woolley [3], Stefan Scherer[1], Mohammad Soleymani[1]**
[1]Institute for Creative Technologies, University of Southern California, Los Angeles, CA, USA
[2]Faculty of Information Technology, Monash University, Melbourne, VIC, Australia
[3]VA Hospital San Francisco, University of California San Francisco, San Francisco, CA, USA
{ltavabi,ttran}@ict.usc.edu, kalin.stefanov@monash.edu,
Brian.Borsari@va.gov, josh.woolley@ucsf.edu,
{scherer,soleymani}@ict.usc.edu

## Abstract

Analysis of client and therapist behavior in counseling sessions can provide helpful insights for assessing the quality of the session and consequently, the client's behavioral outcome. In this paper, we study the automatic classification of standardized behavior codes (i.e. annotations) used for assessment of psychotherapy sessions in Motivational Interviewing (MI). We develop models and examine the classification of client behaviors throughout MI sessions, comparing the performance by models trained on large pretrained embeddings (RoBERTa) versus interpretable and expert-selected features (LIWC). Our best performing model using the pretrained RoBERTa embeddings beats the baseline model, achieving an F1 score of 0.66 in the subject-independent 3-class classification. Through statistical analysis on the classification results, we identify prominent LIWC features that may not have been captured by the model using pretrained embeddings. Although classification using LIWC features underperforms RoBERTa, our findings motivate the future direction of incorporating auxiliary tasks in the classification of MI codes.

## 1 Introduction

Motivational Interviewing (MI) is a psychotherapy treatment style for resolving ambivalence toward a problem such as alcohol or substance abuse. MI approaches focus on eliciting clients' own intrinsic reasons for changing their behavior toward the desired outcome. MI commonly leverages a behavioral coding (annotation) system, Motivational Interviewing Skills Code (MISC) (Miller et al., 2003), which human annotators follow for coding both client's and therapist's utterance-level intentions and behaviors. These codes have shown to be effective means of assessing the quality of the session, training therapists, and estimating clients' behavioral outcomes (Lundahl et al., 2010; Diclemente et al., 2017; Magill et al., 2018). Due to the high cost and labor-intensive procedure of manually annotating utterance-level behaviors, existing efforts have worked on automatic coding of the MI behaviors. The client utterances throughout the MI session are categorized based on their expressed attitude toward change of behavior: (1) Change Talk (CT): willing to change, (2) Sustain Talk (ST): resisting to change, and (3) Follow/Neutral (FN): other talk unrelated to change. An example conversation between a therapist (T) and a client (C) is shown below.

- T: [...] you talked about drinking about 7 times a week [...] Does that sound about right, or?
- C: I don't know so much any, like 5, probably like, the most 4 now, in the middle of the week I try to just kinda do work, **(CT)**
- C: I mean, like I would **(ST)**
- C: but, but getting up's worse, it's like being tired, not so much hungover just feeling uhh, class. [...] **(CT)**
- T: When you do drink, how much would you say, would you say the ten's about accurate?
- C: About around ten, maybe less, maybe more, depends like, I don't really count or anything but, it's probably around ten or so. **(FN)**

Previous work in MI literature mainly approached automatic classification of behavior codes in MI by modeling utterance-level representations. Aswamenakul et al. (2018) trained a logistic regression model using both interpretable linguistic features (LIWC) and GloVe embeddings, finding that Sustain Talk is associated with positive attitude towards drinking, and the opposite for Change Talk. To account for dialog context, Can et al. (2015) formulated the task as a sequence labeling problem, and trained a Conditional Random Field (CRF) to predict MI codes. More recent approaches leveraged advances in neural networks, using standard recurrent neural networks (RNNs) (Xiao et al., 2016; Ewbank et al., 2020; Gibson



*Proceedings of the Seventh Workshop on Computational Linguistics and Clinical Psychology*, pages 110–115
June 11, 2021. 

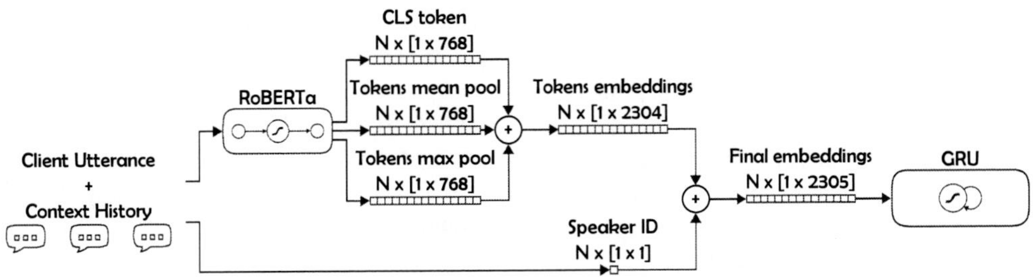


Figure 1: Utterance representation from RoBERTa embeddings.

et al., 2016; Huang et al., 2018) or hierarchical encoders with attention (Cao et al., 2019). In addition to context modeling, Tavabi et al. (2020) leveraged pretrained contextualized embeddings (Devlin et al., 2019) and incorporated the speech modality to classify MI codes, beating the previous baseline of Aswamenakul et al. (2018) on a similar dataset. The most gain seemed to come from powerful pretrained embeddings, as with many other NLP tasks. However, it is unclear what these BERT-like embeddings learn, as they are not as interpretable as the psycholinguistically motivated features (LIWC).

In this paper, we study the quality of automatic MI coding models in an attempt to understand what distinguishes language patterns in Change Talk, Sustain Talk, and Follow/Neutral. We develop a system for classifying clients' utterance-level MI codes by modeling the client's utterance and the preceding context history from both the client and the therapist. We compare the effectiveness and interpretability between contextualized pretrained embeddings and hand-crafted features, by training classifiers using (1) pretrained RoBERTa embeddings (Liu et al., 2019), (2) an interpretable and dictionary-based feature set, Linguistic Inquiry Word Count (LIWC) (Pennebaker et al., 2001). Our best-performing model outperforms the baseline model from previous work on the same dataset (Tavabi et al., 2020), reaching F1=0.66 from F1=0.63.

In examining misclassifications by both models, we identify features that are significant across classes. Our findings suggest that large pretrained embeddings like RoBERTa, despite their high representation power, might not necessarily capture all the salient features that are important in distinguishing the classes. We identified prominent features that are statistically significant across classes on the entire dataset, as well as the misclassified samples. Theses findings suggest that our systems might benefit from fine-tuning pretrained embeddings, adding auxiliary tasks (e.g sentiment classification), and better context modeling.

## 2 Data

We use two clinical datasets (Borsari et al., 2015) collected in college campuses from real MI sessions with students having alcohol-related problems. The data consists of transcripts and audio recordings from the client-therapist in-session dialogues. The sessions are manually transcribed, and labelled per utterance using MISC codes. The dataset includes 219 sessions for 219 clients, consisting of about 93k client and therapist utterances; the client-therapist distribution of utterances is 0.44-0.54. The dataset is highly imbalanced, with a class distribution of [0.13, 0.59, 0.28] for [Sustain Talk, Follow/Neutral, Change Talk]. In addition to the in-session text and speech data, the dataset consists of session-level measures regarding clients' behavioral changes toward the desired outcome. Additional metadata includes session-level global metrics such as therapist empathy, MI spirit, and client engagement.

## 3 Methodology

### 3.1 Embeddings and Feature sets

*Pretrained RoBERTa Embeddings.* RoBERTa (Liu et al., 2019) is an improved representation based on BERT (Devlin et al., 2019). RoBERTa differs from BERT in several aspects: removal of the Next Sentence Prediction objective, introduction of dynamic masking, pretrained on a larger dataset with larger mini-batches and longer sequences. These changes can improve the representations on our data, especially since dialogue utterances in psychotherapy can consist of very long sequences. Our preliminary experiments for fine-tuning both BERT and RoBERTa on our task showed that RoBERTa performed better. We therefore select

RoBERTa to obtain utterance representations.

*Interpretable LIWC Features.* LIWC (Pennebaker et al., 2001) is a dictionary-based tool that assigns scores in psychologically meaningful categories including social and affective processes, based on words in a text input. It was developed by experts in social psychology and linguistics, and provides a mechanism for gaining interpretable and explainable insights in the text input. Given our focus domain of clinical psychology, where domain knowledge is highly valuable, we select the psychologically-motivated LIWC feature set as a natural point of comparison.

### 3.2 Classification Model

For classifying the clients' MI codes, we learn the client utterance representation using features described in 3.1, as well as the preceding history from both the client and therapist. The input window includes the current utterance, and history context. Specifically, the input window consists of a total of 3 or more turn changes across speakers, where each turn consists of one or more consecutive utterances per speaker. In the beginning of the session, where the history context is shorter than the specified threshold, the context history consists of those limited preceding utterances. The size of the context window was selected empirically among 3, 4 or 5 turn changes.

Our input samples contain between 6 and 28 utterances depending on the dynamic of the dialogue, e.g. an example input could be [T C T T T C C T C], where T denotes Therapist's utterance and C denotes Client's. The motivation for using the entire window of context and final utterance is that the encoding by our recurrent neural network (RNN) would carry more information from the final utterance and closer context, while retaining relevant information from the beginning of the window. We also investigated encoding the current utterance separate from the context using a linear layer, but did not see improvements in the classification results.

For RoBERTa embeddings, each utterance representation is the concatenation of (1) CLS token (2) mean pooling of the tokens from the last hidden state (3) max pooling of the tokens from the last hidden state. Figure 1 illustrates this process. For LIWC representations, the features are already extracted on the utterance level. Additionally, for both RoBERTa and LIWC representations, we add a binary dimension for each utterance to indicate the speaker. The history context representation for both RoBERTa and LIWC is obtained by concatenating the utterance-level representation vectors into a 2d matrix. These inputs are then fed into a unidirectional GRU, and the last hidden state is used for the last classification layer.

## 4 Results and Discussions

For training, we use a 5-fold subject-independent cross validation. 10% of the train data from each fold is randomly selected in stratified fashion, and held out as the validation set. We optimize the network using AdamW (Loshchilov and Hutter, 2019), with a learning rate of $10^{-4}$ and batch size of 32. We train our model for 25 epochs with early stopping after 10 epochs, and select the model with the highest macro F1 on the validation set. To handle class imbalance, we use a cross-entropy loss with a weight vector inversely proportional to the number of samples in each class. The GRU hidden dimension is 256 and 32 when running on RoBERTa and LIWC representations, respectively.

We compare our work to the best performing model from previous work (Tavabi et al., 2020), trained on the same dataset and under the same evaluation protocol. Briefly, this baseline model differs from our current model in several aspects: BERT embeddings were used as input; the representation vector for the current client utterance is fed into a linear layer. The client and therapist utterances within the context window are separated, mean-pooled and fed individually to two different linear layers. The output encodings from the three linear layers are merged and fed into another linear layer before being passed to the classification layer.

We perform statistical analysis to identify prominent LIWC features across pairs of classes, as well as misclassified samples from each classifier. Since the classifiers encode context, we incorporate the context in the statistical analysis by averaging the feature vectors along utterances within the input window.

### 4.1 Classifier Performance

The classification results are shown in Table 1. The model trained using RoBERTa outperforms the model trained on LIWC features, in addition to beating the baseline model in (Tavabi et al., 2020) with F1-macro=0.66. Improved results over the baseline model are likely due to the following: 1) The previous linear model encodes the client and therapist utterances from the context history separately, therefore potentially missing information from the dyadic interaction. 2) The RNN in our current model temporally encodes the dyadic interaction window. 3) Using RoBERTa embeddings improved over BERT embeddings, as RoBERTa was trained on larger datasets and on longer sequences, making them more powerful representations.

| | Features | | Baseline |
|---|---|---|---|
| | LIWC | RoBERTa | |
| ST | 0.41 | 0.50 | 0.46 |
| FN | 0.78 | 0.84 | 0.81 |
| CT | 0.56 | 0.64 | 0.63 |
| All (macro) | 0.58 | **0.66** | 0.63 |
| All (micro) | 0.65 | **0.74** | 0.71 |

Table 1: F1-Score Classification Results

The results from other work on classifying client codes in MI range from F1-macro=0.44 (Can et al., 2015) to F1-macro=0.54 (Cao et al., 2019) on different datasets. Aswamenakul et al. (2018), who used a similar dataset to our work, reached F1-macro=0.57. Huang et al. (2018) obtained F1-macro=0.70 by using (ground truth) labels from prior utterances as the model input and domain adaptation for theme shifts throughout the session.

The F1 scores show that Sustain Talk, the minority class, is consistently the hardest to classify and Follow/Neutral, the majority class, the easiest. This is similar to findings from previous work in literature, e.g. (Can et al., 2015) and remains a challenge in automated MI coding. Using approaches like upsampling toward a more balanced dataset will be part of our future work. In order for these systems to be deployable in the clinical setting, the standard we adhere to is guided by a range developed by biostaticians in the field, which indicates values higher than 0.75 to be "excellent" (Cicchetti, 1994). Therefore, despite the good results, there is much room for improvement before such systems can be autonomously utilized in real-world MI sessions.

### 4.2 Error Analysis

Figure 2 shows the confusion matrices from classification results by the model using LIWC features vs. RoBERTa embeddings. Comparing between classes, Sustain Talk gets misclassified about equally as Follow/Neutral and Change Talk by RoBERTa but it is much more often misclassified as Change Talk by LIWC. On the other hand, Change Talk is more often misclassified as Follow/Neutral by RoBERTa, but misclassified as Sustain Talk by LIWC.

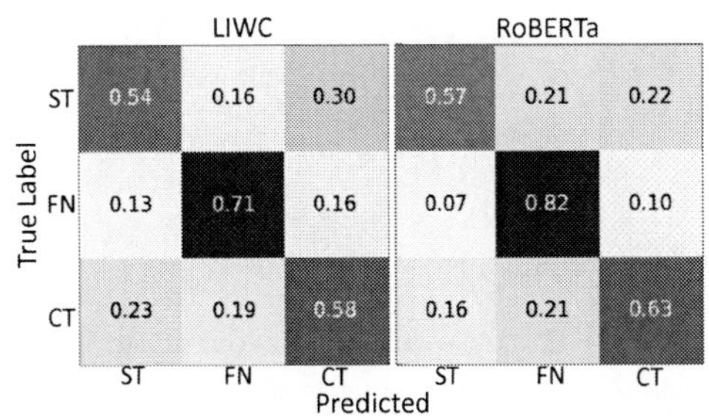


Figure 2: Confusion matrices (normalized by true labels) of classification results by LIWC (left) and RoBERTa (right) features.

Of the wrongly classified utterances by LIWC, 47% were correctly classified by RoBERTa. Of the RoBERTa misclassifications (11k utterances), about 30% were correctly classified by LIWC. Some examples of these cases are presented in Figure 3, which seem to be associated with certain key words related to salient features (Section 4.3).

T: What varies your drinking?
C: Money, (CT → ST)
C: if I have work to do I won't drink. (CT)
T: Okay.
... ...
C: Anxious thing is kinda like I don't have control, like I, I'm shaky and stuff like that. (CT)
T: Ok. Is your heart racing faster or, and, and that type of thing?
C: No, it's not really anxious, it's kinda just like a ... (CT → ST)
T: It's more shaky?
C: It's like agitated, kind of. (CT → ST)

Figure 3: Example dialog with correct and incorrect classifications. T=therapist; C=client; red (true → predicted) denotes misclassification by RoBERTa but correctly classified by LIWC; blue (true label) denotes correct classification by both models.

When both RoBERTa and LIWC misclassified, they give the same wrong prediction on 70% of those utterances. Some anecdotal examples of such cases are shown in Figure 4, most seem to be highly context-dependent, suggesting that better modeling of context would potentially be useful.

T: Oh, ok, so the summer you usually drink a little more
C: Yeah. (**FN**)
T: and then when you get to school, it's...
C: Kinda cut down a little bit. (**CT**)
T: I see, because of like, school and classes and stuff.
C: Yeah. (**CT** → **FN**)
T: And working on the weekends.
C: Yeah. (**CT** → **FN**)

Figure 4: Example dialog with correct and incorrect classifications. T=therapist; C=client. blue (true label) denotes correct classification by our models, red (true → predicted) denotes misclassification by both models.

We also experimented with simple concatenation of RoBERTa and LIWC features, but did not find significant improvements over the RoBERTa-only model. Better models for combining RoBERTa and LIWC features might improve our results, which will be part of future work.

### 4.3 Salient Features

Statistical analysis on LIWC features across the classes can help identify the salient features distinguishing the classes, therefore can signal important information picked up by the LIWC classifier. We used hierarchical Analysis of Variance (ANOVA), with talk types nested under sessions to account for individual differences, to find linguistic features that are significantly different across MI codes. To further examine the statistical significance across pairs of classes, we performed a Tukey post hoc test. We found the following features to be the most statistically different features across all the pairs of classes: 'WPS' (mean words per sentence), 'informal', 'assent' (e.g. agree, ok, yes), 'analytic.' Additionally, 'AllPunc' (use of punctuations) and 'function' (use of pronouns) were prominent features that were significantly distinguishing Follow/Neutral from the other classes.

We further looked into samples where RoBERTa representations might be limited (i.e. misclassified), while LIWC features were correct in the classification. Using ANOVA, we found the most prominent features in such samples across the 3 classes: 'swear' (6.06), 'money' (5.29), 'anger' (2.24), 'death' (2.19), and 'affiliation' (2.00), where numbers in parentheses denote F-statistic from hierarchical ANOVA. This is consistent with our error analysis in Section 4.2, as shown in Figure 3. The mean scores of the 'swear,' 'money,' and 'anger' categories are higher for Change Talk compared to other classes. We hypothesize that 'swear' and 'anger' in Change Talk may represent anger toward oneself regarding drinking behavior. Words in the 'money' category might be related to the high cost of alcohol (especially with college-age clients), which can be motivation for behavior change. The Change Talk samples misclassified by the RoBERTa model may indicate the model's failure to capture such patterns.

## 5 Conclusion

We developed models for the classification of clients' MI codes. We experimented with pretrained RoBERTa embeddings and interpretable LIWC features as our model inputs, where the RoBERTa model outperformed the baseline from previous work, reaching F1=0.66. Through statistical analysis, we investigated prominent LIWC features that are significantly different across pairs of classes. We further looked into misclassified samples across the classifiers, and identified prominent features that may have not been captured by the RoBERTa model. This finding motivates the use of auxiliary tasks like sentiment and affect prediction, in addition to fine-tuning the model with domain-specific data and better context modeling.

With this work, we aim to develop systems for enhancing effective communication in MI, which can potentially generalize to other types of therapy approaches. Identifying patterns of change language can lead to MI strategies that will assist clinicians with treatment, while facilitating efficient means for training new therapists. These steps contribute to the long-term goal of providing cost- and time- effective evaluation of treatment fidelity, education of new therapists, and ultimately broadening access to lower-cost clinical resources for the general population.

## Acknowledgments

This work was supported by NIAAA grants R01 AA027225, R01 AA017427 and R01 AA12518. The content is the responsibility of the authors and does not necessarily represent the official views of the NIAAA, NIH, Dept. of Veterans Affairs, or the US Government. We thank the clients and therapists for their audiotapes to be used in this work, and the anonymous reviewers for their feedback.

## References


Chanuwas Aswamenakul, Lixing Liu, Kate B Carey, Joshua Woolley, Stefan Scherer, and Brian Borsari. 2018. Multimodal Analysis of Client Behavioral Change Coding in Motivational Interviewing. In *Proceedings of the 20th ACM International Conference on Multimodal Interaction*. 356–360.

Brian Borsari, Timothy R Apodaca, Kristina M Jackson, Nadine R Mastroleo, Molly Magill, Nancy P Barnett, and Kate B Carey. 2015. In-session processes of brief motivational interventions in two trials with mandated college students. *Journal of consulting and clinical psychology* 83, 1 (2015), 56.

Doğan Can, David C Atkins, and Shrikanth S Narayanan. 2015. A dialog act tagging approach to behavioral coding: A case study of addiction counseling conversations. In *Sixteenth Annual Conference of the International Speech Communication Association*.

Jie Cao, Michael Tanana, Zac E Imel, Eric Poitras, David C Atkins, and Vivek Srikumar. 2019. Observing dialogue in therapy: Categorizing and forecasting behavioral codes. *arXiv preprint arXiv:1907.00326* (2019).

Domenic V Cicchetti. 1994. Guidelines, criteria, and rules of thumb for evaluating normed and standardized assessment instruments in psychology. *Psychological assessment* 6, 4 (1994), 284.

Jacob Devlin, Ming-Wei Chang, Kenton Lee, and Kristina Toutanova. 2019. BERT: Pre-training of Deep Bidirectional Transformers for Language Understanding. In *Proc. NAACL*. 4171–4186.

C. Diclemente, Catherine M Corno, Meagan M. Graydon, Alicia E Wiprovnick, and Daniel J. Knoblach. 2017. Motivational Interviewing, Enhancement, and Brief Interventions Over the Last Decade: A Review of Reviews of Efficacy and Effectiveness. *Psychology of Addictive Behaviors* 31, 862–887.

MP Ewbank, R Cummins, V Tablan, A Catarino, S Buchholz, and AD Blackwell. 2020. Understanding the relationship between patient language and outcomes in internet-enabled cognitive behavioural therapy: A deep learning approach to automatic coding of session transcripts. *Psychotherapy Research* (2020), 1–13.

James Gibson, Dogan Can, Bo Xiao, Zac E Imel, David C Atkins, Panayiotis Georgiou, and Shrikanth Narayanan. 2016. A deep learning approach to modeling empathy in addiction counseling. *Commitment* 111 (2016), 21.

Xiaolei Huang, Lixing Liu, Kate Carey, Joshua Woolley, Stefan Scherer, and Brian Borsari. 2018. Modeling temporality of human intentions by domain adaptation. In *Proceedings of the 2018 Conference on Empirical Methods in Natural Language Processing*. 696–701.

Yinhan Liu, Myle Ott, Naman Goyal, Jingfei Du, Mandar Joshi, Danqi Chen, Omer Levy, Mike Lewis, Luke Zettlemoyer, and Veselin Stoyanov. 2019. Roberta: A robustly optimized bert pretraining approach. *arXiv preprint arXiv:1907.11692*.

Ilya Loshchilov and Frank Hutter. 2019. Decoupled Weight Decay Regularization. In *International Conference on Learning Representations*. `https://openreview.net/forum?id=Bkg6RiCqY7`

Brad W Lundahl, Chelsea Kunz, Cynthia Brownell, Derrik Tollefson, and Brian L Burke. 2010. A meta-analysis of motivational interviewing: Twenty-five years of empirical studies. *Research on social work practice* 20, 2 (2010), 137–160.

Molly Magill, Timothy R Apodaca, Brian Borsari, Jacques Gaume, Ariel Hoadley, Rebecca EF Gordon, J Scott Tonigan, and Theresa Moyers. 2018. A meta-analysis of motivational interviewing process: Technical, relational, and conditional process models of change. *Journal of consulting and clinical psychology* 86, 2 (2018), 140.

William R Miller, Theresa B Moyers, Denise Ernst, and Paul Amrhein. 2003. Manual for the motivational interviewing skill code (MISC). *Unpublished manuscript. Albuquerque: Center on Alcoholism, Substance Abuse and Addictions, University of New Mexico* (2003).

James W Pennebaker, Martha E Francis, and Roger J Booth. 2001. Linguistic inquiry and word count: LIWC 2001. *Mahway: Lawrence Erlbaum Associates* 71, 2001 (2001), 2001.

Leili Tavabi, Kalin Stefanov, Larry Zhang, Brian Borsari, Joshua D Woolley, Stefan Scherer, and Mohammad Soleymani. 2020. Multimodal Automatic Coding of Client Behavior in Motivational Interviewing. In *Proceedings of the 2020 International Conference on Multimodal Interaction*. 406–413.

Bo Xiao, Dogan Can, James Gibson, Zac E Imel, David C Atkins, Panayiotis G Georgiou, and Shrikanth S Narayanan. 2016. Behavioral Coding of Therapist Language in Addiction Counseling Using Recurrent Neural Networks.. In *Interspeech*. 908–912.

# Automatic Detection and Prediction of Psychiatric Hospitalizations From Social Media Posts

**Zhengping Jiang**
Computer Science Dept.
Columbia University
zj2265@columbia.edu

**Jonathan Zomick**
Psychology Dept.
Hofstra University
jzomick1@pride.hofstra.edu

**Sarah Ita Levitan**
Computer Science Dept.
Hunter College, CUNY
sarah.levitan@hunter.cuny.edu

**Mark Serper**
Psychology Dept.
Hofstra University
Mark.R.Serper@hofstra.edu

**Julia Hirschberg**
Computer Science Dept.
Columbia University
julia@cs.columbia.edu

## Abstract

We address the problem of predicting psychiatric hospitalizations using linguistic features drawn from social media posts. We formulate this novel task and develop an approach to automatically extract time spans of self-reported psychiatric hospitalizations. Using this dataset, we build predictive models of psychiatric hospitalization, comparing feature sets, user vs. post classification, and comparing model performance using a varying time window of posts. Our best model achieves an F1 of .718 using 7 days of posts. Our results suggest that this is a useful framework for collecting hospitalization data, and that social media data can be leveraged to predict acute psychiatric crises before they occur, potentially saving lives and improving outcomes for individuals with mental illness.

## 1 Introduction

Every year, approximately 1% of adults in the United States are hospitalized for psychiatric reasons, including increased suicidality and psychosis (Elflein, 2020). With the global COVID-19 pandemic, hospitalizations due to suicidality are projected to increase substantially (John et al., 2020), and there is already evidence of the adverse impact of the pandemic on the mental health of individuals around the world (Cullen et al., 2020). Psychiatric hospitalizations typically result from crises among individuals struggling with suicidality and mental illness. The present study aims to predict psychiatric hospitalization due to increased suicidality or a psychotic break before it occurs.

There are several motivations for this research goal. Improving our ability to better predict psychiatric hospitalization helps enable the identification of early warning signs of these crises before they fully develop. Early detection and prediction of acute psychiatric crises is essential for lowering mortality rates and improving overall outcomes for individuals suffering with mental illness. Further, psychiatric hospitalizations place a tremendous burden on limited hospital resources, and involve steep costs for patients as well as taxpayers (Stensland et al., 2012; Owens et al., 2019).

Typically, prediction of psychiatric hospitalization has relied on rich and personalized clinical information for a particular patient. This requirement has limited the size of available datasets, and has also limited the possibility of reaching and helping potential patients who do not have a well-documented psychiatric medical history. In this work we circumvent this limitation by leveraging social media data to train and evaluate predictive models of psychiatric hospitalization. This is a necessary step towards the ultimate goal of predicting behavioral and cognitive changes that often lead to hospitalization. There is a rich literature of computer scientists, psychologists, and psychiatrists taking advantage of the vast amount of social media data – which includes language data of posts and comments, as well as meta-information such as preferences, engagement patterns, and group membership – to gain insights about mental states and behaviors of people with psychiatric disorders.



*Proceedings of the Seventh Workshop on Computational Linguistics and Clinical Psychology*, pages 116–121
June 11, 2021. 

Building on this successful line of research, we detect engagement patterns combined with self-disclosures to identify potential periods of psychiatric hospitalization. We compile a dataset of these periods, or time spans, along with the posts preceding those periods, and conduct machine learning experiments to automatically predict whether a post precedes a hospitalization or not. Our results suggest that this is a potentially useful approach for predicting psychiatric hospitalizations before they occur. This can enable clinicians to mitigate and hopefully prevent a psychotic break or suicide attempt, helping to save patients' lives and improve outcomes.

The rest of this paper is organized as follows: Section 2 reviews related work and Section 3 describes our novel data collection approach. Section 4 presents our experiments to predict psychiatric hospitalizations, and Section 5 provides analyses of the data and the learned models to gain further insights about the dataset and our results. We conclude in Section 6 and discuss ideas for future work.

## 2 Related Work

Research over the past decade has supported and validated the use of computational linguistics techniques applied to social media data for predicting and detecting mental illness across a broad range of psychiatric conditions (Guntuku et al., 2017; Wongkoblap et al., 2017). To date, linguistic indicators of psychopathology have been identified for a wide range of psychiatric conditions (Zomick et al., 2019; Coppersmith et al., 2015; Birnbaum et al., 2017; Huang et al., 2017; De Choudhury et al., 2013; Shen and Rudzicz, 2017). Recent work has also looked at detecting and predicting suicidality using linguistic features from social media posts (Du et al., 2018; Coppersmith et al., 2018; Zirikly et al., 2019).

While the majority of past research has compared specific psychiatric conditions with healthy control groups, more recent work has begun analyzing and identifying unique differences and discriminators among psychiatric conditions (Jiang et al., 2020; Cohan et al., 2018a; Coppersmith et al., 2015). As this area progresses, we have begun to investigate whether this technology can be used beyond detection of mental illness for detecting severity of symptomatology and prediction of acute psychiatric episodes that result in hospitalization. This would benefit patients by alerting clinicians to worsening symptoms, allowing for early intervention care and potential mitigation. Relatedly, advancements in machine learning techniques have led to the development of advanced models for predicting psychiatric crises such as increased suicidality and psychotic episodes using a multimodal approach based on clinical data (Koutsouleris et al., 2021). However, to date, these studies have relied exclusively on clinical data and medical data. To our knowledge, this is the first study to leverage a large dataset of publicly available social media posts for predicting psychiatric hospitalization.

## 3 Data Collection

In this section we describe the pipeline components of our dataset construction process, in the order in which they are applied.[1] Table 1 presents the overall statistics of our dataset.

| Candidates | TI | SC | #Posts |
|---|---|---|---|
| 95,904 | 318 | 128 | 7,077 |

Table 1: Overall dataset statistics, where **Candidates** are the total number of users we examined, **TI** corresponds to number of users from which we extracted hospitalization identification with time-span information and **SC** corresponds to the number of users having posts collected for the 21 days directly before the refined hospitalization span. **#Posts** are number of posts from these spans in total.

### 3.1 Candidate Collection

We begin data collection by identifying candidate Reddit users who may be at risk for a psychiatric hospitalization. We focus on two user groups: those that self-identify with a psychiatric disorder, and those that self-identify with suicidal ideation or attempted suicide. To identify such users, we leverage subreddits, or forums on Reddit dedicated to specific topics. Following Shing et al. (2018), we collect posts from the r/SuicideWatch (SW) subreddit, and following (Cohan et al., 2018b; Jiang et al., 2020) we collect posts from subreddits related to 8 different mental health conditions: obsessive compulsive disorder (OCD), schizophrenia (SZ), borderline personality disorder (BPD), post-traumatic stress disorder (PTSD), eating disorder (ED), major depression dis-order (MDD), general

[1] This study received IRB approval and all human subjects protection guidelines were followed.

anxiety disorder (GAD) and bipolar disorder. We then use regular expression matching to extract self-identification statements from these posts to form our candidate user pool. Our data collection methods yield 69,682 candidates for suicidal risk and 35,606 candidates for mental health conditions.

### 3.2 Hospitalization Time Span Identification

After identifying nearly 100k candidate Reddit users at risk for psychiatric hospitalization, we designed an approach to identify users from that pool that have been hospitalized for psychiatric reasons. While previous work has shown that regular expression matching alone is able to create high precision mental health datasets (Coppersmith et al., 2014; Cohan et al., 2018b; Jiang et al., 2020), it is far more difficult to automatically construct a dataset with more fine-grained information. MacAvaney et al. (2018) created a dataset of self-disclosures of depression on Reddit, which includes manually annotated temporal information about the diagnosis date. In our case, it is important to not only identify users that self-disclose psychiatric hospitalizations, but also to pinpoint the time span of the hospital stay. There are several challenges associated with this task: First, we need to ensure that the correct time span is identified when a user mentions multiple events in a single post, and avoid identifying a time span that is not associated with the identified hospitalization instance. Second, there are various ways an adverbial phrase of time could be attached to a predicate, making regular expression design difficult. A third challenge is that some time-related words having other common synsets (e.g. "May").

We address the above mentioned problems by (1) sentence-tokenizing the posts and performing all our matching at sentence-level; and (2) running a state-of-the-art semantic role labeling model first to identify the likely span for regular expression matching. Specifically, we only parse the [ARG-TMP] temporal field related to the hospitalization event, identified by the pre-trained SRL model (Shi and Lin, 2019) provided by AllenNLP (Gardner et al., 2018). When the identification is precise to date level we allow $\pm 7$ days of flexibility. In total, we extracted 72 hospitalization time spans from the SuicideWatch user group, and 349 time-spans from the psychiatric disorders user group. A clinical psychologist trainee manually reviewed all 421 spans and found that 69.12% of them were clearly correctly identified and relevant hospitalizations, while the other time-spans were not incorrect but simply lacked enough context in the post for confident labeling. This validates our proposed time-span identification approach, and suggests that further context (e.g. other posts in the same thread) may be useful to improve time-span identification.

### 3.3 Span Refinement

We observe that the most common duration of the span identified is one month, and it is desirable to have hospitalization time identified on a more fine-grained scale. For example, a user might mention that they were hospitalized "last June," without providing specific start and end dates of their hospital stay. Coppersmith et al. (2017); Coppersmith et al. (2018) shows that social media provides information in the "clinical whitespace." Inspired by them, we further identify rare media blackout periods in the previously found plausible hospitalization span, and use them as a proxy to a ground truth hospitalization period. To do this, we fit an exponential distribution on users' social media posting activity, and define a rare media blackout period as the time span of inactivity where the occurrence probability is less than a certain threshold $r$. This process also provides us with other benefits, as we are able to characterize irregularities like throw-away accounts. Figure 1 is an example of such irregularities, where the user became significantly more active after the identified span; therefore we hypothesize that most of their posts would be related to their mental health condition and perhaps their hospitalization experience. In contrast, Figure 2 is an example of users who actively use their social media before and after the hospitalization blackout. We believe these users and their posts are potentially more useful for research, because they include posts on a wide range of topics over long periods of time, both before and after a psychiatric hospitalization. However, in this paper we make no further use of the features other than to select posts that directly precede a blackout period. When multiple rare media blackout periods are found for an identified span, we empirically select the one with the longest overlap with the span.

## 4 Prediction of Psychiatric Hospitalization

Having collected a dataset of proposed hospitalization spans and preceding posts, we use our col-

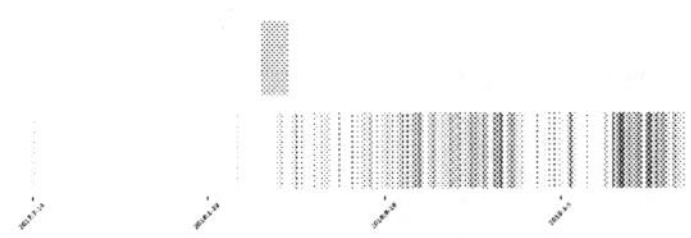

Figure 1: Irregular Reddit activity plots (green), where the user is significantly more active (darker) after the plausible hospitalization span (red).

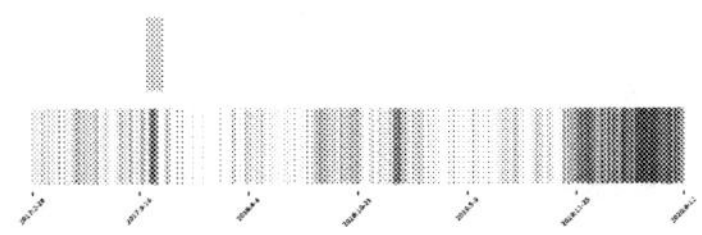

Figure 2: Regular Reddit activity plots, where the user generally post smoothly with some minor irregularity around the plausible hospitalization span (red).

lected dataset to build predictive models of psychiatric hospitalizations. We experiment with two different task formulations: post-level prediction and user-level prediction. Post-level prediction involves a binary classification for each post, determining whether the post is followed by valid hospitalization span or not. User-level prediction classifies a group of posts from a user in a given time window to predict whether the user will be hospitalized. In order to train classification models, we first need to select negative samples as a control group for our experiments. We describe our methods of pairing negative samples in subsection 4.1. We experiment with three set of features: unigram, bigram[2] and LIWC (Pennebaker et al., 2007, 2015) features. We perform hyper-parameter grid search to optimize performance. For all features we use the Naive-Bayes classifier, as it has been found to perform well on small datasets (NG and Jordan, 2002). We pre-process the text by lower-casing all input posts and, following the guidelines of (Benton et al., 2017), we de-identify posts by anonymizing URLs and replacing usernames with randomly generated strings.

### 4.1 Pairing Negative Samples

To form a challenging prediction task, we compile negative samples for classification by selecting control users from the same candidate pool that the target hospitalization group was selected from. The control users are those who do not have associated hospitalization time spans, but did have similar media blackout periods (described in subsection 3.3). We group spans by number of post before the span in a prescribed time window of length $d$ days. For each positive span we randomly sample a span from the negative span pool that has a similar number of posts, creating a balanced classification task. Note that we expect this task to be difficult because the control users either self-identified with mental health conditions or posted in the SW subreddit. For post-level classification, we use the same set of posts sampled on the user-level.

### 4.2 Classification

Table 2 shows mean F-1 scores from cross-validation on both user-level and post-level tasks. In all experiments, we set the span selection probability threshold $t = 0.1$. For user-level and post-level performance comparison, we set the inclusion number of days to $d = 21$.

| | **1-gram** | **1,2-gram** | **LIWC** |
|---|---|---|---|
| **user-level** | 0.687 | 0.698 | 0.655 |
| **post-level** | 0.601 | 0.622 | 0.584 |

Table 2: Experiment result in F-1, with different features on both tasks.

The best performance of 0.698 F1 is obtained using bigrams for the user-level task. In general, user-level classification results in better F-1 scores, indicating that more context is likely crucial to success in psychiatric hospitalization prediction. N-gram features outperform LIWC features for both tasks, and adding bigram features perform better than unigrams alone. Overall, the model performance with a small amount of data is promising, well above a 50% random baseline.

### 4.3 Performance Over Time

We again run experiments for user-level classification with another more strictly paired control group that satisfies the pairing constraints mentioned in subsection 4.1 for $d \in \{1, 7, 14, 21\}$. Table 3 shows the performance change as the window length increases. The results suggest that using a wider context is useful in predicting hospitalization blackouts, and the best performance was obtained using unigrams extracted from 7 days of posts.

## 5 Lexical Analysis

Figure 3 shows the list of most predictive words for the unigram model. We see that many words correspond to time duration (e.g. "week", "month"),

[2]Due to the size of our dataset, we set a minimum document count of 5 for bigram features.

| $d$ (days) | 1-gram | 1,2-gram |
|---|---|---|
| **1** | 0.678 | 0.676 |
| **7** | 0.718 | 0.695 |
| **14** | 0.697 | 0.692 |
| **21** | 0.708 | 0.706 |

Table 3: F-1 performance with different features on different window lengths

medical professions (e.g., "med", "doctor", "hospital") and conversation (e.g., "sorry", "thanks"). We hypothesize that these may correspond to users' frequent online posts seeking advice and describing conditions. Indeed we observe some posts conforming to this pattern through manual examinations.

care, come, person, taken, stuff, able, hear, weeks, &, definitely, bit, let, doctor, does, makes, point, home, tell, times, sorry, family, months, hope, little, use, yeah, sleep, maybe, best, new, post, told, night, probably, voices, went, great, isn, meds, bot, moderator, school, days, thought, week, doesn, trying, started, working, used, mom, message, thank, long, doing, hospital, having, try, hard, love, year, thanks, bad, getting, actually, pretty, sure, thing, help, better, years, life, ll, need, said, right, say, didn, work, way, did, make, lot, day, got, things, url, want, going, feel, good, think, people, time, know, ve, really, don, like, just

Figure 3: The top 100 most predictive words for the hospitalized group by the uni-gram model.

## 6 Conclusion and Future Work

We present a novel social media data collection method for identifying hospitalization time spans and design a novel classification task for predicting psychiatric hospitalizations. We experiment with multiple linguistic feature sets and task formulations, including user-level and post-level classification, as well as varying the time window of posts used. Our results suggest that this is a useful framework for collecting data related to psychiatric hospitalization, and that social media data can be leveraged to predict psychiatric crises before they occur. In our ongoing and future work, we plan to conduct further analysis of the language of pre-hospitalization posts to gain insights about linguistic patterns and changes that occur as the user experiences a psychiatric crisis. We also plan to improve the data collection process to achieve better precision and to expand to a larger scale. We hope that an improved understanding of the linguistic cues that precede psychiatric hospitalizations, as well as improvements in automatic prediction of hospitalizations, will enable interventions that can potentially save lives and improve outcomes for individuals with mental illness.

## References


Adrian Benton, Glen Coppersmith, and Mark Dredze. 2017. Ethical research protocols for social media health research. In *Proceedings of the First ACL Workshop on Ethics in Natural Language Processing*, pages 94–102.

Michael L Birnbaum, Sindhu Kiranmai Ernala, Asra F Rizvi, Munmun De Choudhury, and John M Kane. 2017. A collaborative approach to identifying social media markers of schizophrenia by employing machine learning and clinical appraisals. *Journal of medical Internet research*, 19(8):e289.

Arman Cohan, Bart Desmet, Andrew Yates, Luca Soldaini, Sean MacAvaney, and Nazli Goharian. 2018a. Smhd: a large-scale resource for exploring online language usage for multiple mental health conditions. *arXiv preprint arXiv:1806.05258*.

Arman Cohan, Bart Desmet, Andrew Yates, Luca Soldaini, Sean MacAvaney, and Nazli Goharian. 2018b. SMHD: a large-scale resource for exploring online language usage for multiple mental health conditions. In *Proceedings of the 27th International Conference on Computational Linguistics*, pages 1485–1497, Santa Fe, New Mexico, USA. Association for Computational Linguistics.

G. Coppersmith, C. Hilland, O. Frieder, and R. Leary. 2017. Scalable mental health analysis in the clinical whitespace via natural language processing. In *2017 IEEE EMBS International Conference on Biomedical Health Informatics (BHI)*, pages 393–396.

Glen Coppersmith, Mark Dredze, Craig Harman, and Kristy Hollingshead. 2015. From adhd to sad: Analyzing the language of mental health on twitter through self-reported diagnoses. In *Proceedings of the 2nd Workshop on Computational Linguistics and Clinical Psychology: From Linguistic Signal to Clinical Reality*, pages 1–10.

Glen Coppersmith, Craig Harman, and Mark Dredze. 2014. Measuring post traumatic stress disorder in twitter. In *Proceedings of the International AAAI Conference on Web and Social Media*, volume 8.

Glen Coppersmith, Ryan Leary, Patrick Crutchley, and Alex Fine. 2018. Natural language processing of social media as screening for suicide risk. *Biomedical informatics insights*, 10:1178222618792860.

W Cullen, G Gulati, and BD Kelly. 2020. Mental health in the covid-19 pandemic. *QJM: An International Journal of Medicine*, 113(5):311–312.

Munmun De Choudhury, Michael Gamon, Scott Counts, and Eric Horvitz. 2013. Predicting depression via social media. In *Seventh international AAAI conference on weblogs and social media*.

Jingcheng Du, Yaoyun Zhang, Jianhong Luo, Yuxi Jia, Qiang Wei, Cui Tao, and Hua Xu. 2018. Extracting psychiatric stressors for suicide from social media using deep learning. *BMC medical informatics and decision making*, 18(2):77–87.

John Elflein. 2020. Mental health service use in the past year among u.s. adults from 2002 to 2019, by type of care. `https://www.statista.com/statistics/252316/type-of-mental-health-service-used-by-us-adults-since-2002/`. Accessed: 2021-03-15.

Matt Gardner, Joel Grus, Mark Neumann, Oyvind Tafjord, Pradeep Dasigi, Nelson Liu, Matthew Peters, Michael Schmitz, and Luke Zettlemoyer. 2018. Allennlp: A deep semantic natural language processing platform. *arXiv preprint arXiv:1803.07640*.

Sharath Chandra Guntuku, David B Yaden, Margaret L Kern, Lyle H Ungar, and Johannes C Eichstaedt. 2017. Detecting depression and mental illness on social media: an integrative review. *Current Opinion in Behavioral Sciences*, 18:43–49.

Yen-Hao Huang, Lin-Hung Wei, and Yi-Shin Chen. 2017. Detection of the prodromal phase of bipolar disorder from psychological and phonological aspects in social media. *arXiv preprint arXiv:1712.09183*.

Zhengping Jiang, Sarah Ita Levitan, Jonathan Zomick, and Julia Hirschberg. 2020. Detection of mental health from Reddit via deep contextualized representations. In *Proceedings of the 11th International Workshop on Health Text Mining and Information Analysis*, pages 147–156, Online. Association for Computational Linguistics.

Ann John, Jane Pirkis, David Gunnell, Louis Appleby, and Jacqui Morrissey. 2020. Trends in suicide during the covid-19 pandemic.

Nikolaos Koutsouleris, Dominic B Dwyer, Franziska Degenhardt, Carlo Maj, Maria Fernanda Urquijo-Castro, Rachele Sanfelici, David Popovic, Oemer Oeztuerk, Shalaila S Haas, Johanna Weiske, et al. 2021. Multimodal machine learning workflows for prediction of psychosis in patients with clinical high-risk syndromes and recent-onset depression. *JAMA psychiatry*, 78(2):195–209.

Sean MacAvaney, Bart Desmet, Arman Cohan, Luca Soldaini, Andrew Yates, Ayah Zirikly, and Nazli Goharian. 2018. Rsdd-time: Temporal annotation of self-reported mental health diagnoses. *arXiv preprint arXiv:1806.07916*.

Andrew NG and Michael I Jordan. 2002. On discriminative vs. generative classifiers: A comparison of logistic regression and naive bayes. *Advances in neural information processing systems*, 14:841.

Pamela L Owens, Kathryn R Fingar, Kimberly W McDermott, Pradip K Muhuri, and Kevin C Heslin. 2019. Inpatient stays involving mental and substance use disorders, 2016: Statistical brief# 249. *Healthcare cost and utilization project (HCUP) statistical briefs*.

James W Pennebaker, Roger J Booth, and Martha E Francis. 2007. Linguistic inquiry and word count: Liwc [computer software]. *Austin, TX: liwc. net*, 135.

James W Pennebaker, Ryan L Boyd, Kayla Jordan, and Kate Blackburn. 2015. The development and psychometric properties of LIWC2015. Technical report, TX: University of Texas at Austin.

Judy Hanwen Shen and Frank Rudzicz. 2017. Detecting anxiety through reddit. In *Proceedings of the Fourth Workshop on Computational Linguistics and Clinical Psychology—From Linguistic Signal to Clinical Reality*, pages 58–65.

Peng Shi and Jimmy Lin. 2019. Simple bert models for relation extraction and semantic role labeling. *arXiv preprint arXiv:1904.05255*.

Han-Chin Shing, Suraj Nair, Ayah Zirikly, Meir Friedenberg, Hal Daumé III, and Philip Resnik. 2018. Expert, crowdsourced, and machine assessment of suicide risk via online postings. In *Proceedings of the Fifth Workshop on Computational Linguistics and Clinical Psychology: From Keyboard to Clinic*, pages 25–36, New Orleans, LA. Association for Computational Linguistics.

Michael Stensland, Peter R Watson, and Kyle L Grazier. 2012. An examination of costs, charges, and payments for inpatient psychiatric treatment in community hospitals. *Psychiatric Services*, 63(7):666–671.

Akkapon Wongkoblap, Miguel A Vadillo, and Vasa Curcin. 2017. Researching mental health disorders in the era of social media: systematic review. *Journal of medical Internet research*, 19(6):e228.

Ayah Zirikly, Philip Resnik, Ozlem Uzuner, and Kristy Hollingshead. 2019. Clpsych 2019 shared task: Predicting the degree of suicide risk in reddit posts. In *Proceedings of the sixth workshop on computational linguistics and clinical psychology*, pages 24–33.

Jonathan Zomick, Sarah Ita Levitan, and Mark Serper. 2019. Linguistic analysis of schizophrenia in reddit posts. In *Proceedings of the Sixth Workshop on Computational Linguistics and Clinical Psychology*, pages 74–83.

# Automatic Identification of Ruptures in Transcribed Psychotherapy Sessions

**Adam Tsakalidis[1,2], Dana Atzil-Slonim[3], Asaf Polakovski, Natalie Shapira[3], Rivka Tuval-Mashiach[3], Maria Liakata[1,2,4]**

[1] Queen Mary University of London, London, United Kingdom
[2] The Alan Turing Institute, London, United Kingdom
[3] Bar Ilan University, Ramat Gan, Israel
[4] University of Warwick, Coventry, United Kingdom
a.tsakalidis;m.liakata@qmul.ac.uk,dana.atzil@biu.ac.il

## Abstract

We present the first work on automatically capturing alliance rupture in transcribed therapy sessions, trained on the text and self-reported rupture scores from both therapists and clients. Our NLP baseline outperforms a strong majority baseline by a large margin and captures client reported ruptures unidentified by therapists in 40% of such cases.

## 1 Introduction

The client-therapist relationship within a psychotherapy treatment ('therapeutic alliance') is considered a powerful predictor of therapy success across treatment modalities and disorders (Flückiger et al., 2018; Norcross and Lambert, 2019). Conversely, when a tension or a breakdown (*rupture*) occurs in the therapeutic alliance, it can often lead to unilateral termination of the treatment by the client or to poor psychotherapy outcomes (Eubanks et al., 2018). However, when alliance ruptures are recognised they can become meaningful therapeutic events (Chen et al., 2018). Indeed, alliance ruptures have been found to be beneficial to the therapeutic process and outcome when they are recognized and followed by repair of the rupture (Stevens et al., 2007; Stiles et al., 2004) and to hinder the process or outcome of therapy when they go unrecognized (Chen et al., 2018).

**Challenges in capturing alliance rupture**: Most studies have explored alliance ruptures using self-reports at relatively low time resolution (once each session, typically weekly). However, ruptures may occur at higher time resolutions within a session (Coutinho et al., 2014). In addition, standardized subjective measures have critical shortcomings, including the extent of participants' self-insights, willingness to complete questionnaires, and the restricted choice of responses (Kazdin, 2016). Recent studies have used within-session coding tools to detect ruptures moment-by-moment during a session, yielding important insights into the within-session processes that lead to ruptures (e.g., (Eubanks et al., 2015). These insights have been used to train therapists to recognize ruptures when they happen (Eubanks-Carter et al., 2015). However, since observational human-coding is very labor intensive and expensive, these studies have focused on a small number of therapeutic components in a small sample of clients and at limited time points.

**Benefits of capturing alliance rupture from text** originating from the transcribed dialogue between therapist and client during therapy sessions include:

- Detecting alliance rupture even when therapists or clients are unaware of it. This would allow signaling the rupture to therapists and help them acknowledge it. Such information may be used alongside existing monitoring tools to inform therapists about meaningful instances of alliance rupture that went unrecognized.
- Subtler and more implicit content associated with a rupture would be captured, increasing our understanding of the specific moments and reasons for it.
- Alliance rupture would be captured in a cost-effective manner.

**Contributions**: To the best of our knowledge there is no work on capturing alliance rupture automatically from transcribed therapist or client utterances. Recently Goldberg et al. (2020) used 1,235 transcribed recorded sessions with client reported alliance to automatically predict per session alliance using the text from both therapist and client. They used four variants of a linear regression model with linguistic features from either the therapist or client. Their best performing model was only 0.02 more accurate than a baseline predicting the average alliance rating. They also provided a list of unigrams which correlate most with high and low alliance scores respectively.



*Proceedings of the Seventh Workshop on Computational Linguistics and Clinical Psychology*, pages 122–128
June 11, 2021. 

Here we make the following contributions:

- We present the first work on automatically capturing alliance rupture (rather than alliance) trained on transcribed therapy sessions and self-reported rupture scores.
- We provide a detailed description of the dataset creation.
- We provide strong NLP baselines which outperform majority baselines by a large margin. Moreover we have an original privacy preservation setting whereby the data given to the NLP researchers was in encrypted format, facilitating the collaboration of NLP researchers with clinicians and companies with strong privacy concerns.
- We provide a qualitative analysis of examples where our NLP baselines capture client reported ruptures unrecognised by the therapist.

## 2 Dataset Description

**Clients**: were sampled from a pool of clients receiving individual psychotherapy at a university training outpatient clinic. Data were collected between Aug'14-Aug'16 as part of the clinic's regular practice of monitoring clients' progress. From an initial sample of 180 consented clients 34 (18.9%) dropped out. Clients were selected according to two criteria: (a) treatment duration of at least 15 sessions and (b) availability of full data, including audio recordings and session-by-session questionnaires. Clients were also excluded based on the M.I.N.I. 6.0 (Sheehan et al., 1998) if they were diagnosed as severely disturbed. The data of 68 (37.8%) clients who met the inclusion criteria were transcribed, for a total of 873 transcribed sessions. Clients were above the age of 18 ($\mu_{age}$=39.06, SD=13.67, range 20–77), the majority of whom were women (58.9%). 53.5% had at least a bachelor's degree, 53.5% reported being single, 8.9% were in a committed relationship, 23.2% were married and 14.2% were divorced or widowed. Clients' diagnoses were established based on the Mini International Neuropsychiatric Diagnostic Interview for Axis I DSM-IV diagnoses (Sheehan et al., 1998). 22.9% of the clients had a single diagnosis, 20.0% had two and 25.7% had three or more. The most common diagnoses were comorbid anxiety and affective disorders (25.7%), followed by other comorbid disorders (17.1%), anxiety disorders (14.3%), and affective disorders (5.7%).

**Therapists and Therapy**: Clients were treated by 52 therapists at various stages of their clinical training. Clients were assigned to therapists in an ecologically valid manner based on therapist availability and caseload. 42 therapists treated one client each; eight treated two clients. Each therapist received one hour of individual supervision biweekly and four hours of group supervision on a weekly basis. All therapy sessions were audiotaped for supervision by senior clinicians. Supervision focused heavily on reviewing audiotaped case material and technical interventions designed to facilitate the appropriate use of therapist interventions. Individual psychotherapy consisted of once- or twice-weekly sessions. The language of therapy was Modern Hebrew (MH). The dominant approach in the clinic includes a short-term psychodynamic psychotherapy treatment model (e.g., (Blagys and Hilsenroth, 2000; Shedler, 2010; Summers and Barber, 2009). On average, treatment length was 37 sessions (SD=23.99, range=18–157). Treatment was open- ended in length, but given that psychotherapy was provided by clinical trainees at a university-based outpatient community clinic, treatment duration was often restricted to 9 months.

**Instruments and Procedure**: Clients and/or therapists responded to several scales during the treatment, including the Outcome Rating Scale (Miller et al., 2003) and the Post-Session Questionnaire (PSQ). In this work, we focus specifically on the *alliance ruptures*. Alliance ruptures were assessed after each session with one question to the therapist and client: "Did you experience any tension, misunderstanding, conflict or disagreement in the relationship with your client/therapist?". This item is answered subjectively on a 5-point Likert scale from 1 ('not at all') to 5 ('constantly') by the two involved entities separately. Following (Muran et al., 2009), a rupture was defined as any rating higher than 1 on the scale. The PSQ has been widely used in psychotherapy research and demonstrates sound psychometric properties, including predictive validity with a variety of process indices such as the Working Alliance Inventory (Tracey and Kokotovic, 1989). Here the PSQ mean score was 2.06 (SD=1.43).

**Transcription**: Due to the high associated cost manual transcriptions were conducted alternately (sessions 2, 4, 6, etc.). In cases where material was incomplete (e.g., questionnaire or poor recording quality), the following session was transcribed instead. The transcriber team was composed of seven graduate students in the University's psychology department. The transcribers went through a one day training workshop which included how to handle private/sensitive information; monthly meetings were held throughout the transcription process to supervise the quality of their work. The transcription protocol followed general guidelines, as described in (Mergenthaler and Stinson, 1992; Albert et al., 2013). The word forms, the form of commentaries and the use of punctuation were kept as close as possible to the speech presentation. Everything was transcribed, including word fragments as well as syllables or fillers (e.g., "ums", "ahs", "you know"). The transcripts included elisions, mispronunciations, slang, grammatical errors, non-verbal sounds (e.g., laughs, cry, sighs), and background noises. The rules were limited in number and simple and the format used several symbols to indicate comments (e.g. '[...]' to indicate the correct form when the actual utterance was mispronounced).

There were 873 transcripts in total (the mean transcribed sessions per client was 12.56; SD=4.93). The transcriptions totaled about four million words over 150,000 talk turns (i.e., switching between speakers). On average, there were 5800 words in a session, of which 4538 (78%; SD=1409.62; range 416-8176) were client utterances and 1266 (22%; SD=674.99; range 160-6048) were therapist utterances.

**Text Processing & Privacy**: In morphologically rich languages such as Hebrew, each token may have multiple different morphological analyses where only one is pertinent to the context. To tackle this, we used the YAP parser (More and Tsarfaty, 2016), which performs a lexicon-based morphological analysis followed by joint morpho-syntactic disambiguation and dependency parsing. Finally, to work in a privacy preserving manner due to the sensitive nature of our data, we replaced each word with a token ID. We further used a separate mapping of the token IDs to indices in a dictionary of word vectors to share the data within our team for our experiments. The word vectors were also rotated, as an additional security step.

## 3 Experiments

**Task**: We define the problem of capturing rupture alliance as a binary classification task. In particular, we aim at identifying whether a rupture occurred within a session, given the language used by the therapist and/or the client during that session. The presence or absence of rupture is defined via the self-assessed questionnaire, which is completed by each of the client and therapist. We treat their responses as two separate tasks: (a) Client's Rupture (CR) prediction and (b) Therapist's Rupture (TR) prediction, where in each task the goal is to predict the corresponding self-reported outcome given the transcribed session as input.

| Task | Client's Rupture [CR] | | | Therapist's Rupture [TR] | | |
|---|---|---|---|---|---|---|
| Features | Client | Therapist | Both | Client | Therapist | Both |
| Majority | 59.00 | 59.00 | 59.00 | 37.50 | 37.50 | 37.50 |
| LogReg | **61.90** | 61.30 | 58.80 | 45.60 | 46.60 | **46.70** |

Table 1: F-score for the two binary classification tasks.

**Dataset**: Since some of the transcriptions were not associated with alliance rupture labels, the final dataset used in our experiments consists of 849 transcribed sessions from 68 clients. Due to missing labels, the two tasks also have a different number of instances. There were 821 sessions for CR and 829 for the TR task. The distribution of the labels for the two tasks differs: for TR there is a balance between rupture *vs* no-rupture labels (48% *vs* 52%); the same does not hold for CR (23% *vs* 77%).

**Experimental Setting**: The input to our classifier in the text from a transcribed therapy session. We represent each session via dense word vectors consisting of: (a) the client's text, (b) the therapist's text and (c) both of them in concatenation. The vectors were obtained by training a skip-gram model (Mikolov et al., 2013) on a large collection of tweets in Hebrew. With each word represented as a 100-dim vector, we represent each session by averaging the dimensions of words used by either the client, therapist or both during the session.

We train a Logistic Regression for our two tasks, CR and TR. We perform a leave-one-client-out cross validation (68 folds) to avoid any potential bias in our evaluation (DeMasi et al., 2017; Tsakalidis et al., 2018; Harrigian et al., 2020). This way we can assess the model's ability to generalise in previously unseen clients. For each task, we experiment with the three types of representations discussed above. For evaluation we use the macro-averaged F-score between the two classes, averaged across all folds. We contrast performance against the majority (no-rupture) classifier to get some first insights into the difficulty of the tasks.

**Results**: Table 1 shows the macro-average F-score achieved in the two tasks, averaged across all clients (folds). The performance on the CR task is higher compared to the TR task due to the imbalanced nature of our dataset. However, there is only a minor relative improvement of 4.9% in CR over the majority baseline (52.8% over a completely random classifier) compared to the 24.5% in TR. This large difference between the two tasks is attributed to the fact that therapists are trained to recognise ruptures and are more likely to report ruptures than miss a potential rupture. This makes the dataset more balanced in terms of rupture and non-rupture labels.

Next, we examine the performance on the 801 sessions were we have reports on rupture by both the therapist and the client.In particular, we are interested in inspecting cases of sessions where the client indicated that there was a rupture, but the therapist missed it. Therefore, we treat the label provided by the Client ('rupture') as our ground truth and test our models' performance based on them, when leveraging both of the Client's and the Therapist's text. Overall, there were 72 such cases (9%), as shown in Table 2. Logistic Regression trained for the TR task successfully identified 29 (40%) of these cases. This encouraging finding suggests that incorporating NLP methods for detecting such cases – which is of particular importance for therapists – could act as a tool too assist with rupture detection to improve psychotherapy treatment. On the other extreme combination of labelling shown in Table 2 (i.e., in 341 cases which both the client and the therapist reported as "no rupture"), there were 205 (60%) sessions that have been correctly classified by *both* of the CR and TR models jointly, while there were only 10 of these cases (3%) that were jointly misclassified by the two models. Overall, by considering only the rather "clear" 274 sessions (i) which have been given the same ground-truth label by both client and therapist and (ii) for which the CR/TR models agree on their prediction, the (%) macro-average F1-score is 70.9% (accuracy 83.6%). This suggests that the task of predicting rupture alliance by analysing the language used within a psychotherapy session is indeed feasible. However, there is plenty of room for improvement both in terms of language representation as well as modelling.

Finally, we inspect the language used within rupture *vs* non-rupture sessions. We are particularly interested the sessions that were labelled as 'rupture' by the client only (see Table 2) and also correctly

| | | Therapist | |
|---|---|---|---|
| | | No rupture | Rupture |
| Client | No rupture | 341 | 280 |
| | Rupture | 72 | 108 |

Table 2: Distribution of labels in the 801 sessions that were labelled by both entities (Therapist, Client).

identified by our model (40%). We find that most such cases were withdrawal ruptures (see example in Table 3a). The literature on ruptures highlights two main subtypes: withdrawal and confrontational ruptures (Eubanks et al., 2018). In withdrawal ruptures (see example in Table 3b), the client moves away from the therapist and the work of therapy, e.g. by avoiding the therapist's questions or by hiding their dissatisfaction with therapy by being overly appeasing. In confrontational ruptures (see example in Table 3c), the client moves against the therapist by expressing anger or dissatisfaction with the therapist or treatment, or by trying to apply pressure on the therapist. It seems that it was easier for therapists in our sample to identify the occurrence of confrontational ruptures, which may be more apparent in the client's behavior than the withdrawal ruptures. The latter may be more subtle and less emotionally charged. This finding is in line with other qualitative studies showing that therapists tend to better recognize confrontational ruptures (Hill, 2010). It also highlights the importance of using automated methods to capture ruptures that are challenging for therapists to capture.

## 4 Conclusion and Future Work

In this work we focused on the task of automatically predicting alliance rupture between a therapist and a client from the language of therapy sessions. We collected and transcribed sessions between clients and their therapists, conducted in Hebrew. We also obtained self-reported rupture labels for sessions by clients and therapists, used in clinical psychotherapy research. We tested baseline models leveraging the language used within a session to predict the occurrence of alliance rupture based on the perception of both the therapist and the client. We yield good performance and showcase the potential for using NLP for aiding therapists in identifying rupture during psychotherapy sessions. In the future we plan to build on our initial findings by incorporating contextual language models (Chriqui and Yahav, 2021; Devlin et al., 2019) and

| I had to pick up my kind from his music lessons and I was busy and I asked my husband if he could take the child and he said he was busy and that I was the one who should give up. |
|---|
| Why do you think this is happening? |
| I always have to run from one thing to another. He's busy with his own affairs. But what did you ask? I'm not in focus. |
| No, no, it's okay, please continue. |
| I feel like I was unlucky in life. Yesterday I needed his help, but he is never there to help or hug me. I just don't have anyone who can do that for me. It's hard. I need someone who can support me. I never had such support in my life. I tried to get closer to him, but I feel that I am the only one whose needs are dismissed. He never gives up his needs. I feel so tired of all that. I have no desire to do anything. |
| We talked in the last session about your difficulties to bring your needs. But last time you also said that you felt closer to him, didn't you? |
| Yes, I should try to get closer to him, I don't know, maybe I am wrong. |
| How is it for you with other people? |
| I don't know. |

(a) Example of part of a session that was labelled by the Client as 'rupture', but not from the Therapist. Logistic Regression trained for the TR task predicted that there is a 'rupture'.

| I think I should be an employee instead of a boss. That pressure... I can't stand it. I'm not good at it. When a client comes I'm at the height of my enthusiasm, I have a lot of ideas on what to do, and I make plans & invest a lot of thought, I want it to be perfect, but something stops me, I cannot do it the way I want. |
|---|
| You are afraid of disappointing. |
| Yes, exactly. I invest too much time in planning and then something stops me from doing it. I want it to be perfect and I'm working on the planning and I'm getting exhausted. I feel so much pressure to implement the plan & then I just become lazy and unable to actually do it. Maybe if I was an employee then I would have cared less & the job would have been easier. |
| Sounds like there is a lot of pressure, also around the thought of finding another job. |
| No, it's not about finding another job. |
| But you also said - I feel that .. I have lots of strength and lots of motivation and I have many ideas, and suddenly when it comes to execution I can not find them. |
| There is some kind of fatigue, laziness, I feel I do not have the strength, not the physical strength, the mental strength. |
| Something stops you. Lets try to understand what it is. |
| I tend to postpone everything. |
| What do you postpone here? |
| Everything. |
| What do you postpone here, in treatment? |
| Nothing specific. I just tend to postpone everything. |

(b) **Withdrawal rupture:** A translated snippet of a session where the client reported a 'rupture', but not the therapist. Logistic Regression trained for the TR task predicted that there is a 'rupture', agreeing with the client.

| It's cold in here. |
|---|
| Cold? |
| Um .. this is, I'm coming here and the feelings are really .. confused, turbulent. I had a really completely confused week, I had a very very hard time at the end of the previous session. |
| Mmm.. |
| It made me tense, and I was thinking if this form of treatment is good for me or if it's doing me any harm. I was looking for answers. I don't know if going deeper into things is good for me or if the right way for me is the opposite – to let go. |
| Mmmm |
| And I met again that person I have worked with last summer. He is helping me to raise my self-confidence. Sometimes that's what I need when I feel confused and unstable. |
| I hear you. I also thought a lot about the hard things you talked about in the previous meeting. |
| I felt overwhelmed and confused after the session. |
| Let's try to talk about what was it that you needed from me last time and that you felt that I did not provide. |

(c) **Confrontational rupture:** This snippet of a translated transcribed session that was labelled by both client and therapist as a 'rupture'.

Table 3: Examples of alliance rupture

by developing models that can perform this task in a sequential and temporally sensitive manner. Finally, a limitation of our work stems from the fact that the clients and therapists come from the same background, both linguistically and culturally. Confirming our findings via analysing data from therapy sessions across different backgrounds is an important future direction.

## Acknowledgements

This work was supported by a UKRI/EPSRC Turing AI Fellowship to Maria Liakata (grant EP/V030302/1) and the The Alan Turing Institute (grant EP/N510129/1).

## References


Aviad Albert, Brian MacWhinney, Bracha Nir, and Shuly Wintner. 2013. The hebrew childes corpus: transcription and morphological analysis. *Language resources and evaluation*, 47(4):973–1005.

Matthew D Blagys and Mark J Hilsenroth. 2000. Distinctive features of short-term psychodynamic-interpersonal psychotherapy: A review of the comparative psychotherapy process literature. *Clinical psychology: Science and practice*, 7(2):167–188.

Roei Chen, Dana Atzil-Slonim, Eran Bar-Kalifa, Ilanit Hasson-Ohayon, and Eshkol Refaeli. 2018. Therapists' recognition of alliance ruptures as a moderator of change in alliance and symptoms. *Psychotherapy Research*, 28(4):560–570.

Avihay Chriqui and Inbal Yahav. 2021. Hebert & hebemo: a hebrew bert model and a tool for polarity analysis and emotion recognition. *arXiv preprint arXiv:2102.01909*.

Joana Coutinho, Eugénia Ribeiro, Catarina Fernandes, Inês Sousa, and Jeremy D Safran. 2014. The development of the therapeutic alliance and the emergence of alliance ruptures.[el desarrollo de la alianza terapéutica y la aparición de rupturas en la alianza]. *Anales de Psicología/Annals of Psychology*, 30(3):985–994.

Orianna DeMasi, Konrad Kording, and Benjamin Recht. 2017. Meaningless comparisons lead to false optimism in medical machine learning. *PloS one*, 12(9):e0184604.

Jacob Devlin, Ming-Wei Chang, Kenton Lee, and Kristina Toutanova. 2019. Bert: Pre-training of deep bidirectional transformers for language understanding. In *Proceedings of the 2019 Conference of the North American Chapter of the Association for Computational Linguistics: Human Language Technologies, Volume 1 (Long and Short Papers)*, pages 4171–4186.

Catherine F Eubanks, J Christopher Muran, and Jeremy D Safran. 2015. Rupture resolution rating system (3rs): Manual. *Unpublished manuscript, Mount Sinai-Beth Israel Medical Center, New York*.

Catherine F Eubanks, J Christopher Muran, and Jeremy D Safran. 2018. Alliance rupture repair: A meta-analysis. *Psychotherapy*, 55(4):508.

Catherine Eubanks-Carter, J Christopher Muran, and Jeremy D Safran. 2015. Alliance-focused training. *Psychotherapy*, 52(2):169.

Christoph Flückiger, AC Del Re, Bruce E Wampold, and Adam O Horvath. 2018. The alliance in adult psychotherapy: A meta-analytic synthesis. *Psychotherapy*, 55(4):316.

Simon B Goldberg, Nikolaos Flemotomos, Victor R Martinez, Michael J Tanana, Patty B Kuo, Brian T Pace, Jennifer L Villatte, Panayiotis G Georgiou, Jake Van Epps, Zac E Imel, et al. 2020. Machine learning and natural language processing in psychotherapy research: Alliance as example use case. *Journal of counseling psychology*, 67(4):438.

Keith Harrigian, Carlos Aguirre, and Mark Dredze. 2020. Do models of mental health based on social media data generalize? In *Proceedings of the 2020 Conference on Empirical Methods in Natural Language Processing: Findings*, pages 3774–3788.

Clara E Hill. 2010. Qualitative studies of negative experiences in psychotherapy.

Alan E Kazdin. 2016. *Methodological issues and strategies in clinical research*. American Psychological Association.

Erhard Mergenthaler and Charles Stinson. 1992. Psychotherapy transcription standards. *Psychotherapy research*, 2(2):125–142.

Tomas Mikolov, Ilya Sutskever, Kai Chen, Greg Corrado, and Jeffrey Dean. 2013. Distributed representations of words and phrases and their compositionality. In *Proceedings of the 26th International Conference on Neural Information Processing Systems-Volume 2*, pages 3111–3119.

Scott D Miller, BL Duncan, J Brown, JA Sparks, and DA Claud. 2003. The outcome rating scale: A preliminary study of the reliability, validity, and feasibility of a brief visual analog measure. *Journal of brief Therapy*, 2(2):91–100.

Amir More and Reut Tsarfaty. 2016. Data-driven morphological analysis and disambiguation for morphologically rich languages and universal dependencies. In *Proceedings of COLING 2016, the 26th international conference on computational linguistics: Technical papers*, pages 337–348.

J Christopher Muran, Jeremy D Safran, Bernard S Gorman, Lisa Wallner Samstag, Catherine Eubanks-Carter, and Arnold Winston. 2009. The relationship of early alliance ruptures and their resolution to process and outcome in three time-limited psychotherapies for personality disorders. *Psychotherapy: Theory, Research, Practice, Training*, 46(2):233.

John C Norcross and Michael J Lambert. 2019. *Psychotherapy relationships that work: Volume 1: Evidence-based therapist contributions*. Oxford University Press.

Jonathan Shedler. 2010. The efficacy of psychodynamic psychotherapy. *American psychologist*, 65(2):98.

David V Sheehan, Yves Lecrubier, K Harnett Sheehan, Patricia Amorim, Juris Janavs, Emmanuelle Weiller, Thierry Hergueta, Roxy Baker, and Geoffrey C Dunbar. 1998. The mini-international neuropsychiatric interview (mini): the development and validation of a structured diagnostic psychiatric interview for dsm-iv and icd-10. *The Journal of clinical psychiatry*.

Christopher L Stevens, J Christopher Muran, Jeremy D Safran, Bernard S Gorman, and Arnold Winston. 2007. Levels and patterns of the therapeutic alliance in brief psychotherapy. *American journal of psychotherapy*, 61(2):109–129.

William B Stiles, Meredith J Glick, Katerine Osatuke, Gillian E Hardy, David A Shapiro, Roxane Agnew-Davies, Anne Rees, and Michael Barkham. 2004. Patterns of alliance development and the rupture-repair hypothesis: Are productive relationships u-shaped or v-shaped? *Journal of Counseling Psychology*, 51(1):81.

R. F. Summers and J. P. Barber. 2009. *Psychodynamic therapy: A guide to evidence-based practice*. New York and London: Guilford Press.

Terence J Tracey and Anna M Kokotovic. 1989. Factor structure of the working alliance inventory. *Psychological Assessment: A journal of consulting and clinical psychology*, 1(3):207.

Adam Tsakalidis, Maria Liakata, Theo Damoulas, and Alexandra I Cristea. 2018. Can we assess mental health through social media and smart devices? addressing bias in methodology and evaluation. In *Joint European Conference on Machine Learning and Knowledge Discovery in Databases*, pages 407–423. Springer.

# Automated coherence measures fail to index thought disorder in individuals at risk for psychosis

**Kasia Hitczenko**[1], **Henry R. Cowan**[2], **Vijay A. Mittal**[2,3,4,5,6], **Matthew Goldrick**[1,6]
Departments of Linguistics[1], Psychology[2], and Psychiatry[3]
Institute for Policy Research[4], Medical Social Sciences[5],
Institute for Innovations in Developmental Sciences[6]
Northwestern University
kasia.hitczenko@northwestern.edu

## Abstract

Thought disorder – linguistic disturbances including incoherence and derailment of topic – is seen in individuals both with and at risk for psychosis. Methods from computational linguistics have increasingly sought to quantify thought disorder to detect group differences between clinical populations and healthy controls. While previous work has been quite successful at these classification tasks, the lack of interpretability of the computational metrics has made it unclear whether they are in fact measuring thought disorder. In this paper, we dive into these measures to try to better understand what they reflect. While we find group differences between at-risk and healthy control populations, we also find that the measures mostly do not correlate with existing measures of thought disorder symptoms (what they are intended to measure), but rather correlate with surface properties of the speech (e.g., sentence length) and sociodemographic properties of the speaker (e.g., race). These results highlight the importance of considering interpretability front and center as the field continues to grow. Ethical use of computational measures like those studied here – especially in the high-stakes context of clinical care – requires us to devote substantial attention to potential biases in our measures.

## 1 Introduction

Individuals with psychosis exhibit language disturbances, often referred to as thought disorder. At the discourse level, this includes poverty of speech (low quantities of speech), poverty of speech content (vague, repetitive speech), as well as the focus of this work: incoherence and derailment (slow but steady loss of topic; e.g., 'I always liked geography. My last teacher in that subject was Professor August A. He was a man with black eyes. I also like black eyes. There are also blue and grey eyes and other sorts.') (Andreasen, 1986; Bleuler, 1950; Kuperberg, 2010). These symptoms are used to diagnose psychotic disorders and are thought to have predictive clinical value (Andreasen, 1979, 1986; Andreasen and Grove, 1986; First, 1997; Roche et al., 2016; Wilcox et al., 2012). Similar, but attenuated, symptoms are observed in individuals who do not have psychosis, but who meet criteria for being at clinical high-risk for psychosis (CHR). In this population, the presence of these linguistic symptoms predicts later transition to psychosis (Bearden et al., 2011; Demjaha et al., 2017; Perkins et al., 2015).

However, despite the clinical value of these measures, these symptoms have generally been evaluated via self-report and/or overall clinician impressions, which may capture only the most extreme disturbances. Manual annotations of specific linguistic features may allow for more nuanced measures; however, they are time-intensive and infeasible to apply on a wide scale. As a result, these linguistic measures, despite their clinical value, have been underused in the field.

There is a growing body of literature trying to automatically quantify these linguistic differences using methods from computational linguistics, both in psychosis (Elvevåg et al., 2007; Iter et al., 2018; Just et al., 2019; Hitczenko et al., 2020) and CHR populations (Bedi et al., 2015; Corcoran et al., 2018; Gupta et al., 2018; Corcoran et al., 2020). This work has been quite successful, replicating group differences between patient and healthy populations and accurately categorizing individuals into appropriate groups. However, much of the focus of this work has been on separating groups, and there has been less of a focus on relating these metrics to symptoms. Work examining this relationship has sometimes found correlations between these computational metrics and relevant symptoms, but has often failed to find such relationships.

In order for these measures to be useful clinically, it is important to establish their construct validity:



*Proceedings of the Seventh Workshop on Computational Linguistics and Clinical Psychology*, pages 129–150
June 11, 2021. ©2021 Association for Computational Linguistics

Do they relate to relevant symptoms? Or, do their instead reflect other linguistic/demographic factors? Establishing trust for a system's predictions is particularly important in the clinical/medical setting where these systems could have substantial consequences (Ribeiro et al., 2016). This is especially true as the machine learning systems that these metrics rely on are known to exhibit potentially harmful biases in other domains (Bolukbasi et al., 2016; Caliskan et al., 2017; Koenecke et al., 2020).

In this paper, we dive into measures utilized in previous work to try to understand what they reflect. Following this work, we use a suite of models to quantify incoherence and derailment on speech produced by the CHR vs. HC groups (individuals who meet criteria for being at high-risk for psychosis vs. healthy controls). We examine group differences, finding significant differences using a subset of measures (at uncorrected $\alpha = .05$). We then critique these measures to determine if they reflect the target thought disorder symptoms – and fail to find specific correlations. Finally, we consider what these measures *do* reflect, finding that they partially reflect surface properties of the speech (sentence length) and sociodemographic properties of the speaker. These results highlight the need to consider the interpretability of these measures as the field continues to grow.

## 2 A Note on Terminology

Past work applying computational methods to study thought disorder in psychosis has used the words 'incoherence' or 'tangentiality' to describe their object of study, which has focused on the cohesion between sentences. However, this terminology is somewhat misaligned with the terminology discussed in the original thought disorder literature, which uses 'incoherence' to describe a lack of cohesion *within* sentences and 'tangentiality' for cases where participants give an off-topic response to a question (Andreasen, 1986). In this paper, we follow the naming conventions of past computational work in this area. We will refer to methods measuring the cohesion between neighboring sentences as 'coherence measures' and methods measuring how much a text drifts off topic as 'tangentiality measures'. However, it is very important to note that these methods better relate to derailment as defined in Andreasen (1986), as they measure how much a participant shifts topics between sentences.

| | CHR | HC |
|---|---|---|
| **Sociodemographics** | | |
| Age | 21.0(2.3) | 21.6(3.2) |
| Sex | | |
| Female | 47% | 71% |
| Male | 53% | 29% |
| Education Level | 14.4(2.1) | 14.6(2.2) |
| Racial Identity | | |
| First Nations | 0% | 2% |
| East Asian | 9% | 7% |
| Southeast Asian | 0% | 5% |
| South Asian | 6% | 2% |
| Black | 37% | 17% |
| Central/South American | 11% | 2% |
| West/Central Asia and ME | 0% | 2% |
| White | 31% | 51% |
| Interracial | 6% | 10% |
| Ethnicity | | |
| Hispanic | 23% | 12% |
| Not Hispanic | 77% | 88% |
| WRAT Score | 108(15) | 118(13) |
| **Speech Samples** | | |
| Sentence Length | 29.2(6.5) | 30.8(10.1) |
| Lexical Diversity | 0.70(0.04) | 0.71(0.03) |
| Response Length | 295(169) | 275(121) |

Table 1: Summary of participant and speech sample measures. ME = Middle East.

## 3 Methods

### 3.1 Participants

Speech samples were obtained from 77 participants aged 16-30: (a) 36 who met criteria for being at clinical high-risk for psychosis, and (b) 41 healthy controls. Participants were recruited from the larger Chicago, Illinois area through newspaper, transit, and Craigslist ads, e-mail postings, flyers, and community professional referrals. The Structured Interview for Prodromal Syndromes (SIPS) was used to determine the CHR vs. HC status of the participant (Miller et al., 1999) and to assess symptomatology. The Structured Clinical Interview for the DSM (First, 1997) was used to rule out Axis I psychotic disorder diagnoses within both groups.

Written informed consent was obtained from all participants. Data collection took place in a research lab setting and was approved by the institutional review board at Northwestern University.

### 3.2 Participant Measures

We obtained self-reported demographic information from participants (including age, sex, education level, and racial identity). In addition, participants completed the Word Reading subtest of the fourth edition of the Wide Range Achievement Test (WRAT) (Wilkinson and Robertson, 2006), which is a measure of scholastic achievement, strongly associated with general intelligence (Johnstone et al., 1996). As described, symptom severity was measured using the SIPS clinical interview. Our analyses focused on the following symptom items: P5 ("disorganized communication") (range 0-6), N5 ("ideational richness") (0-6), and D2 ("bizarre thinking") (0-6), in addition to the positive symptoms subscale total (0-30), the negative symptoms subscale total (0-36), and the disorganized symptoms subscale total (0-24) (see Miller et al. (1999), McGlashan et al. (2001), and Appendix A for more details about the SIPS).

### 3.3 Speech Measures

#### 3.3.1 Speech Elicitation

Participants were prompted to describe (1) a challenge they had overcome, (2) a self-defining memory, (3) a turning-point memory, and (4) an unusual memory (see Appendix B for full prompts). Their responses were professionally transcribed. For the CHR group, responses were 275 words long on average (range: 111-835 words), while for the HC group, responses were 255 words long on average (range: 98-559 words). We analyze the first full uninterrupted response participants provided and remove the following filler words: *um, uh, you know, I mean, okay, so, actually basically, right, yeah* as in Iter et al. (2018) (see Appendix E for analyses with filler words included). We analyzed each participant's four responses separately before averaging them to obtain a mean coherence and a mean tangentiality score for each individual.

#### 3.3.2 Automated Coherence/Tangentiality Measures

We obtain a measure of **coherence**, using the same word embedding methods used in past work on both psychosis and CHR populations (Bedi et al., 2015; Corcoran et al., 2018). At a high-level, this measure represents how similar, on average, the adjacent sentences in each participant's speech samples are to one another. If their sentences tend to be dissimilar to one another, then this is taken as evidence of incoherence.

To do this, we represent each word in the speech sample as a vector (using one of three pre-trained word embedding models e.g., word2vec), and combine the vectors of the words in a sentence (using one of 4 methods e.g., by averaging the word vectors) to obtain a vector for each sentence. We then calculate the cosine similarity between each pair of adjacent sentences, and average these, to obtain one coherence score per speech sample. We average across speech samples to obtain one overall score per participant.

We also obtain a measure of **tangentiality** as in Elvevåg et al. (2007) and Iter et al. (2018). At a high-level, this measure represents how quickly the topic of the speech sample changes. To do this, with sentence-level vectors in hand, we calculate the cosine similarity between the first sentence of a speech sample and each subsequent sentence (i.e., sentence 1 vs. sentence 2, sentence 1 vs. sentence 3, and so forth). We then fit a linear regression model to these values, treating the sentence number as the independent variable and the similarity score against the first sentence as the dependent variable. We use the slope of this line as the tangentiality measure. As with coherence, we obtain one measure for each speech sample, which we average within participants to obtain one overall tangentiality score per participant.

We follow Iter et al. (2018) in deciding which embedding models to use to obtain the sentence-level vector representations needed for these measures. We use either LSA (Landauer et al., 1998), GLoVE (Pennington et al., 2014), or word2vec (Mikolov et al., 2013) to obtain word-level vectors.[1] For sentence embedding methods, we simply average the vectors of all of the words in the sentence (**Mean(All)**), or use one of three methods that puts more weight on the content words of the sentence. **Mean(Content)** averages only the content word vectors of the sentence. **TF-IDF** divides each word's embedding by its frequency (operationalized as the number of times it occurs in a large corpus, like Wikipedia), essentially calculating a weighted average where more frequent words (e.g.,'the') are given less weight (Lintean et al., 2010). **SIF** also computes a weighted average for each sentence, but then removes the projection of the first principal component of the singular value decomposition of the sentence embedding matrix, which removes "semantically meaningless directions" (Arora et al., 2017). Finally, we use sent2vec, which works similarly to word2vec but on the sentence level: it directly learns sentence representations that predict neighboring sentences (Pagliardini et al., 2017). Using these methods, we obtain one coherence score per participant for

[1] We focus on LSA, GLoVE, word2vec, and sent2vec in the main text to align with past work, but Appendix D shows that results are qualitatively similar for the more modern and contextualized ELMo and BERT embeddings.

each combination of sentence and word embedding models, plus one for sent2vec (13 total). We refer the reader to Corcoran et al. (2018), Iter et al. (2018), and Hitczenko et al. (2020) for more details on embedding models.

#### 3.3.3 Other Speech Measures

In addition to automated coherence and tangentiality, we calculated the average sentence length (number of words per sentence) for each participant as well as a measure of each participant's lexical diversity. For lexical diversity, we used the moving average type-to-token ratio (MATTR) with a window of 50 words (Covington and McFall, 2010), which calculates the word type to word token ratio over each overlapping window of 50 words, and then averages them to obtain one overall measure of lexical diversity.

### 3.4 Analyses and Predictions

First, we ask whether there are group differences in coherence and tangentiality between the CHR and HC groups by running two sample t-tests as in past work. We expect to observe significant differences between the groups, with the HC group being more coherent and less tangential than the CHR group.

Second, we ask whether these automated scores correlate with item scores on the SIPS clinical interview related to disorganized speech or thought disorder, as well as with overall symptomatology measured by the SIPS. Where tested, past work has reported mixed findings, with some seeing correlations between automated measures and symptom severity (Just et al., 2019), but many not (Corcoran et al., 2018; Iter et al., 2018). As these automated measures are intended to measure thought disorder, we expect to find that worse symptom severity (i.e. higher symptom scores) is associated with worse coherence scores (i.e. lower coherence scores), especially for P5 ("disorganized communication").

Finally, we ask whether these automated linguistic scores relate to other linguistic properties of the speech (i.e., sentence length and lexical diversity) as well as sociodemographic factors of the individuals speaking (i.e., scholastic achievement/general intelligence, education, race, etc.). We calculate correlations for continuous measures and compare groups for discrete measures.

## 4 Results

### 4.1 Question 1: Are there CHR vs HC group differences in coherence/tangentiality?

As shown in Table 2, we find significant differences in coherence between the CHR and HC groups in 3 out of 13 of the methods we report (see Appendix C.1 for difference plots). However, it is important to note that these differences may be spurious based on multiple comparisons; with a Bonferonni correction ($\alpha$ = .004), these differences no longer reach significance. In 6 out of the remaining 10 methods, the healthy controls have numerically, but non-significantly, greater coherence scores than the CHR group. In the remaining 4 methods, the groups show near identical scores.

For tangentiality, we do not find any significant differences in tangentiality between the CHR and HC groups (Table 3). As a result, we do not conduct additional analyses of this measure.

These results suggest that these automated measures of thought disorder are very sensitive to the particular method used to derive it. Notably, previous work has not found any particular method to be consistently successful in separating groups. One of the methods where we find a significant difference is also successful in Just et al. (2019), who find significant coherence differences using TF-IDF GLoVE and no significant differences in tangentiality. However, Iter et al. (2018) only found differences in coherence using SIF word2vec, while other papers (Bedi et al., 2015; Elvevåg et al., 2007; Corcoran et al., 2018) have found significant differences using LSA Mean(All).

Overall, while we do not find group differences in tangentiality, we do find the predicted group differences in coherence between CHR and HC in a subset of cases. However, more work needs to be done to understand whether these are meaningful effects and what they reflect. To this end, for the remainder of the paper, we ask whether these automated linguistic methods of coherence relate to symptoms or other linguistic/sociodemographic factors. For these analyses, we zoom in on the sentence/word embedding models that separate CHR from HC groups. We present GLoVE Mean(Content) analyses in the main text; all other analyses are presented in Appendix C.

| Sentence | Word | CHR mean | HC mean | CHR sd | HC sd | T-stat | P-value |
|---|---|---|---|---|---|---|---|
| Mean | LSA | 0.58 | 0.60 | 0.07 | 0.06 | -1.12 | 0.13 |
| (All) | word2vec | 0.79 | 0.80 | 0.04 | 0.04 | -0.94 | 0.18 |
| | **GLoVE** | **0.92** | **0.93** | **0.02** | **0.02** | **-1.89** | **0.03** |
| Mean | LSA | 0.31 | 0.30 | 0.07 | 0.06 | 0.77 | 0.77 |
| (Content) | word2vec | 0.63 | 0.65 | 0.05 | 0.06 | -1.17 | 0.12 |
| | **GLoVE** | **0.81** | **0.82** | **0.04** | **0.03** | **-1.74** | **0.04** |
| TF-IDF | LSA | 0.42 | 0.44 | 0.07 | 0.07 | -1.05 | 0.15 |
| | word2vec | 0.75 | 0.76 | 0.04 | 0.05 | -0.71 | 0.24 |
| | **GLoVE** | **0.87** | **0.89** | **0.03** | **0.02** | **-2.14** | **0.02** |
| SIF | LSA | 0.08 | 0.08 | 0.09 | 0.07 | 0.23 | 0.59 |
| | word2vec | 0.03 | 0.02 | 0.06 | 0.06 | 0.96 | 0.83 |
| | GLoVE | 0.05 | 0.04 | 0.06 | 0.06 | 1.08 | 0.86 |
| sent2vec | sent2vec | 0.47 | 0.48 | 0.04 | 0.05 | -1.21 | 0.11 |

Table 2: Coherence results. We see a significant difference between groups in 3/13 methods (in bold), though these differences are no longer significant using the Bonferroni correction for multiple comparisons ($\alpha = 0.004$).

| Sentence | Word | CHR mean | HC mean | CHR sd | HC sd | T-stat | P-value |
|---|---|---|---|---|---|---|---|
| Mean | LSA | -0.007 | -0.018 | 0.03 | 0.05 | 1.18 | 0.88 |
| (All) | word2vec | -0.007 | -0.01 | 0.02 | 0.02 | 0.68 | 0.75 |
| | GLoVE | -0.002 | -0.004 | 0.01 | 0.01 | 0.94 | 0.83 |
| Mean | LSA | -0.017 | -0.013 | 0.04 | 0.03 | -0.5 | 0.31 |
| (Content) | word2vec | -0.013 | -0.016 | 0.02 | 0.03 | 0.57 | 0.72 |
| | GLoVE | -0.007 | -0.01 | 0.02 | 0.02 | 0.76 | 0.78 |
| TF-IDF | LSA | -0.011 | -0.017 | 0.04 | 0.04 | 0.59 | 0.72 |
| | word2vec | -0.008 | -0.011 | 0.02 | 0.03 | 0.52 | 0.70 |
| | GLoVE | -0.004 | -0.006 | 0.01 | 0.02 | 0.8 | 0.79 |
| SIF | LSA | -0.02 | -0.029 | 0.07 | 0.07 | 0.59 | 0.72 |
| | word2vec | -0.029 | -0.039 | 0.05 | 0.08 | 0.66 | 0.74 |
| | GLoVE | -0.034 | -0.04 | 0.06 | 0.07 | 0.41 | 0.66 |
| sent2vec | sent2vec | -0.013 | -0.011 | 0.03 | 0.03 | -0.34 | 0.37 |

Table 3: Tangentiality results. We observe no significant differences between the CHR vs. HC groups.

### 4.2 Question 2: Do automated coherence scores correlate with symptoms?

Do lower coherence scores (within the CHR group) relate to worse thought disorder? We examine this using symptoms in the SIPS that are related to thought disorder. As shown in Figure 1, we find generally poor correlations. The computational measures intended to measure thought disorder do not show any correlation with currently used clinical interviews measuring thought disorder in the CHR group. This result adds to a growing but mixed literature on the relationship between automated linguistic measures and the symptoms they are intended to measure.

Of past work that has reported correlations, Corcoran et al. (2018) and Iter et al. (2018) found no correlation between coherence scores and clinical interview symptoms, while Just et al. (2019) found their coherence measures did correlate negatively with symptom severity as measured by the Scale for the Assessment of Negative Symptoms (Andreasen, 1989). Bedi et al. (2015) included coherence in a canonical correlation identifying the maximal correlation between a linear combination of 3 linguistic features – coherence, maximal word phrase length, and number of determiners – and a linear combination of the positive and negative SIPS subscales. They found an overall positive correlation, but it's unclear what role coherence played in driving this correlation. Taken together, our results and previous results suggests that coherence scores are not reliably related to clinical measures of thought

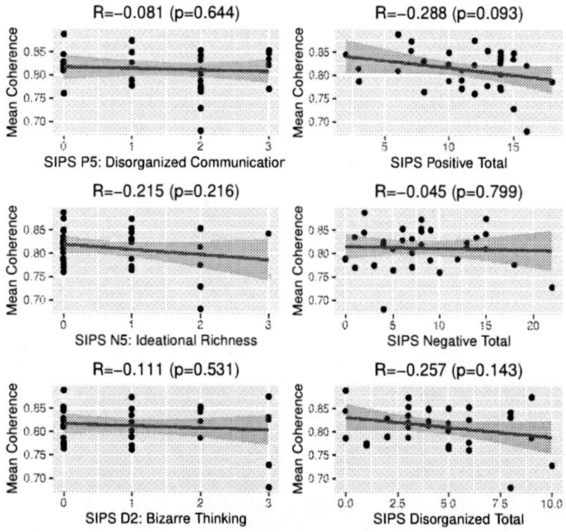


Figure 1: Correlation between mean coherence scores and relevant SIPS subitems and SIPS symptom totals. The lines show the estimated linear regression models and the shading shows 95% confidence intervals. Each point represents one participant.

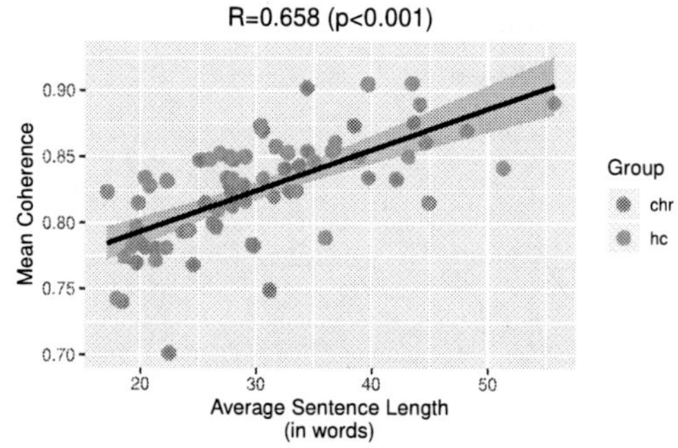


Figure 2: Correlation between mean coherence scores and average sentence length. The line show the estimated linear regression model and the shading shows 95% confidence intervals. Each point represents one participant, colored by CHR status.

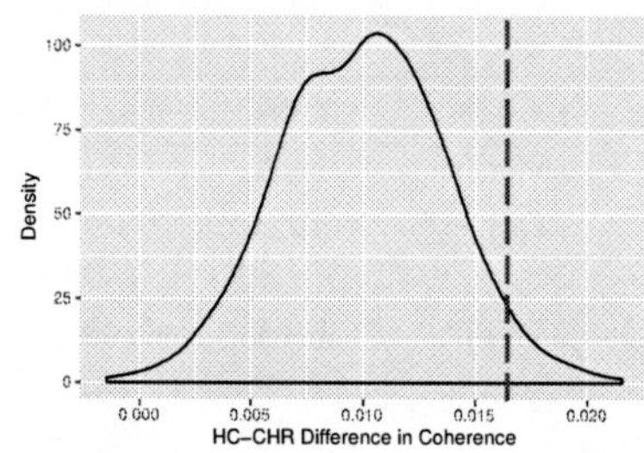


Figure 3: Length-only baseline distribution of HC-CHR differences in coherence (1000 samples). The vertical line shows the location of the true difference in this distribution.

disorders; however, a high-powered investigation is warranted.

### 4.3 Question 3: Do automated coherence scores correlate with linguistic features of speech samples or sociodemographic factors of the speaker?

If these measures are not capturing thought disorder symptoms, what are they measuring? To examine this issue, we examine the relationship of these computational measures to surface linguistic features of the speech samples and sociodemographic factors of the speakers. We focus on three features that show a significant relationship to this 'coherence' measure – sentence length, a measure of general intelligence, and racial identity of the speaker – and report non-significant correlations in Appendix C.

#### 4.3.1 Sentence length

We find a significant positive correlation between average sentence length and automated measures of coherence: that is, longer sentences are measured as more coherent ($r$ (75)=0.66; p<0.001) all else being equal (Figure 2).

This raises the possibility that the observed CHR-HC difference simply reflects differences in average sentence length (CHR mean: 29 words/sentence; HC mean: 31 words/sentence). To test for this possibility, we calculated the distribution of group differences predicted by a length-only baseline. In particular, we use a Monte Carlo method to compare the group differences in coherence scores against a surface-only baseline based on sentence length. We estimate this baseline by randomly replacing each word in our corpus – generating random word strings matching the length of our participants' productions. We then recalculate the group difference, providing an estimate of the difference in coherence scores predicted to occur by differences in sentence length alone. This procedure is repeated 1000 times to estimate the distribution of baseline differences. If the difference in coherence scores is based on the content of what participants are saying, then the observed difference should lie at the extreme tail of this baseline distribution.

As shown in Figure 3, only 3.9% of the runs had a more extreme HC-CHR difference than observed in the original participant data (shown with the blue dotted line), suggesting that there is something in the linguistic content that is contributing to the difference observed above and beyond the sentence lengths. However, we also note that the baseline difference is always greater than zero. Even though we completely randomized the content of the speech in both groups, the sentence length differences observed between the groups still resulted in greater coherence for the HC group, suggesting that sentence length plays a large role in the observed outcomes. Group differences can be obtained without considering any of the linguistic content spoken by participants. This is not a good property for this measure.

#### 4.3.2 WRAT scores

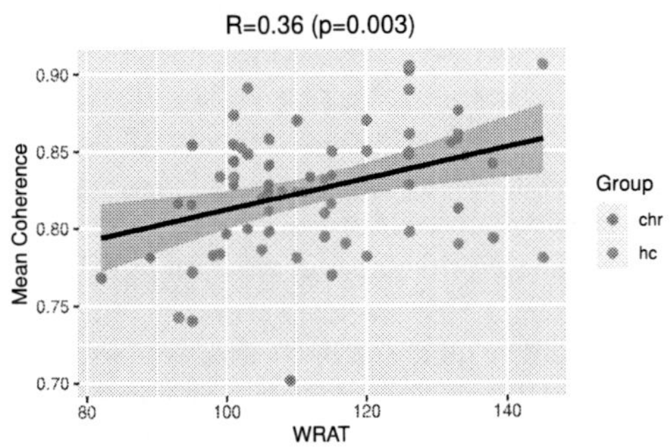


Figure 4: WRAT vs. Coherence Scores. The line shows the estimated linear regression model and the shading shows 95% confidence intervals. Each point represents one participant, colored by CHR status.

Next we observe in Figure 4 that higher coherence is associated with higher scores on the WRAT, a measure of scholastic achievement, associated with general intelligence ($r(75)=0.36$; $p<0.001$). Those with higher WRAT scores tend to produce more coherent speech (though it could also be that they tend to produce longer sentences). As with sentence length, we cannot make conclusions about causality here. However, this finding again reduces our confidence in the use of this computational measure as an index of thought disorder. Future work utilizing this coherence measure must control for the correlation with WRAT.

#### 4.3.3 Race

Finally, as shown in Figure 5, coherence scores may be correlated with racial identity. In our sample, Black speakers' speech was measured as less 'coherent' than that of White speakers' (all else being equal). However, it is critical to note that these analyses were based on a small numbers of participants (including just 7 Black participants

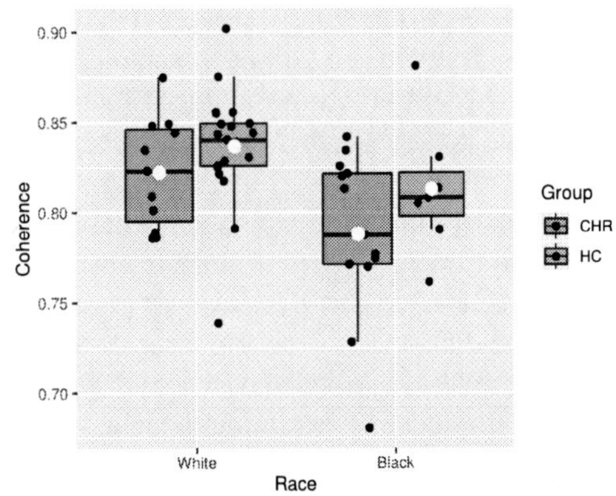


Figure 5: Coherence scores by race and clinical group. Each black point represents the mean coherence score of one individual grouped by their race (Black vs. White) and clinical status (HC vs. CHR). The four white dots represent the mean value for each group.

in the healthy control group), and this warrants a high-powered study directly investigating the relationship between coherence models and racial identity.

Nonetheless, this is a troubling finding that calls for a deeper dive into understanding what factors these computational measures are sensitive to before they can be used clinically. In particular, this result parallels other findings from the computational world - e.g., that ASR systems and computer vision systems work less well for Black individuals than White individuals (Koenecke et al., 2020; Buolamwini and Gebru, 2018). As the field develops, it is crucial to place analyses such as these front and center to ensure that this does not become another domain that perpetuates existing systemic biases.

#### 4.3.4 Relationship between effects

In summary, we observed relationships between automated coherence scores and (1) average sentence length, (2) intelligence/achievement scores as measured by the WRAT, and (3) racial identity. To get a better understanding of these effects and their interrelationship, we fit a linear model predicting average coherence scores from average sentence length, WRAT score, and race. We found that coherence scores were significantly higher for participants with longer average sentences ($\hat{\beta} = 0.001$, $p = 0.009$), but found no other significant effects – suggesting that the relationships between coherence and racial identity as well as scholastic achievement reflected correlations of these factors with sentence length. Indeed, White speakers produced longer sentences than Black speakers (White mean: 32 words, Black mean: 26 words) and individuals with higher WRAT scores produced longer sentences and passages than those with lower WRAT scores ($r = 0.3$; $p = 0.01$).

Overall, the findings in this section make clear that there is more work to be done to ensure that group differences reported in this body of literature reflect the differences in thought disorder they are meant to reflect, especially given non-correlations with SIPS symptoms measuring thought disorder. Of all of the factors, including thought disorder symptoms, sentence length was the factor that most correlated with coherence scores. Our results not only suggest that these measures may not be measuring what we think they are, but that this could have harmful downstream consequences (e.g., predicting lower coherence scores for Black speakers than White speakers).

## 5 Discussion

We tested methods of quantifying coherence and tangentiality, applying them to speech samples produced by individuals at clinical high-risk for psychosis. We found group differences between the CHR and HC groups for a subset of the tested methods (3 out of 13, significant only at uncorrected $\alpha = .05$). Surprisingly, we did not find significant correlations with items from clinical interviews that measure thought disorder (i.e. what these measures are meant to capture). In order for these measures to be useful clinically, it is important to show construct validity – that the measures actually index what they are meant to, rather than other features of the speech/speaker. This is especially true as the methods we use here have been shown to exhibit potentially harmful biases in other work. To this end, our final exploratory analyses were designed to better understand what these measures *are* capturing. We found correlations with sentence length, WRAT scores, and race, which suggests that these methods partially reflect properties that these measures are *not* intended to measure. These results suggest that there is substantial and careful work that needs to be done for these methods to be useful clinically.

### 5.1 Group differences are sensitive to the methods used and vary across papers

Replicating past work, we find group differences in coherence between the CHR vs. HC groups. However, as in past work using multiple word/sentence embedding methods, we find this difference in a subset of cases, suggesting this finding is sensitive to the particular method used. We fail to find group differences in tangentiality between CHR vs. HC groups. While these results overlap with those of one paper (Just et al., 2019), they do not overlap with other work (Corcoran et al., 2018; Bedi et al., 2015; Elvevåg et al., 2007; Iter et al., 2018) (and there is substantial variation within these papers as well). We offer two possible factors underlying these diverging findings. First, each paper has made different methodological decisions. Research differs in: the kinds of speech samples collected (shorter vs. longer length, individuals with vs. at-risk for psychosis); the analysis methods (some researchers remove fillers but others do not); and modeling decisions (some compare similarity between sentences, while others compare similarity between windows of words of length N), and so forth. These differences could easily give rise to differences across studies. Second, the true effect size could be quite small to begin with, especially in the CHR group who displays attenuated symptoms, and we know there is substantial heterogeneity between individuals. Some healthy individuals show linguistic disturbances, while some individuals with psychosis do not show any or show disturbances of almost opposite nature (e.g., perseveration, staying fixed on a single topic) (Andreasen, 1979). The substantial heterogeneity and differing sample sizes observed could also give rise to substantial differences between studies.

Overall, while past work has highlighted successes in the important goal of establishing differences between groups, it is critical to acknowledge where this line of work has fallen short: small changes in the particular methods used can substantially change the outcome, and which methods are successful varies unpredictably between studies. Moving forward, it may be useful to better align the methodological, analytical, and modeling choices across studies to better understand what gives rise to these differences. Due to the heterogeneity observed, it may also be worth focusing less on group differences and more on symptoms and outcome measures. In addition, as these methods continue to develop, it may be easier to accurately and more transparently evaluate their performance, by testing them on speech samples that are known to contain vs. not contain the particular studied linguistic disturbances. This shift in focus may allow us to gain a better understanding of what these measures reflect and how they can be useful on an individual basis.

### 5.2 Lack of correlations with SIPS thought disorder symptoms

We did not find correlations with the SIPS items that are thought to measure disorganized language and thought disorder. We note that is possible that, with 36 CHR participants, we did not have sufficient power to detect existing correlations with SIPS symptoms. However, this null finding adds to a growing literature of inconsistent findings, with some past work finding correlations with thought disorder and/or other clinical symptoms, but other past work failing to find these same correlations. This underscores the importance of doing careful work to establish construct validity with automated measures. Rigorous testing is needed to verify that novel measures relate to the properties of speech and cognition that they are intended to index.

### 5.3 Coherence scores correlate with sentence length and speaker sociodemographics

Perhaps most troublingly, we find that the differences in coherence between groups partially reflect irrelevant surface properties of the speech and sociodemographic qualities of the speakers. In fact, the single factor that best correlated with these measures was the length of the sentence. On the one hand, this raises concern that we are not measuring what we think we are. On the other hand, due to the fact that other factors (e.g., racial identity, achievement and intelligence, as measured by the WRAT) correlate with differences in average sentence length, this could have downstream harmful consequences (e.g., rating Black speakers as less coherent than White speakers due to differences unrelated to coherence). Overall, these results provide evidence that there is substantial work to be done to understand what these measures reflect to a degree where they can be used clinically.

### 5.4 Ethics and Broader Impacts Statement

Ethical use of computational measures like those studied here – especially in the high-stakes context of clinical care – requires us to devote substantial attention to potential biases in our measures. To that end, we recommend that future researchers in this area conduct and report analyses examining relations to symptoms, as well as the linguistic and sociodemographic factors studied here. This will allow us to gain a better understanding of what these measures reflect, and make sure that they are developed to be equally useful for all. To this end, we provide all of our code to hopefully facilitate these crucial cross-study comparisons.[2]

### 5.5 Conclusion

Linguistic disturbances characterize psychosis, yet they have been understudied in the field, largely due to how time-intensive it is to obtain meaningful and reliable measures of them. Automated linguistic methods have the potential to transform the scale at which we can study and identify these linguistic disturbances. However, with this strength come some downsides that the field must address: these methods are less transparent and can be harder to interpret. Facing these challenges head-on will allow us to develop a stronger, more ethical practice in this important and promising area of research.

## Acknowledgements

This work was supported by the National Institutes of Health (NIH R21 MH119677 and T32 NS047987) and the Canadian Institutes of Health Research (CIHR DFS-152268). We thank Rachel Ostrand and Ewan Dunbar for their helpful suggestions and feedback on this work.

## References

Nancy C Andreasen. 1979. Thought, language, and communication disorders: II. Diagnostic significance. *Archives of General Psychiatry*, 36(12):1325–1330.

Nancy C Andreasen. 1986. Scale for the assessment of thought, language, and communication (TLC). *Schizophrenia Bulletin*, 12(3):473.

Nancy C Andreasen. 1989. The scale for the assessment of negative symptoms (SANS): Conceptual and theoretical foundations. *The British Journal of Psychiatry*, 155(S7):49–52.

Nancy C Andreasen and William M Grove. 1986. Thought, language, and communication in schizophrenia: Diagnosis and prognosis. *Schizophrenia Bulletin*, 12(3):348–359.

Sanjeev Arora, Yingyu Liang, and Tengyu Ma. 2017. A simple but tough-to-beat baseline for sentence embeddings. In *5th International Conference on Learning Representations, ICLR 2017*.

[2] github.com/khitczenko/chr_coherence

Carrie E Bearden, Keng Nei Wu, Rochelle Caplan, and Tyrone D Cannon. 2011. Thought disorder and communication deviance as predictors of outcome in youth at clinical high risk for psychosis. *Journal of the American Academy of Child & Adolescent Psychiatry*, 50(7):669–680.

Gillinder Bedi, Facundo Carrillo, Guillermo A Cecchi, Diego Fernández Slezak, Mariano Sigman, Natália B Mota, Sidarta Ribeiro, Daniel C Javitt, Mauro Copelli, and Cheryl M Corcoran. 2015. Automated analysis of free speech predicts psychosis onset in high-risk youths. *npj Schizophrenia*, 1(1):1–7.

Eugen Bleuler. 1950. Dementia praecox or the group of schizophrenias. *International Universities Press*.

Tolga Bolukbasi, Kai-Wei Chang, James Zou, Venkatesh Saligrama, and Adam Kalai. 2016. Man is to computer programmer as woman is to homemaker? Debiasing word embeddings. In *Proceedings of the 30th International Conference on Neural Information Processing Systems (NeurIPS)*, pages 4356–4364.

Joy Buolamwini and Timnit Gebru. 2018. Gender shades: Intersectional accuracy disparities in commercial gender classification. In *Conference on Fairness, Accountability and Transparency*, pages 77–91.

Aylin Caliskan, Joanna J Bryson, and Arvind Narayanan. 2017. Semantics derived automatically from language corpora contain human-like biases. *Science*, 356(6334):183–186.

Cheryl M Corcoran, Facundo Carrillo, Diego Fernández-Slezak, Gillinder Bedi, Casimir Klim, Daniel C Javitt, Carrie E Bearden, and Guillermo A Cecchi. 2018. Prediction of psychosis across protocols and risk cohorts using automated language analysis. *World Psychiatry*, 17(1):67–75.

Cheryl M Corcoran, Vijay A Mittal, Carrie E Bearden, Raquel E Gur, Kasia Hitczenko, Zarina Bilgrami, Aleksandar Savic, Guillermo A Cecchi, and Phillip Wolff. 2020. Language as a biomarker for psychosis: A natural language processing approach. *Schizophrenia Research*, 226:158–166.

Michael A Covington and Joe D McFall. 2010. Cutting the gordian knot: The moving-average type–token ratio (MATTR). *Journal of Quantitative Linguistics*, 17(2):94–100.

Arsime Demjaha, Sara Weinstein, Daniel Stahl, Fern Day, Lucia Valmaggia, Grazia Rutigliano, Andrea De Micheli, Paolo Fusar-Poli, and Philip McGuire. 2017. Formal thought disorder in people at ultra-high risk of psychosis. *BJPsych Open*, 3(4):165–170.

Brita Elvevåg, Peter W Foltz, Daniel R Weinberger, and Terry E Goldberg. 2007. Quantifying incoherence in speech: An automated methodology and novel application to schizophrenia. *Schizophrenia Research*, 93(1-3):304–316.

Michael B First. 1997. Structured Clinical Interview for DSM-IV Axis I disorders. *Biometrics Research Department*.

Tina Gupta, Susan J Hespos, William S Horton, and Vijay A Mittal. 2018. Automated analysis of written narratives reveals abnormalities in referential cohesion in youth at ultra high risk for psychosis. *Schizophrenia Research*, 192:82–88.

Kasia Hitczenko, Vijay A Mittal, and Matthew Goldrick. 2020. Understanding language abnormalities and associated clinical markers in psychosis: The promise of computational methods. *Schizophrenia Bulletin*.

Dan Iter, Jong Yoon, and Dan Jurafsky. 2018. Automatic detection of incoherent speech for diagnosing schizophrenia. In *Proceedings of the Fifth Workshop on Computational Linguistics and Clinical Psychology: From Keyboard to Clinic*, pages 136–146.

Brick Johnstone, Charles D Callahan, Cynthia J Kapila, and Dawn E Bouman. 1996. The comparability of the WRAT-R reading test and NAART as estimates of premorbid intelligence in neurologically impaired patients. *Archives of Clinical Neuropsychology*, 11(6):513–519.

Sandra Just, Erik Haegert, Nora Kořánová, Anna-Lena Bröcker, Ivan Nenchev, Jakob Funcke, Christiane Montag, and Manfred Stede. 2019. Coherence models in schizophrenia. In *Proceedings of the Sixth Workshop on Computational Linguistics and Clinical Psychology*, pages 126–136.

Allison Koenecke, Andrew Nam, Emily Lake, Joe Nudell, Minnie Quartey, Zion Mengesha, Connor Toups, John R Rickford, Dan Jurafsky, and Sharad Goel. 2020. Racial disparities in automated speech recognition. *Proceedings of the National Academy of Sciences*, 117(14):7684–7689.

Gina R Kuperberg. 2010. Language in schizophrenia part 1: An introduction. *Language and Linguistics Compass*, 4(8):576–589.

Thomas K Landauer, Peter W Foltz, and Darrell Laham. 1998. An introduction to latent semantic analysis. *Discourse Processes*, 25(2-3):259–284.

Mihai Lintean, Cristian Moldovan, Vasile Rus, and Danielle McNamara. 2010. The role of local and global weighting in assessing the semantic similarity of texts using latent semantic analysis. In *Twenty-Third International FLAIRS Conference*.

Thomas H McGlashan, Barbara C Walsh, Scott W Woods, J Addington, K Cadenhead, T Cannon, and E Walker. 2001. Structured Interview for Psychosis-risk Syndromes. *New Haven, CT: Yale School of Medicine*.

Tomas Mikolov, Kai Chen, Greg Corrado, and Jeffrey Dean. 2013. Efficient estimation of word representations in vector space. *arXiv preprint arXiv:1301.3781*.

Tandy J Miller, Thomas H McGlashan, Scott W Woods, Kelly Stein, Naomi Driesen, Cheryl M Corcoran, Ralph Hoffman, and Larry Davidson. 1999. Symptom assessment in schizophrenic prodromal states. *Psychiatric Quarterly*, 70(4):273–287.

Matteo Pagliardini, Prakhar Gupta, and Martin Jaggi. 2017. Unsupervised learning of sentence embeddings using compositional n-gram features. *arXiv preprint arXiv:1703.02507*.

Jeffrey Pennington, Richard Socher, and Christopher D Manning. 2014. GLoVE: Global vectors for word representation. In *Proceedings of the 2014 Conference on Empirical Methods in Natural Language Processing (EMNLP)*, pages 1532–1543.

Diana O Perkins, Clark D Jeffries, Barbara A Cornblatt, Scott W Woods, Jean Addington, Carrie E Bearden, Kristin S Cadenhead, Tyrone D Cannon, Robert Heinssen, Daniel H Mathalon, et al. 2015. Severity of thought disorder predicts psychosis in persons at clinical high-risk. *Schizophrenia Research*, 169(1-3):169–177.

Marco Tulio Ribeiro, Sameer Singh, and Carlos Guestrin. 2016. Model-agnostic interpretability of machine learning. *arXiv preprint arXiv:1606.05386*.

Eric Roche, John Lyne, Brian O'Donoghue, Ricardo Segurado, Caragh Behan, Laoise Renwick, Felicity Fanning, Kevin Madigan, and Mary Clarke. 2016. The prognostic value of formal thought disorder following first episode psychosis. *Schizophrenia Research*, 178(1-3):29–34.

James Wilcox, George Winokur, and Ming Tsuang. 2012. Predictive value of thought disorder in new-onset psychosis. *Comprehensive Psychiatry*, 53(6):674–678.

GS Wilkinson and GJ Robertson. 2006. Wide Range Achievement Test 4 Professional Manual. *Psychological Assessment Resources*.

## A More Information on the Structured Interview for Psychosis-Risk Syndromes

The SIPS is a clinical interview administered by experienced clinicians that is used to classify individuals as being at clinical high-risk for psychosis. It consists of 19 symptoms that are grouped into four symptom classes: 5 positive (P) symptoms, 6 negative (N) symptoms, 4 disorganized (D) symptoms, and 4 general (G) symptoms. Patients are rated along each of the 19 individual symptoms (scores for each individual symptom range from 0, least severe, to 6, most severe). The scores on the individual symptoms within each of the four classes are totaled to get total positive (range 0-30), negative (range 0-36), disorganized (range 0-24), and general (range 0-24) symptom scores. Our analyses focus on items P5 ("Disorganized Communication"), N5 ("Ideational Richness"), and D2 ("Bizarre Thoughts"), as well as the positive, negative, and disorganized symptom totals, as described below. We refer readers to Miller et al. (1999) and McGlashan et al. (2001) for more information about the SIPS.

**Positive Symptoms [0-30]**: There are five positive symptoms: P1 (Unusual Thought Content/Delusional Ideas), P2 (Suspiciousness/Persecutory Ideas), P3 (Gradiose Ideas), P4 (Perceptual Abnormalities/Hallucinations), and P5 (Disorganized Communication).

**P5- Disorganized Communication [0-6]**: The types of inquiries used to establish the score include:

- Do people ever tell you that they can't understand you? Do people ever seem to have difficulty understanding you?
- Are you aware of any ongoing difficulties getting your point across, such as finding yourself rambling or going off track when you talk?
- Do you ever completely lose your train of thought or speech, like suddenly blanking out?

**Negative Symptoms [0-36]**: There are six negative symptoms: N1 (Social Anhedonia), N2 (Avolition), N3 (Expression of Emotion), N4 (Experience of Emotions and Self), N5 (Ideational Richness), and N6 (Occupational Functioning).

**N5- Ideational Richness [0-6]**: The types of inquiries used to establish the score include:

- Do you sometimes find it hard to understand what people are trying to tell you because you don't understand what they mean?
- Do people more and more use words that you don't understand?

**Disorganized Symptoms [0-24]**: There are four disorganized symptoms: D1 (Odd Behavior or Appearance), D2 (Bizarre Thinking), D3 (Trouble with Focus and Attention), and D4 (Impairment in Personal Hygiene). In our analyses, we use the total disorganized score (range: 0-24), as well as the D2 item (bizarre thinking).

**D2- Bizarre Thinking [0-6]**: The types of inquiries used to establish the score include:

- Do people ever say your ideas are unusual or that the way you think is strange or illogical?

**General Symptoms [0-24]**: We do not include these symptoms in our analyses, but there are four general symptoms: G1 (Sleep Disturbance), G2 (Dysphoric Mood), G3 (Motor Disturbances), and G4 (Impaired Tolerance to Normal Stress).

## B Complete Question Prompts

- **Challenge**: Looking back over your life, what do you think is the single greatest challenge you have ever faced? Tell me the story of that challenge, what it is or was, how did the challenge or problem develop, and how did you address or deal with the challenge or problem?

- **Self-Defining**: A self-defining memory is a scene or an episode from your life that was very important for how you see yourself. This would be something that happened at least one year ago that you have thought about many times since it happened so that the memory of it is clear and familiar to you. This scene or episode helps you know who you are as a person. You might even tell this story to a friend if you wanted to help them understand you better. I'd like you to take a moment to think of a self-defining memory like this and then tell me the story of that memory and specifically what happened, when and where it happened, and who was involved?

- **Turning Point**: In most people's lives we experience episodes that change the direction of our lives or change how we see ourselves in some important way. We call those memories turning points. Looking back over your life, there may be a few key moments that stand out as turning points or episodes that marked an important change in you or your life story. I'd like you to identify a particular memory that you see as a turning point in your life and then tell me the story about that turning point: what happened, when and where it happened, and who was involved?

- **Unusual**: Next I'll ask you about an unusual experience that you might have had. Any unusual, strange or profound things that are hard to explain, for example, some coincidences, supernatural events, seeing visions of spirits, feeling like you're the center of attention, like you have special powers, or like one of your dreams had really happened. These experiences might be difficult to explain and might feel like the world is not as it seems or like your mind is playing tricks on you in some way. Take a moment to think of an unusual experience like this and then tell me the story of that experience: what happened, when and where it happened, and who was involved?

# C Additional Analyses: Participant's Main Response with Fillers Removed

The main text reports results from running the participant's first main response, with fillers removed. This section provides additional analyses that were omitted from the main text, including correlations for all three models that were found to be significant (GLoVE TF-IDF, GLoVE Mean(All), and GLoVE Mean(Content)), as well as non-significant correlations (e.g., for age and education).

## C.1 Group Differences

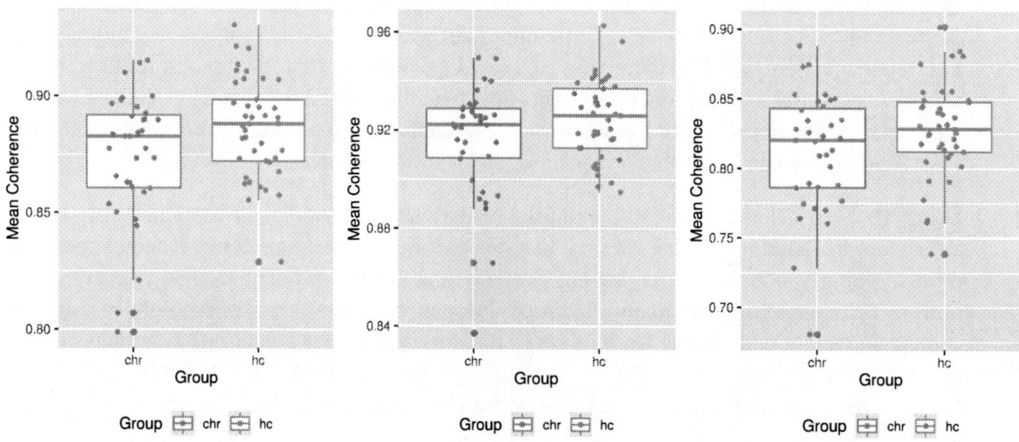


Figure 6: Coherence scores by group for each of the three methods that yield significant differences between the CHR and HC groups: GLoVE TF-IDF, GLoVE Mean(All), and GLoVE Mean(Content).

## C.2 Correlations with thought disorder symptoms

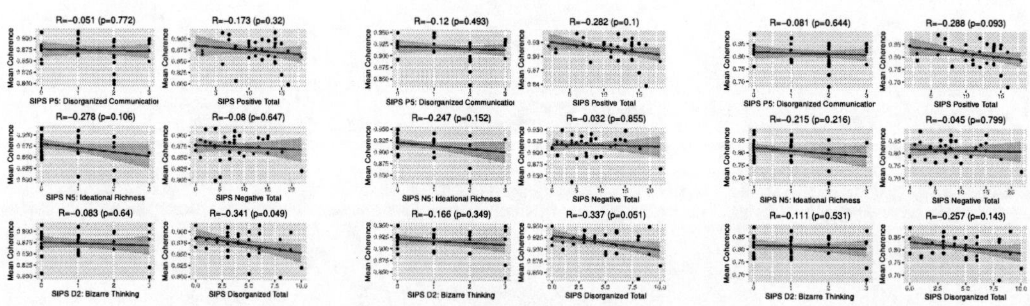


Figure 7: Correlations between coherence scores and SIPS symptoms for methods that yielded significant results (from left to right: GLoVE TF-IDF, GLoVE Mean(All), GLoVE Mean(Content)). Most correlations are not significant with one exception: GLoVE TF-IDF coherence scores correlate negatively with SIPS Total Disorganized Scores ($r$ = -0.34, p = 0.049).

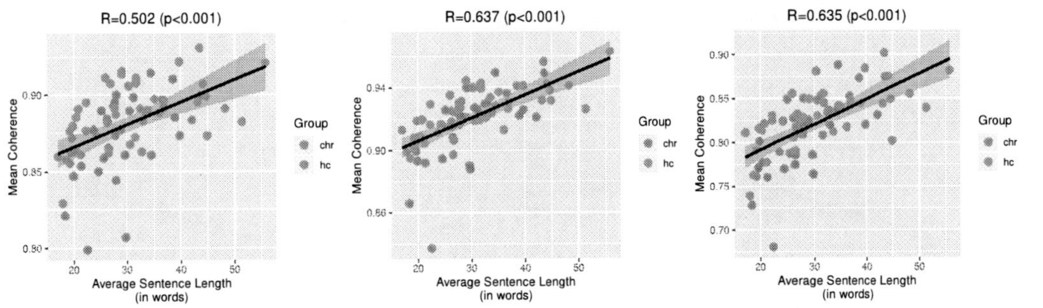


Figure 8: In all three cases (L-to-R: GLoVE TF-IDF, GLoVE Mean(All), GLoVE Mean(Content)), we observe significant positive correlations between average sentence length and average coherence with correlation coefficients ranging from 0.5 to 0.64.

### C.3 Correlations with sentence length

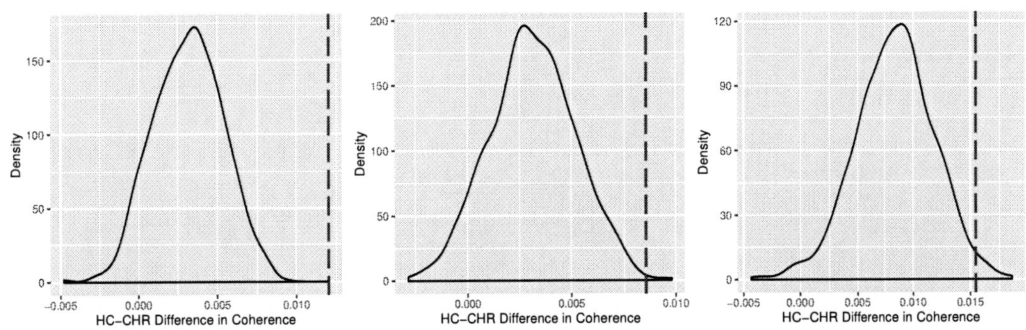


Figure 9: For each of the three significant methods (L-to-R: GLoVE TF-IDF, GLoVE Mean(All), GLoVE Mean(Content)), we randomly replace words and recalculate the coherence scores 1000 times. This graph shows the distribution of HC-CHR differences over these 1000 runs. For all three graphs, the vast majority of the differences are positive, meaning that the HC group scores as more coherent than the CHR group despite complete randomization of words. Nonetheless, the true difference (shown in the blue dotted line) is more extreme than most (GLoVE Mean(All), GLoVE Mean(Content)) or all (GLoVE TF-IDF) of the 1000 differences, suggesting that the coherence measures are partially based on the content of the speech.

### C.4 Correlations with lexical diversity

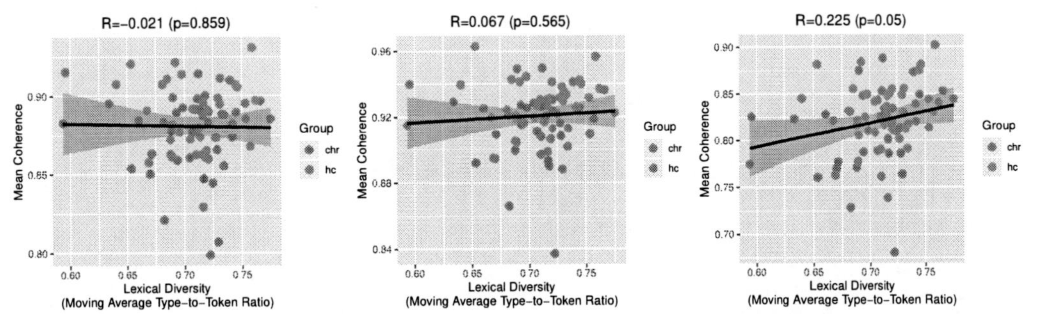


Figure 10: Left-to-right: GLoVE TF-IDF, GLoVE Mean(All), GLoVE Mean(Content). Lexical diversity, as measured by MATTR, does not correlate with coherence scores, though the correlation approaches significance for GLoVE Mean(Content), such that greater lexical diversity is associated with greater average coherence. As these automated measures calculate similarity between sentences, we might expect that repeating words would be associated with greater coherence scores. However, we do not observe this effect.

## C.5 Correlations with Scholastic Achievement and Intelligence (WRAT)

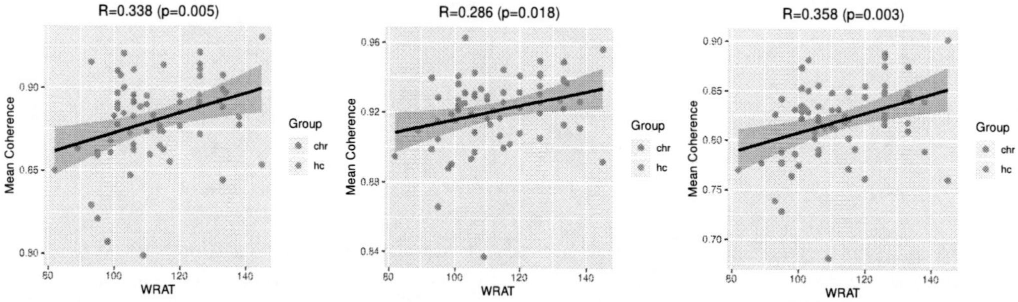


Figure 11: Left-to-right: GLoVE TF-IDF, GLoVE Mean(All), GLoVE Mean(Content). In all three methods, WRAT scores correlate positively with coherence scores, such that greater coherence is associated with higher WRAT scores. Correlation coefficients range from 0.29 to 0.36.

## C.6 Race

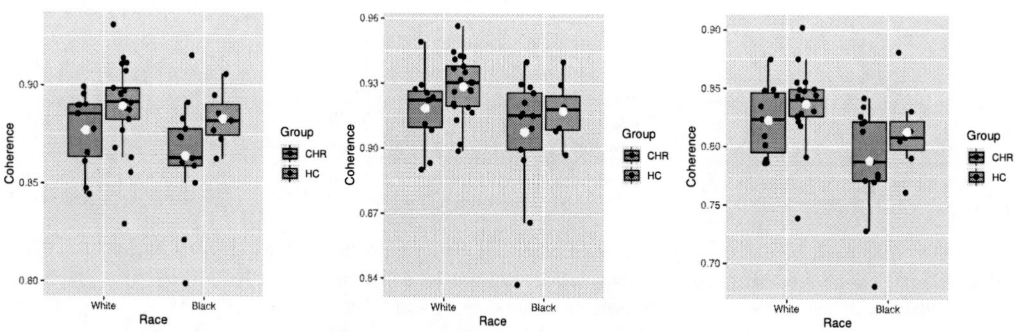


Figure 12: Left-to-right: GLoVE TF-IDF, GLoVE Mean(All), GLoVE Mean(Content). Coherence scores by CHR status (HC vs. CHR) and racial identity (Black vs. White). Across the three methods, these automated measures rate Black speakers as less coherent than White speakers.

## C.7 Age

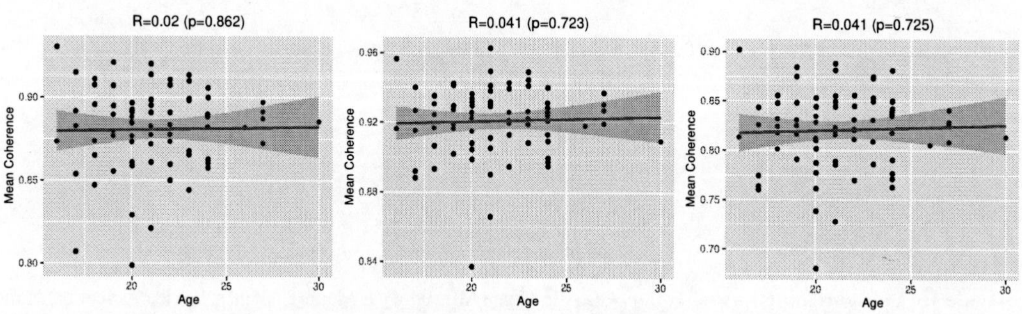


Figure 13: Left-to-right: GLoVE TF-IDF, GLoVE Mean(All), GLoVE Mean(Content). As expected, we find no correlation between age and coherence scores, although we note that this relationship has been observed in past work with older individuals scoring as more coherent (Corcoran et al., 2018).

### C.8 Education

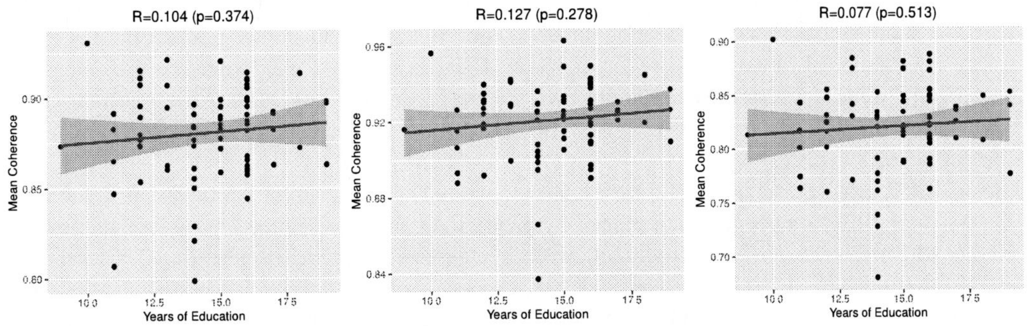


Figure 14: Left-to-right: GLoVE TF-IDF, GLoVE Mean(All), GLoVE Mean(Content). Finally, as expected, we find no correlation between level of education and coherence scores.

## D Contextualized Word Embeddings (Fillers Removed)

To align with past work, the main text reports results using word2vec, GLoVE, and LSA embeddings. Here, we show similar results for the more modern, contextualized embeddings from BERT and ELMo.

The analyses for ELMo mirror those for word2vec, GLoVE, and LSA: once we have word embeddings from ELMo, we obtain sentence embeddings by averaging all of the words (Mean(All)) or just the content words (Mean(Content)) or using TF-IDF or SIF weights, which both essentially give more weight to more content-bearing words.

For BERT, however, we used a different approach, taking advantage of in-built features of the model. In particular, BERT embeddings are trained by giving the model two sentences and having the model predict whether or not one immediately followed the other (Next Sentence Prediction). That means that given a first sentence and a second sentence, we can obtain a score for how likely it is that the second sentence directly follows the first one. We used this to obtain coherence scores for each participant's speech sample, with the idea that more coherent passages will have adjacent sentences that are more predictive of one another. We obtained BERT embeddings for each word in the participant's speech. Then, directly from these embeddings, for each pair of adjacent sentences in a speech sample, we obtained the model's score for how likely it was that the second sentence followed the first sentence (BERT Next Sentence Prediction). We averaged these scores within speech samples to obtain one coherence score for each speech sample (which, in turn, were averaged to obtain one coherence score per participant).

| Word | Sentence | CHR mean | HC mean | CHR sd | HC sd | T-stat | P-value |
|---|---|---|---|---|---|---|---|
| ELMo | Mean (All) | 0.71 | 0.72 | 0.03 | 0.03 | -0.86 | 0.20 |
| | Mean (Content) | 0.62 | 0.63 | 0.05 | 0.05 | -0.68 | 0.25 |
| | TF-IDF | 0.69 | 0.70 | 0.03 | 0.04 | -0.85 | 0.20 |
| | SIF | 0.02 | 0.01 | 0.05 | 0.05 | 0.98 | 0.84 |
| BERT | n/a | 0.977 | 0.983 | 0.05 | 0.05 | -1.07 | 0.14 |

Table 4: Coherence results, using ELMo embeddings. We find no significant differences between groups.

Although we found no significant differences between groups, we checked whether these embeddings also exhibited the same crucial problem of being correlated with sentence length and found that they did (Figure 15). The effect is reduced using BERT, as many sentence pairs are predicted to be adjacent with scores approaching 1; however, we still observe a significant correlation between average sentence length and mean coherence, finding that participants who produce shorter sentences are relatively more likely to have lower coherence scores.

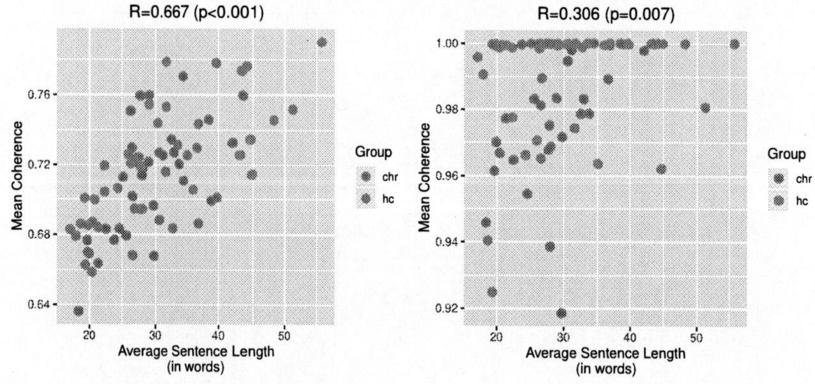


Figure 15: Left-to-right: ELMo (Mean(All)), BERT (Next Sentence Prediction). In both cases, we see a correlation between automated coherence scores and sentence length.

## E Including Filler Words

In the main text, we report findings from analyzing the participants' first uninterrupted response removing filler words as in Iter et al. (2018). Here, we report results from the same speech samples, but with fillers included.

### E.1 Group Differences

We test for group differences between the CHR and HC groups. As in the main text, we find significant differences in coherence for a subset of the methods used (here 2/13: GLoVE Mean(All) is no longer significant), but no significant differences in tangentiality. For the remainder of the analyses, we focus on the two methods that yielded significant differences between groups: coherence as measured by GLoVE TF-IDF and GLoVE Mean(Content).

| Sentence | Word | CHR mean | HC mean | CHR sd | HC sd | T-stat | P-value |
|---|---|---|---|---|---|---|---|
| Mean (All) | LSA | 0.57 | 0.59 | 0.08 | 0.07 | -1.16 | 0.12 |
| | word2vec | 0.80 | 0.81 | 0.03 | 0.04 | -0.98 | 0.17 |
| | GLoVE | 0.91 | 0.92 | 0.03 | 0.02 | -1.55 | 0.06 |
| Mean (Content) | LSA | 0.32 | 0.31 | 0.07 | 0.07 | 0.43 | 0.66 |
| | word2vec | 0.64 | 0.66 | 0.05 | 0.06 | -1.38 | 0.09 |
| | **GLoVE** | **0.82** | **0.83** | **0.04** | **0.04** | **-1.81** | **0.04** |
| TF-IDF | LSA | 0.42 | 0.44 | 0.07 | 0.08 | -1.35 | 0.09 |
| | word2vec | 0.78 | 0.78 | 0.04 | 0.05 | -0.38 | 0.35 |
| | **GLoVE** | **0.88** | **0.89** | **0.03** | **0.03** | **-1.75** | **0.04** |
| SIF | LSA | 0.10 | 0.10 | 0.09 | 0.07 | 0.18 | 0.57 |
| | word2vec | 0.05 | 0.04 | 0.05 | 0.06 | 1.37 | 0.91 |
| | GLoVE | 0.07 | 0.06 | 0.06 | 0.08 | 0.96 | 0.83 |
| sent2vec | sent2vec | 0.47 | 0.48 | 0.05 | 0.05 | -1.37 | 0.09 |

Table 5: Coherence results. We see a significant difference between groups in 2/13 methods (GLoVE TF-IDF and GLoVE Mean(Content)), though these differences are no longer significant using the Bonferroni correction for multiple comparisons.

| Sentence | Word | CHR mean | HC mean | CHR sd | HC sd | T-stat | P-value |
|---|---|---|---|---|---|---|---|
| Mean(All) | LSA | -0.015 | -0.022 | 0.04 | 0.05 | 0.73 | 0.77 |
| | word2vec | -0.006 | -0.01 | 0.02 | 0.02 | 0.88 | 0.81 |
| | GLoVE | -0.004 | -0.004 | 0.02 | 0.01 | 0.1 | 0.54 |
| SIF | LSA | -0.026 | -0.027 | 0.06 | 0.06 | 0.07 | 0.53 |
| | word2vec | -0.032 | -0.038 | 0.08 | 0.07 | 0.34 | 0.63 |
| | GLoVE | -0.038 | -0.037 | 0.08 | 0.06 | -0.02 | 0.49 |
| TF-IDF | LSA | -0.016 | -0.019 | 0.04 | 0.04 | 0.35 | 0.64 |
| | word2vec | -0.005 | -0.009 | 0.02 | 0.02 | 0.96 | 0.83 |
| | GLoVE | -0.004 | -0.006 | 0.01 | 0.01 | 0.5 | 0.69 |
| Mean(Content) | LSA | -0.02 | -0.014 | 0.04 | 0.03 | -0.69 | 0.25 |
| | word2vec | -0.011 | -0.015 | 0.02 | 0.03 | 0.69 | 0.75 |
| | GLoVE | -0.007 | -0.009 | 0.02 | 0.02 | 0.37 | 0.64 |
| sent2vec | sent2vec | -0.017 | -0.011 | 0.03 | 0.03 | -0.84 | 0.20 |

Table 6: Tangentiality results. As in the main text, we observe no significant differences between groups.

## E.2 Correlations with thought disorder symptoms

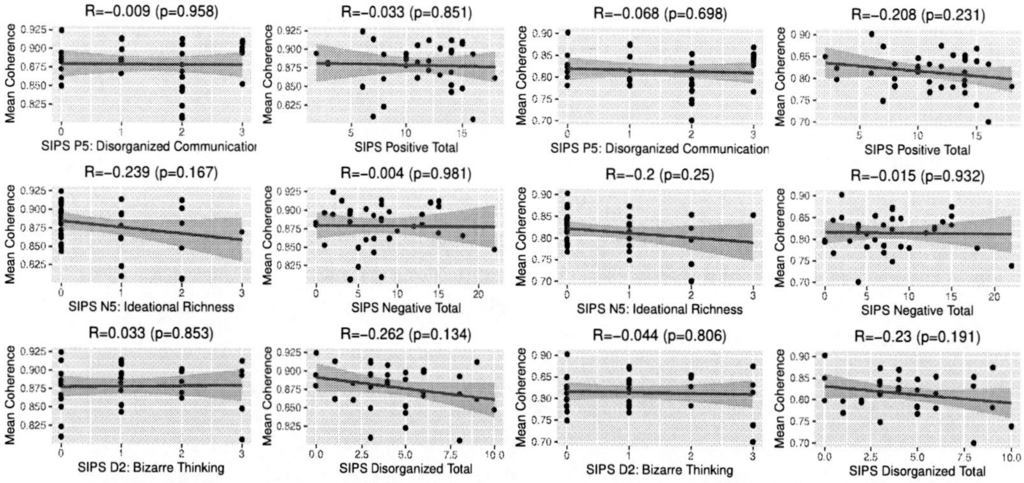


Figure 16: Correlations between coherence scores and SIPS symptoms for methods that yielded significant results (left two columns: GLoVE TF-IDF, right two columns: GLoVE Mean(Content)). As in the main text, we observe no significant correlations between SIPS symptoms and mean coherence.

## E.3 Correlations with sentence length

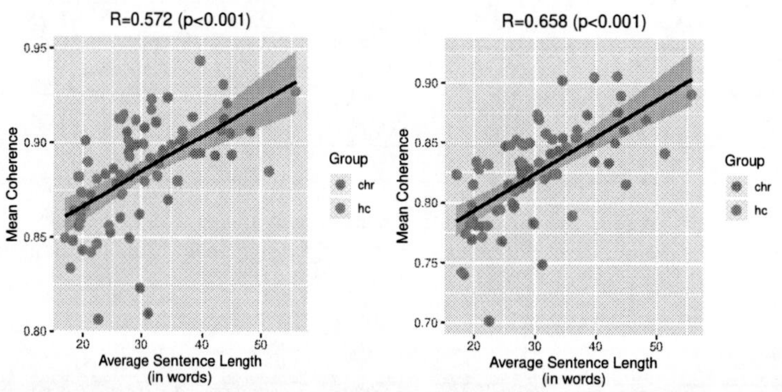


Figure 17: As in the main text, in both cases (L-to-R: GLoVE TF-IDF, GLoVE Mean(Content)), we observe significant positive correlations between average sentence length and average coherence with correlation coefficients.

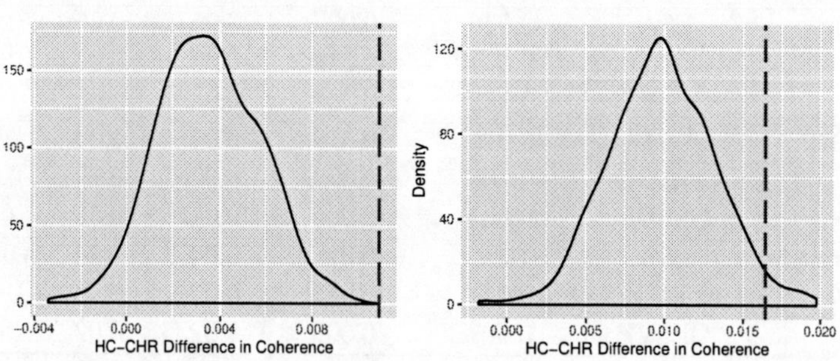


Figure 18: Left: GLoVE TF-IDF, Right: GLoVE Mean(Content). As in the case of removing fillers, in both graphs, the vast majority of the differences are positive, meaning that the HC group scores as more coherent than the CHR group despite complete randomization of words.

## E.4 Correlations with lexical diversity

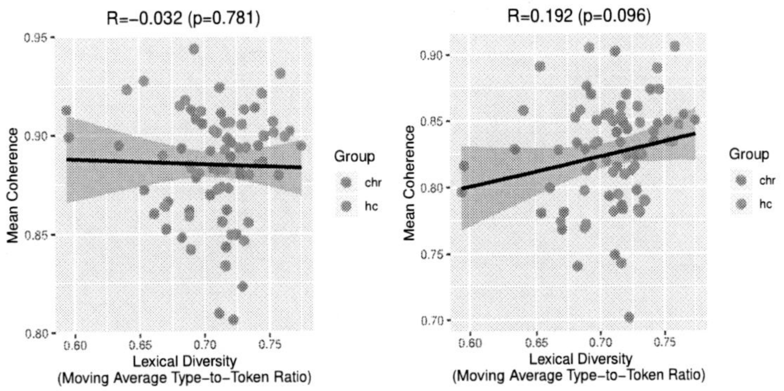


Figure 19: Left-to-right: GLoVE TF-IDF, GLoVE Mean(Content). As in the case of removing fillers, we find no significant correlation between lexical diversity, as measured by the MATTR, and mean coherence scores.

## E.5 Correlations with Scholastic Achievement and Intelligence (WRAT)

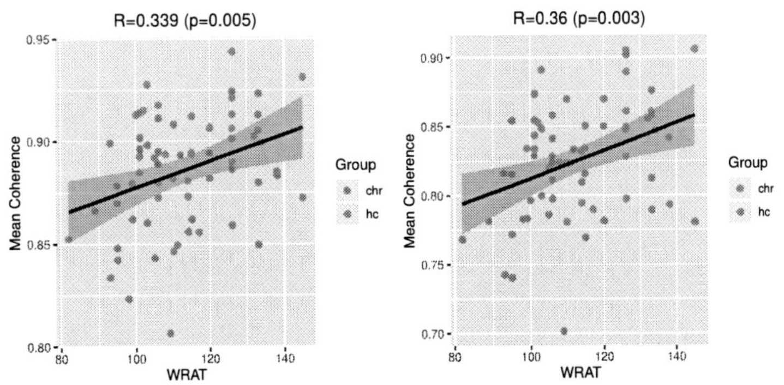


Figure 20: Left-to-right: GLoVE TF-IDF and GLoVE Mean(Content). As in the main text, WRAT scores correlate positively with coherence scores, such that greater coherence is associated with higher WRAT scores (a measure of achievement, associated with intelligence).

## E.6 Race

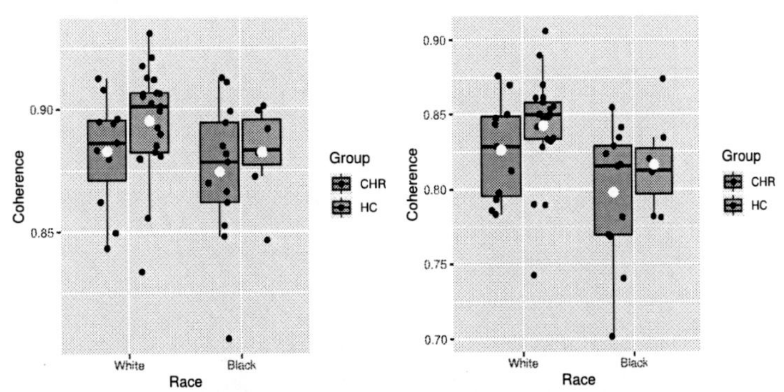


Figure 21: Left-to-right: GLoVE TF-IDF and GLoVE Mean(Content). As in the case of removing fillers, we find that both of these automated methods assign lower coherence scores to Black speakers than White speakers.

### E.7 Age

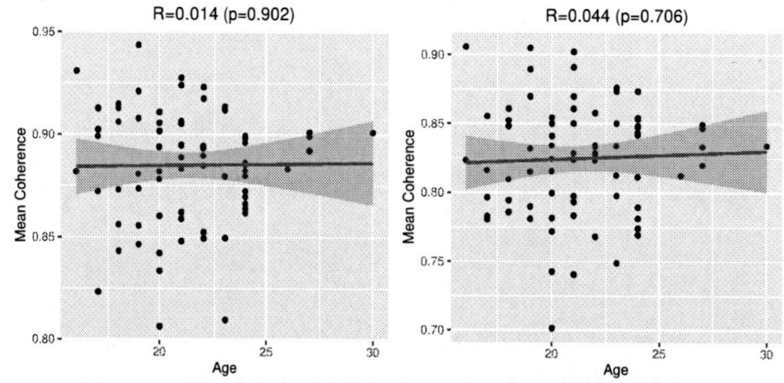


Figure 22: Left-to-right: GLoVE TF-IDF and GLoVE Mean(Content). Using both methods, we find no correlation between age and coherence scores.

# Detecting Cognitive Distortions from Patient-Therapist Interactions

**Sagarika Shreevastava**
Department of Computer Science, and
Department of Linguistics,
University of Colorado, Boulder
sagarika.shreevastava@colorado.edu

**Peter W. Foltz**
Institute of Cognitive Science
University of Colorado, Boulder
peter.foltz@colorado.edu

## Abstract

An important part of Cognitive Behavioral Therapy (CBT) is to recognize and restructure certain negative thinking patterns that are also known as cognitive distortions. This project aims to detect these distortions using natural language processing. We compare and contrast different types of linguistic features as well as different classification algorithms and explore the limitations of applying these techniques on a small dataset. We find that pre-trained Sentence-BERT embeddings to train an SVM classifier yields the best results with an F1-score of 0.79. Lastly, we discuss how this work provides insights into the types of linguistic features that are inherent in cognitive distortions.

## 1 Introduction

Cognitive Behavioral Therapy (CBT) is one of the most common methods of psycho-therapeutic intervention to treat depression or anxiety. Due to the COVID-19 pandemic, mental health issues are on the rise. At the same time, more and more interactions are now held virtually. Furthermore, mental health issues are not limited to the one-hour-per-week window that patients usually get with their therapists. This has led to a growth in the demand for digitally accessible therapy sessions. As mental health care is often inaccessible to people, there is a need for innovative ways to make it more widely available and affordable (Holmlund et al., 2019).

One possible solution is to develop an automated system that could serve by performing some ancillary tasks more efficiently. Towards that, Natural Language Processing (NLP) and Machine learning (ML) algorithms are now gaining widespread popularity and are being implemented in many fields where language is used. While we are far from a chatbot replacing a therapist's nuanced skillset, having easy access to an intelligent support system can help fill in these gaps.

One of the major aspects of CBT is to recognize and restructure certain types of negative thinking patterns. Some established negative thinking patterns are commonly observed in patients dealing with anxiety or depression. These cognitive distortions arise due to errors in reasoning (Beck, 1963). The aim of educating the patient about these distortions during CBT is to equip the patient with the right tools to detect errors in their own thought processes. Once the patient is aware of the error in their reasoning, they can start to work on restructuring how to perceive the same situations in a healthier way.

### 1.1 Cognitive Distortions

The concept of cognitive distortions was first introduced by Beck (1963). There is no definitive number of types of distortions, and the number varies widely in existing literature depending on the level of detail in reasoning considered by the author. For example, the Cognitive Distortion Scale developed by Briere (2000) consists of only five types. In this work, we consider a total of ten types of cognitive distortions that are described below:

1. **Emotional Reasoning**: Believing "I feel that way, so it must be true"
2. **Overgeneralization**: Drawing conclusions with limited and often un negative experience.
3. **Mental Filter**: Focusing only on limited negative aspects and not the excessive positive ones.
4. **Should Statements**: Expecting things or personal behavior should be a certain way.
5. **All or Nothing**: Binary thought pattern. Considering anything short of perfection as a failure.
6. **Mind Reading**: Concluding that others are reacting negatively to you, without any basis in fact.
7. **Fortune Telling**: Predicting that an event will always result in the worst possible outcome.



*Proceedings of the Seventh Workshop on Computational Linguistics and Clinical Psychology*, pages 151–158
June 11, 2021. ©2021 Association for Computational Linguistics

8. **Magnification**: Exaggerating or Catastrophizing the outcome of certain events or behavior.
9. **Personalization**: Holding oneself personally responsible for events beyond one's control.
10. **Labeling**: Attaching labels to oneself or others (ex: "loser", "perfect").

These distortions are based on the 10 types of cognitive distortion defined by Burns and Beck (1999). Some of these distortions are either combined into a super-category, or further divided into sub-categories, and hence the varying number of types of distortions. For example, mind reading and fortune telling are sometimes grouped and considered as a single distortion called Jumping to conclusions.

### 1.2 Problem statement

The first goal of this research project is to detect cognitive distortions from natural language text. This can be done by implementing and comparing different methodologies for binary classification of annotated data, obtained from mental health patients, into Distorted and Non-Distorted thinking. The second goal is to analyze the linguistic implications of classification tasks of different types of distortions.

In particular, this research aims to answer the following questions:

1. Which type of NLP features is more suitable for cognitive distortion detection: semantic or syntactic? Simply put, to compare *what* is said and *how* is it said in the context of this task. And, how important is word order in this context?
2. How well do these NLP features and ML classification algorithms perform this task with a limited-sized dataset?

### 1.3 Related work

Previous work done in this field includes the Stanford Woebot, which is a therapy chatbot (Fitzpatrick et al., 2017). The dialogue decision in Woebot is primarily implemented using decision trees. It functions on concepts based on CBT including the concept of cognitive distortions. However, it only outlines several types of distortions for the user and leaves the user to identify which one applies to their case.

Another study established a mental health ontology based on the principles of CBT using a gated-CNN mechanism (Rojas-Barahona et al., 2018). The model associated certain thinking errors (cognitive distortions) with specific emotions and situations. Their study uses a dataset consisting of about 500k posts taken from a platform that is used for peer-to-peer therapy. The distribution of types of distortion is very similar to our results. These tasks come with annotator agreement issues - their inter-annotator agreement rate was 61%. One possible reason for the low agreement rate given by the authors is the presence of multiple distortions in a single data point.

As there is a lack of publicly available structured data that was curated specifically for the detection of cognitive distortions, datasets from other domains, such as social media data or personal blogs are used instead. One such study was conducted on Tumblr data collected by using selected keywords (Simms et al., 2017). By using the LIWC features (Section 3.3) to train a Decision Tree model to detect the presence of cognitive distortions, they were able to lower the false positive rate to 24% and the false-negative rate to 30.4%.

A similar study was conducted by Shickel et al. (2020) on a crowdsourced dataset and some mental health therapy logs. Their approach was to divide the task into two sub-tasks - first to detect if an entry has a distortion (F1-score of 0.88) and second to classify the type of distortion (F1-score of 0.68). For this study, 15 different classes are considered for the types of distortion. For both of the tasks - logistic regression outperformed more complex deep learning algorithms such as Bi-LSTMs or GRUs. On applying this model to smaller counseling datasets, however, the F1-score dropped down to 0.45.

## 2 Methods and Dataset

One of the most common roadblocks in using Artificial Intelligence for Clinical Psychology is the lack of available data. Most of the datasets that have patients interacting with licensed professionals are confidential and therefore not publicly available.

Here, we use a dataset, named Therapist Q&A, obtained from the crowd-sourced data science repository, Kaggle[1]. The dataset follows a Question and Answer format and the identity of each patient is anonymized, to maintain their privacy.

Each patient entry usually consists of a brief description of their circumstance, symptoms, and

[1]https://www.kaggle.com/arnmaud/therapist-qa

their thoughts. Each of these concerns is then answered by a licensed therapist addressing their issues followed by a suggestion. Since the patient entry is not just a vague request and it provides some insight into the situation as well as their reaction to it, it can be used to detect if they were engaging in any negative thinking patterns.

### 2.1 Annotation of dataset

For the annotation task, we have just focused on the patient's input. One of the key factors in detecting cognitive distortions is context. While the data does give some insight into the situation a patient is in, it should be noted that the description itself is given by the patient themselves. As a result, their version of the situation itself may be distorted.

In this task, we focus on detecting cues in language that would indicate any type of distortion and there was no way to verify the veracity of their statements. Thus each entry is perceived as a viable candidate for cognitive distortion and given one out of 11 labels ('No distortion' and 10 different types of distortions as listed in section 1.1). It is noted that an entry can have multiple types of distortions. However for this project, the annotators were asked to determine a dominant distortion for each of the entries, and an optional secondary distortion if it is too hard to determine a dominant distortion. The decision between dominant or secondary distortion was made based on the severity of each distortion. Since the project aims to detect the presence of these distortions, the severity of distortions was not marked by any quantitative value. They were also asked to flag the sentences that led them to conclude that the reasoning was distorted.

The annotators coded 3000 samples out of which, 39.2% were marked as not distorted, while the remaining were identified to have some type of distortion. The highly subjective nature of this task makes it very hard to achieve a high agreement rate between the annotators. On comparing the dominant distortion of about 730 data points encoded by two annotators, the Inter-Annotator Agreement (IAA) for specific *type* of distortion was 33.7%. Considering the secondary distortion labels as well and computing a more relaxed agreement rate bumped the agreement to $\sim$ 40%. On the other hand, the agreement rate increased to 61% when we focus on distorted versus non-distorted thinking only. The IAA metric used here is the Joint Probability of Agreement. These disagreements were resolved by enabling the annotators to discuss their reasoning and come to a consensus. The types of distortion were found to be evenly distributed across the 10 classes of distortions mentioned earlier (figure 1). The annotated dataset will be made available to the public to encourage similar work in this domain.

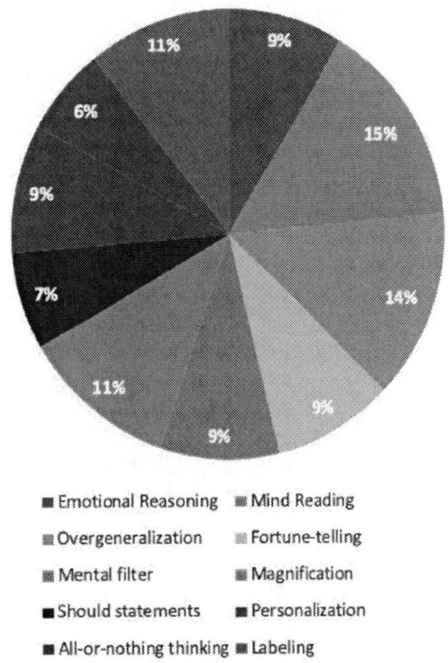


Figure 1: Distribution of the types of Cognitive Distortions in the Kaggle dataset

### 2.2 Experiments

Due to the limited size of the annotated dataset, several machine learning algorithms such as complex deep learning methods were eliminated from the experiments. Finally, the four types of features (Table 1) were tested using the following classification algorithms:

1. Logistic regression
2. Support vector machines
3. Decision trees
4. K- Nearest Neighbors (k = 15)
5. Multi-Layer Perceptron (with a single hidden layer having 100 units)

All of these classification algorithms were implemented with the default hyper-parameter settings using the python package commonly used for ML algorithms, scikit-learn [2].

## 3 Feature Selection

To address the different aspects of language, feature selection was divided into two categories - Semantic and Syntactic features. Two different training approaches were implemented for each of these categories. A brief description of each training method is given below.

| | Bag-of-words approach | Sequential approach |
|---|---|---|
| Semantic | SIF | S-BERT |
| Syntactic | LIWC | POS |

Table 1: Types of linguistic features. Note that LIWC features are not limited to the Syntactic category.

### 3.1 Smooth Inverse Frequency (SIF)

There are multiple ways of encoding Sentence embeddings where the word order does not matter. One of the most common methods is simply using the mean value of all the word embeddings.

Another common approach is to treat these sentences as documents and use TF-IDF (Term Frequency - Inverse Document Frequency) vectors. However, the issue with treating sentences as documents is that sentences usually do not have multiple words repeated.

To address this, smooth inverse frequency (SIF) can be used instead. The SIF method for sentence embeddings improves the performance for textual similarity tasks, beating sequential deep learning models such as RNNs or LSTM (Arora et al., 2016).

Here, the sentence embeddings are generated using the SIF method on pre-trained GloVe embeddings (Pennington et al., 2014) for each word in the sentence.

### 3.2 Sentence-BERT (Bidirectional Encoder Representations from Transformers)

For the sequential semantic representation of these entries, a pre-trained sentence-BERT model was used (Reimers and Gurevych, 2019). To ensure that in this vector space, semantically similar sentences are closer, the authors have used Triplet Objective Function as the loss function. This triplet objective function minimizes the distance between the anchor sentence and a positive sample while maximizing the distance between the anchor sentence and a negative sample.

### 3.3 Linguistic Inquiry and Word Count (LIWC) Features

The linguistic inquiry and word count (LIWC) is a tool used to analyze textual data (Pennebaker et al., 2001). The LIWC program generates about 80 features based on the words used in the text. While we categorize the LIWC features as syntactic in table 1, these features reflect the percentage of words in different categories. A lot of these features are syntactic, such as the count of pronouns, proper nouns, etc. Other categories are psychological, linguistic, cognitive, or other (Tausczik and Pennebaker, 2010).

LIWC features are widely used for conducting linguistic analysis in almost any domain. Specific to mental illness, these features were used to detect the linguistic indicators of Schizophrenia (Zomick et al., 2019), Depression (Jones et al., 2020) and even Cognitive Distortions (Simms et al., 2017).

### 3.4 Parts of Speech (POS) tag embeddings

The main motivation behind using Parts of speech tags was to eliminate any specific Noun or Verb from heavily dominating the classification process. Two entries having the same context can have different distortions. Using POS tags as features have proved to be useful for similar applications, such as detecting depression from text (Morales and Levitan, 2016).

Syntactic features generally do not consider word order as an important aspect. To maintain the impact of word order each word is replaced with its Part-Of-Speech (POS) tag [3] using the pre-trained Spacy language model [4].

These POS tags are then converted to embeddings by similarly training them as word embeddings using Skip-gram word2vec model (Mikolov et al., 2013). This is done to encode POS tag-order in the embeddings. Once each tag has an embedding, these vectors are padded with zeros for normalization.

[2]https://scikit-learn.org

[3]https://universaldependencies.org/docs/u/pos

[4]https://spacy.io/usage/linguistic-featurespos-tagging

## 4 Results and Discussion

### 4.1 Detecting Cognitive Distortion

The task of detecting cognitive distortions is treated as a binary classification problem here. From the F1 scores given in Table 2, we can see that the SVM outperforms all the other candidate algorithms. All types of features were found to be performing best with Support vector machines.

| | SIF | BERT | LIWC | POS | BERT + LIWC |
|---|---|---|---|---|---|
| **Log. reg.** | 0.75 | 0.74 | 0.77 | 0.73 | 0.74 |
| **SVM** | 0.77 | **0.79** | 0.78 | 0.77 | 0.76 |
| **Decision Tree** | 0.65 | 0.67 | 0.67 | 0.66 | 0.64 |
| **k-NN** | 0.74 | 0.75 | 0.76 | 0.75 | 0.75 |
| **MLP** | 0.73 | 0.70 | 0.77 | 0.72 | 0.74 |

Table 2: The F1 scores on testing each type of features mentioned above on a 80-20 training test data split.

SIF embeddings perform very similarly to the sentence BERT embeddings. This indicates that the word order might not give much insight for this task when it comes to the semantic features.

The LIWC features, while comparable, always perform slightly better than the POS tags as features. As the POS tag embeddings have the word order encoded it, whereas LIWC features (be it semantic or syntactic) do not, this reinforces our conclusion that the word order does not contribute much to the classification task.

To get the best of semantic as well as syntactic insights, we tried a hybrid model that combines these features. This method yielded strikingly similar results to the other tests. For example, the combination of best performing semantic as well as syntactic features, i.e. S-BERT with LIWC features, still yields the highest F1-score of 0.76 by using SVM. This result may be because the combined model tends to overgeneralize in training, which in turn results in a slight decrease in performance on the test set.

### 4.2 Detecting the Type of Cognitive Distortion

While the aforementioned results show good performance in detecting the presence of cognitive distortions, detection of the type of distortion fails to yield good results. None of the algorithms mentioned above got a weighted F1-score more than 0.30. This could also be attributed to a poor IAA rate of $\sim 34\%$ which creates an upper bound for the performance in this task. Despite the discouraging classification results, we can draw some meaningful conclusions based on these experiments.

One way to test if semantically similar sentences tend to have the same type of distortion was to use k-Nearest Neighbors (k-NN) on the semantic embeddings using cosine similarity. When applied to the sentence-BERT embeddings, using k-NN on the multi-class classification problem yields 24% accuracy in the best case scenario as shown in figure 2.

Here, 'the count-based k-NN' would simply count the most redundant class of the 'k' nearest neighbors of a new data point, and then classify it as the same distortion. Whereas, the 'probability based' model applies more weight to the entries which were semantically closer to the data point in question. Both of these models perform best at $k \geq 15$. At lower values, however, the count-based model performs slightly better than the probability-based model. So, we can conclude that semantically similar sentences do not have the same cognitive distortions.

Focusing on the syntactic features, if we analyze the behavior of these distortions based on their POS tags we can draw some conclusions about the type of language used for these distortions (figure 3). For example, the distortion "labeling" had a higher probability of having Adjectives, interjections, and Punctuations. The distortion "mind-reading" has a higher probability of having Pronouns, more specifically 3rd-person pronouns. Both of these examples are in accordance with the definition of the respective distortions.

On the other hand, some findings are more unexpected. The expectation with the "should statement"

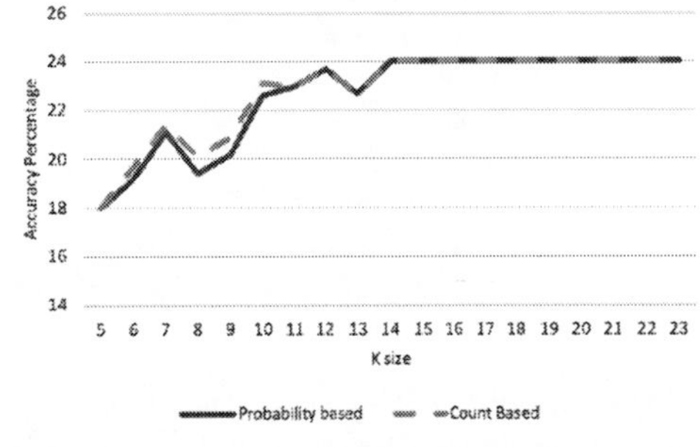


Figure 2: Performance of k-Nearest Neighbors as a multi-class classifier for Cognitive Distortions

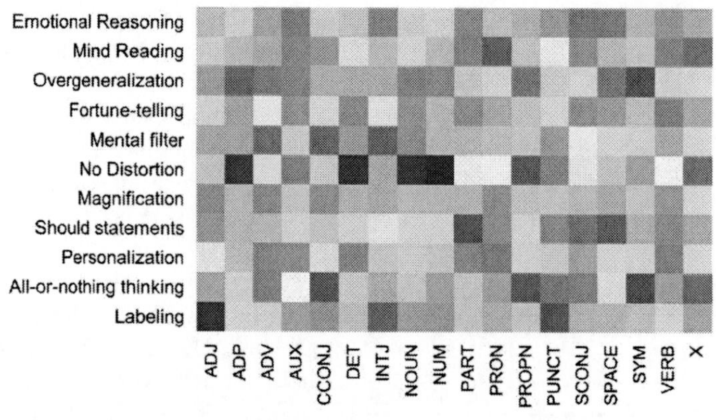


Figure 3: Heatmap showing normalized frequency of each POS tag used in different types of distortion. The darker colors indicates higher than normal frequency and the lighter colors indicate lower than normal frequency for a particular tag in the corresponding distortion.

was to have a higher probability of having auxiliary verbs such as 'should', 'must', 'ought to' etc. However, the results show that should statement have a lower than average probability of having auxiliary verbs. An example of this distortion without using any of the words listed above could be "While others my age are busy with their jobs and life I am just wasting my time".

Unsurprisingly, entries having no cognitive distortions usually behave very differently than the mean behavior of distorted data (and hence the high F1-scores for the binary classification task). This can also be supported by the analysis of the LIWC features, more than 50% of the features do not conform to the patterns exhibited by other distorted entries. In addition to having the lowest score on the LIWC features - 'feel', 'perception',

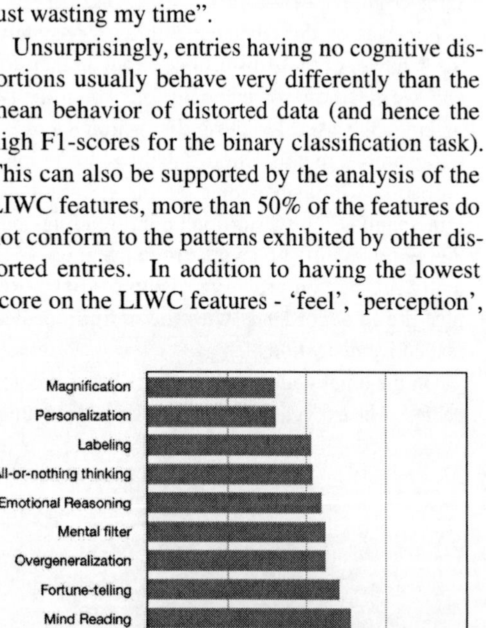


Figure 4: Normalized z-scores calculated for the types of cognitive distortion for each of 93 features of LIWC. Higher magnitude of z-score indicates higher deviation from the norm.

'insight', 'negative emotion', 'risk' and 'reward'; The non-distorted entries also tend to have more Adpositions, Determiners, Nouns, and Numerals, which indicates low subjectivity (Sahu, 2016).

Conducting a similar analysis on the LIWC features, we can conclude that some types of distortions are easier to detect than others. While most of the features of the entries conform to a mean pattern, some of the distortions deflect in behavior for specific features. For example, Fortune-telling distortion has the highest score for 'focus future', emotional reasoning has the highest 'feel' score, and so on.

Figure 4 shows a visual representation of how difficult it is to classify a certain distortion. The x-axis shows the magnitude of deviation (normalized z-score) from the mean behavior. The higher deviation from mean behavior, the easier it would be to classify that label using the LIWC features. This was done by calculating the z-score for each feature to quantify how far is the data point from the mean. The mean behavior here represents the average LIWC features expected from a natural language entry by a patient in the context of this study. This analysis is consistent with the finding that it's easier to detect 'No distortion' than any of the specific types of distortions since the 'no distortion' category shows maximum deviation from the mean behavior.

## 5 Conclusion

In this work, we compare and contrast the performance of five classification algorithms in detecting cognitive distortion.

We find that the task of determining whether or not an input indicates distorted thinking is computationally feasible, wherein semantic as well as syntactic features perform equally well. The order of words was found to have no impact on the results. Entries with cognitive distortions tend to be more subjective than non-distorted entries. The best classification results were obtained by SVM using pre-trained S-BERT embeddings with an F1 score of 0.79.

Regarding the task of identifying the type of distortion, we found that semantically similar entries do not always get categorized as the same distortion. Some of the distortions are easier to classify than others, e.g. 'should statements', 'mind reading', 'fortune-telling' etc. None of the implemented ML techniques obtained an F1 score higher than 0.30 on the classification of each type of cognitive distortion.

A challenging aspect of this research is getting a standardized inter-annotator agreement. One reason for that is a lack of clear distinction in psychology literature itself wherein some of these distortions are sometimes grouped as one. Another reason for this could be the presence of multiple distortions in a single patient entry (Rojas-Barahona et al., 2018).

As with the clinical application of detection algorithms, there are some ethical risks to keep in mind. If the algorithm is implemented as an unregulated flagging system, the false negatives would go undiagnosed and the false positives would be put through an unnecessary position of second-guessing their cognitive capabilities. However, 100% accuracy of classification from a single interaction (as used for training here) may not be needed for such clinical applications. If this were to be implemented in a dialogue system, an ongoing conversation with the participant will serve to make the system more accurate and personalized. As the main goal is to develop effective feedback to help any participants, having less than perfect predictions is still valuable in informing the types of feedback that an automated clinical tool could provide to the participant.

Lastly, we discuss several applications of this work in the mental healthcare sector. It could be used to flag or screen people for referrals to mental health care providers. Likewise, it could also be used in tandem with the diagnosis to establish an estimate of the severity of the anxiety or depression. This approach might also be useful in detecting delusions or paranoia as well as suicide risk in natural language. Lastly, the measure of a patient's distorted thinking can be used as an indicator of remission which can be used to determine which therapy techniques (or therapists, from the perspective of insurance companies) are more effective. In conclusion, this tool can be adapted for applications in mental health screening, diagnosis, and tracking treatment effectiveness.

## 6 Future work

This is an ongoing project with the ultimate goal to implement feedback to support CBT through the detection of cognitive distortions. Our next step is to implement a multi-class classification framework to improve the type of distortion detection accuracy. Once this study is complete, the annotated dataset will be made available to the public to encourage similar work in this domain.

The annotators have also identified and flagged specific parts of sentences wherein the negative thinking patterns were most evident. We can then train a classification model by using algorithms such as IOB (inside-outside-between) type tagging which can pinpoint the errors in a patient's reasoning that give rise to cognitive distortions.

## Acknowledgements

We thank the annotators: Rebecca Lee, Josh Daniels, Changing Yang, and Beilei Xiang; and Prof. Martha Palmer for funding this work. We also acknowledge the support of the Computational Linguistics, Analytics, Search and Informatics (CLASIC) department at the University of Colorado, Boulder in creating an interdisciplinary environment that is critical for research of this nature. Lastly, the critical input from three anonymous is gratefully acknowledged in improving the quality of this paper.

## References

Sanjeev Arora, Yingyu Liang, and Tengyu Ma. 2016. A simple but tough-to-beat baseline for sentence embeddings.

Aaron T Beck. 1963. Thinking and depression: I. idiosyncratic content and cognitive distortions. *Archives of general psychiatry*, 9(4):324–333.

John Briere. 2000. *Cognitive Distortions Scale (CDS) Professional manual.*

David D Burns and Aaron T Beck. 1999. Feeling good: The new mood therapy.

Kathleen Kara Fitzpatrick, Alison Darcy, and Molly Vierhile. 2017. Delivering cognitive behavior therapy to young adults with symptoms of depression and anxiety using a fully automated conversational agent (woebot): a randomized controlled trial. *JMIR mental health*, 4(2):e19.

Terje B Holmlund, Peter W Foltz, Alex S Cohen, Håvard D Johansen, Randi Sigurdsen, Pål Fugelli, Dagfinn Bergsager, Jian Cheng, Jared Bernstein, Elizabeth Rosenfeld, et al. 2019. Moving psychological assessment out of the controlled laboratory setting: Practical challenges. *Psychological assessment*, 31(3):292.

Lauren Stephanie Jones, Emma Anderson, Maria Loades, Rebecca Barnes, and Esther Crawley. 2020. Can linguistic analysis be used to identify whether adolescents with a chronic illness are depressed? *Clinical psychology & psychotherapy*, 27(2):179–192.

Tomas Mikolov, Kai Chen, Greg Corrado, and Jeffrey Dean. 2013. Efficient estimation of word representations in vector space.

Michelle Renee Morales and Rivka Levitan. 2016. Speech vs. text: A comparative analysis of features for depression detection systems. In *2016 IEEE spoken language technology workshop (SLT)*, pages 136–143. IEEE.

James W Pennebaker, Martha E Francis, and Roger J Booth. 2001. Linguistic inquiry and word count: Liwc 2001. *Mahway: Lawrence Erlbaum Associates*, 71(2001):2001.

Jeffrey Pennington, Richard Socher, and Christopher Manning. 2014. GloVe: Global vectors for word representation. In *Proceedings of the 2014 Conference on Empirical Methods in Natural Language Processing (EMNLP)*, pages 1532–1543, Doha, Qatar. Association for Computational Linguistics.

Nils Reimers and Iryna Gurevych. 2019. Sentence-bert: Sentence embeddings using siamese bert-networks. *arXiv preprint arXiv:1908.10084*.

Lina Rojas-Barahona, Bo-Hsiang Tseng, Yinpei Dai, Clare Mansfield, Osman Ramadan, Stefan Ultes, Michael Crawford, and Milica Gasic. 2018. Deep learning for language understanding of mental health concepts derived from cognitive behavioural therapy. *arXiv preprint arXiv:1809.00640*.

Ishan Sahu. 2016. *A study on detecting fact vs non-fact in news articles*. Ph.D. thesis, Indian Statistical Institute, Kolkata.

Benjamin Shickel, Scott Siegel, Martin Heesacker, Sherry Benton, and Parisa Rashidi. 2020. Automatic detection and classification of cognitive distortions in mental health text. In *2020 IEEE 20th International Conference on Bioinformatics and Bioengineering (BIBE)*, pages 275–280. IEEE.

T Simms, C Ramstedt, M Rich, M Richards, T Martinez, and C Giraud-Carrier. 2017. Detecting cognitive distortions through machine learning text analytics. In *2017 IEEE international conference on healthcare informatics (ICHI)*, pages 508–512. IEEE.

Yla R. Tausczik and James W. Pennebaker. 2010. The psychological meaning of words: Liwc and computerized text analysis methods. *Journal of Language and Social Psychology*, 29(1):24–54.

Jonathan Zomick, Sarah Ita Levitan, and Mark Serper. 2019. Linguistic analysis of schizophrenia in Reddit posts. In *Proceedings of the Sixth Workshop on Computational Linguistics and Clinical Psychology*, pages 74–83, Minneapolis, Minnesota. Association for Computational Linguistics.

# Evaluating Automatic Speech Recognition Quality and Its Impact on Counselor Utterance Coding

**Do June Min, Verónica Pérez-Rosas, Rada Mihalcea**
Department of Electrical Engineering and Computer Science
University of Michigan, Ann Arbor, MI, USA
dojmin@umich.edu, vrncapr@umich.edu, mihalcea@umich.edu

## Abstract

Automatic speech recognition (ASR) is a crucial step in many natural language processing (NLP) applications, as often available data consists mainly of raw speech. Since the result of the ASR step is considered as a meaningful, informative input to later steps in the NLP pipeline, it is important to understand the behavior and failure mode of this step. In this work, we analyze the quality of ASR in the psychotherapy domain, using motivational interviewing conversations between therapists and clients. We conduct domain agnostic and domain-relevant evaluations using evaluation metrics and also identify domain-relevant keywords in the ASR output. Moreover, we empirically study the effect of mixing ASR and manual data during the training of a downstream NLP model, and also demonstrate how additional local context can help alleviate the error introduced by noisy ASR transcripts.

## 1 Introduction

Evaluating the quality of psychotherapy is an essential step in assessing the fidelity of treatment and providing feedback to practitioners. In psychotherapy practice, this is usually done through a process called behavioral coding that consists of manually analyzing recordings of therapy conversations and then labeling specific behaviors from participants.

Recent efforts have addressed the automatic analysis and evaluation of psychotherapy quality, including the study of conversational dynamics between therapists and clients, the analysis of empathy and emotional responses, and the automatic assessment of therapist's skills (Althoff et al., 2016; Zhang and Danescu-Niculescu-Mizil, 2020a; Pérez-Rosas et al., 2017).

Most of these research studies have been conducted using small collections of manually transcribed counseling conversations due to the need of an accurate representation of what is being said during the conversation. However, the use of manual transcription restricts the inclusion of a larger number of conversations into the analysis as it is a costly and slow process, making it challenging to apply data hungry machine learning approaches. As an alternative, some studies have explored the use of automatic speech recognition (ASR) systems that are able to quickly transcribe a large number of conversations (Flemotomos et al., 2021). However, there are several open questions regarding the feasibility of using automatic transcriptions in the evaluation of psychotherapy (Miner et al., 2020).

In this work, we study the quality of ASR in counseling conversations and its impact on the task of behavioral coding. We use an existing dataset of behavioral counseling conversations consisting of audio recordings and manual transcriptions as well as annotations of ten behaviors related to therapists' counseling skills. We start by generating automatic transcriptions using a commercially available ASR system (Google, 2020). Using the resulting parallel corpus of manual and ASR transcriptions, we conduct an assessment of the ASR quality using three main approaches. First, we use automatic evaluation metrics such as word error rate (WER) and semantic distance to conduct domain agnostic evaluations of the ASR performance across conversation participants. Second, we conduct a domain-specific examination of the ASR output by identifying domain-relevant keywords using behavioral codes and keywords identified using the Linguistic Inquiry and Word Count (LIWC) (Pennebaker et al., 2001). Finally, we study the effect of the noisy ASR on the downstream behavioral coding task and empirically show that additional local context in the form of neighboring utterances can help alleviate the impact of ASR errors.

We believe that studying the role of ASR systems in the NLP pipeline is an important step to develop and evaluate robust systems for better understanding of counseling dialogues.



*Proceedings of the Seventh Workshop on Computational Linguistics and Clinical Psychology*, pages 159–168
June 11, 2021. ©2021 Association for Computational Linguistics

## 2 Related Work

As the overall accuracy of ASR systems keeps improving, the ability of producing accurate transcriptions of conversational data has enabled the development of NLP applications in health. Particularly, in the psychotherapy domain, where a large fraction of therapy sessions are conducted in spoken language, ASR can help reduce the burden of manual transcription, potentially allowing for large-scale analysis of interactions between counselors and patients.

There have been several efforts on applying NLP on conversation analysis and utterance coding tasks in the psychotherapy domain. NLP was used to evaluate counselor behaviors and strategies (Zhang and Danescu-Niculescu-Mizil, 2020b; Pérez-Rosas et al., 2019; Xiao et al., 2015), or to provide feedback by generating appropriate responses to client utterances (Shen et al., 2020).

While most of previous work was conducted on manual transcriptions, there are only a few cases where automatically generated transcripts have been used, limiting the use of computational methods in psychiatry (Imel et al., 2015). The main reason behind this is the need for reliable ASR systems that are able to produce accurate transcriptions as the error introduced by transcribing words incorrectly can have a great impact on the performance of the overall application.

It has been pointed out by previous research that automatic evaluation metrics such as word error rate alone are not a good indicator of accuracy in speech understanding (Park et al., 2008). Our work is similar to Miner et al. (2020) recent work in that we use both agnostic and domain-relevant approaches to assess ASR systems in the mental health domain. However, we additionally investigate how the ASR error, both domain-agnostic and domain-relevant, propagates through the common NLP pipeline, in training and inference times, and provide an advice for researchers.

Finally, Mani et al. (2020) recently framed post-processing ASR error correction as a machine translation task from noisy transcription to ground truth transcription, and trains a sequence to sequence error correction model. Although this approach can provide a modular solution to mitigate ASR errors in many speech understanding systems, we note that building such a parallel corpus can be prohibitive for many researchers.

| | Average | Std |
|---|---|---|
| Session Length | | |
| Duration (min) | 21.03 | 9.33 |
| Length (words) | 3320.02 | 1494.68 |
| Words Spoken per Session ($n$) | | |
| Therapist | 2002.24 | 1024.63 |
| Client | 1317.77 | 858.25 |

Table 1: Session statistics

## 3 Dataset

### 3.1 Data Source

We evaluate utterances and behavioral codes from 213 counseling sessions compiled by Pérez-Rosas et al. (2016). The sessions were originally drawn from various sources, including two studies on smoking cessation and medication adherence. The full set comprises a total of 97.8 hours of audio with average session duration of 20.8 minutes. All the sessions were manually anonymized to remove identifiable information such as counselor and patient names and references to counseling sites' location. The sessions were transcribed using manual and crowd-sourced methods. The transcription set consist of 707,165 words distributed across 52,658 utterances and 39,637 talk-turns. More detailed statistics on words and utterances per session are provided in Table 1. The average conversation in the dataset has a duration of 21 minutes and a length of 3320 words.

The dataset also includes utterance-level annotations for ten behavioral codes from the Motivational Interviewing Treatment Integrity (MITI) coding scheme, the current gold standard for evaluating MI fidelity. MITI is focused on therapist language only and measures how well the therapist adhered to MI strategies by counting behaviors such as asking questions, using reflective language, seeking collaboration and emphasizing autonomy, among others. The dataset annotations were conducted by annotators with previous MI experience and trained on the use of MITI system. In addition to the MITI coding, our study uses two additional categories for utterances that are not labeled in the original dataset. The first includes therapist's speech that is not labeled under any MITI code (NAT) and the second includes client's utterances (NAC). Table 2 list the different behavioral codes, their count and their average word length.

| Code | Count | Avg Len. |
|---|---|---|
| Question (QUEST) | 6269 | 14.55 |
| Simple reflection (SR) | 2564 | 14.33 |
| Complex reflection (CR) | 3354 | 16.95 |
| Seeking collaboration (SEEK) | 927 | 20.34 |
| Emphasizing autonomy (AUTO) | 170 | 17.68 |
| Affirm (AF) | 550 | 17.71 |
| Confront (CON) | 139 | 12.97 |
| Persuading without permission (PWOP) | 1046 | 20.62 |
| Persuading with permission (NPWP) | 378 | 20.27 |
| Giving Information (NGI) | 1894 | 20.59 |
| Non-coded Therapist (NAT) | 12814 | 10.60 |
| Non-coded Counselor (NAC) | 22553 | 12.63 |

Table 2: Statistics for MITI behaviors coded in the dataset

### 3.2 Preprocessing

**Alignment.** Since the manual transcriptions provided in the dataset consist of transcribed speech without corresponding timestamps, we used forced alignment to automatically align speakers' speech with its corresponding transcription. We used *Gentle* (Ochshorn and Hawkins), a forced speech aligner implemented using the Kaldi toolkit for speech recognition (Povey et al., 2011). Note that this is a necessary step to enable comparisons between manual and automatic transcriptions for the same audio segments.

**Automatic Transcription.** To automatically transcribe each counseling session, we first spliced its audio into smaller segments using the obtained timestamps. Next, we individually transcribed each segment using the Google's Speech-to-Text recognition system (Google, 2020).[1] Again, our choice of transcribing segments rather than full conversations is motivated by the need of comparable units so we can avoid potential misalignment generated by ASR segmentation.

## 4 Domain-agnostic Evaluation

We start by conducting a domain-agnostic evaluation of the automatic transcription process that considers that the accuracy of the ASR system is equally important for all speech in the conversation. To this end, we focus on two automatic evaluation metrics: word error rate and semantic distance. The first one evaluates transcription error at the word-level; the second one aims to evaluate transcription error considering the semantic distance between the ASR output and the ground truth i.e., human transcription.

**Word Error Rate (WER)** We calculate WER using the equation below, where $S, D, I$ each denote the number of substitutions, deletions, and insertions respectively required to make the reference sequence identical to the ASR sequence. $C$ refers to the number of correct words, whereas $N$ is the number of words in the reference.

$$WER = \frac{S+D+I}{S+D+C} = \frac{S+D+I}{N} \quad (1)$$

We use the Python Jiwer package[2] to automatically calculate WER for all conversations in the dataset. Our calculations are done by aggregating transcriptions by the corresponding speaker and averaging across sessions.

**Semantic Distance.** Although recent works show that averaging WERs over large benchmark sets can provide good estimation of model performance (Likhomanenko et al., 2020), there have been criticisms against relying solely on WERs, on the grounds that some important aspects of transcription quality are ignored when focusing on word overlaps (Kong et al., 2016; Szymański et al., 2020). For instance, "This is a cap" and "This is a cat" will have a low score of WER because of the low edit distance between the sentences, while their semantic contents are about two distant concepts (Kim et al., 2021). We use semantic distance to complement WER as semantics play an important role in understanding psychotherapy language and the meaning of a particular utterance could be greatly affected by substitutions done during the ASR process.

More specifically, we measure the difference in semantic content between the ground truth and ASR transcriptions. Our calculations are conducted at the utterance level and aggregated overall all conversations. We define the semantic distance between a manually transcribed utterance $Utt_{MAN}$ and an automatically transcribed utterance $Utt_{ASR}$ as the cosine distance between the sentence embeddings of each utterance:

$$\text{Semantic Distance}(Utt_{MAN}, Utt_{ASR}) = 1 - \frac{emb(Utt_{MAN}) \cdot emb(Utt_{ASR})}{\|emb(Utt_{MAN})\| \|emb(Utt_{ASR})\|} \quad (2)$$

Thus, lower semantic distance between a manual transcription and an ASR transcription would indicate lower degree of transcription error.

[1] We use the Google Cloud speech-to-text enhanced model

[2] https://pypi.org/project/jiwer/

| | $n$ | WER | Semantic Distance |
|---|---|---|---|
| Aggregated | 426 | 0.35± 0.09 | 0.28±0.06 |
| | | Speaker Role | |
| Therapist | 213 | 0.35±0.10 | 0.27±0.07 |
| Client | 213 | 0.40±0.16 | 0.30±0.10 |
| | | Speaker Gender | |
| Female | 344 | 0.35±0.09 | 0.27±0.06 |
| Male | 82 | 0.46±0.17 | 0.34±0.11 |
| | | Therapist Gender | |
| Female | 195 | 0.34±0.09 | 0.27±0.07 |
| Male | 18 | 0.40±0.11 | 0.31±0.05 |
| | | Client Gender | |
| Female | 149 | 0.37±0.14 | 0.28±0.08 |
| Male | 64 | 0.48±0.18 | 0.35±0.11 |

Table 3: WER and Semantic Distance statistics by speaker role and gender for manual and automatic transcriptions. Plus and minus values denote standard deviation.

| Code | WER | Semantic Distance |
|---|---|---|
| AF | 0.36±0.23 | 0.18±0.16 |
| AUTO | 0.34±0.29 | 0.18±0.17 |
| CON | 0.38±0.40 | 0.13±0.12 |
| CR | 0.32±0.14 | 0.18±0.16 |
| NGI | 0.33±0.27 | 0.16±0.15 |
| NPWP | 0.35±0.57 | 0.17±0.16 |
| PWOP | 0.29±0.14 | 0.15±0.14 |
| QUEST | 0.31±0.19 | 0.18±0.17 |
| SEEK | 0.32±0.43 | 0.17±0.15 |
| SR | 0.36±0.19 | 0.20±0.18 |
| NAT | 0.48±0.20 | 0.37±0.26 |
| NAC | 0.40±0.16 | 0.30±0.10 |

Table 4: WER and Semantic Distance statistics for ten MITI codes and non-annotated utterances in the dataset by therapists (NAT) and clients (NAC). Plus and minus values denote standard deviation.

For the $emb(\cdot)$ function we use sentence transformer embeddings (Reimers and Gurevych, 2019). We chose the sentence transformer over alternative methods of sentence embeddings such as BERT or word2vec, since recent research has shown that off-the-shelf transformer models without fine-tuning often lead to representations that perform poorly on semantic similarity tasks (Li et al., 2020).

### 4.1 Results

Table 3 summarizes the results obtained by speaker's role (i.e., therapist, client) and gender (i.e., male, female). Overall, transcription of therapist's speech shows significantly lower error than client speech in terms of WER, but not on semantic distance (two tailed Mann-Whitney U-test, $p < .05$). We also observe significant differences in female and male speech recognaition for both WER and semantic distance ($p < .05$, two tailed Mann-Whitney U-test). The difference between genders is also confirmed when the speaker roles are considered. This result is aligned with previous findings that ASR systems tend to perform better on female speakers due to being more consistent to standard pronunciations than male speakers (Adda-Decker and Lamel, 2005; Goldwater et al., 2008). However, it is important to mention that other work on ASR evaluation have encountered the opposite trend, where transcription of female speakers speech obtained higher WER than of males (Tatman, 2017). A factor that potentially affected our analysis is that due to the unavailability of identity data for speakers in the dataset, we treated each session as featuring a unique set of speakers. This might have been caused by the over-representation of speakers who appear multiple times in the dataset.

## 5 Domain-relevant Evaluation

Although the domain-agnostic evaluation can provide insights into the aggregate performance of an ASR system, a domain informed evaluation can help to better understand the quality of derived transcriptions and its potential impact on downstream tasks. In the counseling domain, incorrect transcription of words or phrases related to emotion, mental state, addiction, or medication can cause more harm than the incorrect transcription of other types of words. Seeking to evaluate the role of domain on ASR quality in our automatically transcribed conversations, we focus on speech that is relevant to counseling quality. To identify such speech, we use the behavioral coding provided in the dataset and also word categories from the Linguistic Inquiry and Word Count (LIWC) lexicon (Pennebaker et al., 2001).

**Behavioral codes.** We measure WER and semantic distance on utterances coded with the ten counselor behaviors included in the dataset and also examined transcription error in uncoded utterances from both, therapists and clients. For WER, we first concatenated all the utterances labeled with a given code in each single conversation, and then averaged the obtained WER across all conversations. Semantic distances for each utterance are averaged over all utterances in the dataset.

**LIWC Categories.** LIWC is a psycholinguistic lexicon that maps words and its stems to a set of categories related to psychological processes. There are 69 predefined categories that cover four high-level topics: psychological processes, personal concerns, linguistic dimensions, and linguistic fillers.

For our analysis, we identify and select a subset of categories from psychological processes and personal concerns as they have been found relevant to psychotherapy conversations. For words in the different categories appearing in the ground truth utterances, we evaluated whether the ASR system was able to correctly transcribed them. We calculate the true positive, false negative, and false positive rates as well the standard metrics of recall and precision.

### 5.1 Results

Table 4 shows the average WER and semantic distance of transcription for behavior codes and also for non-coded ("Non-coded Client", "Non-coded Therpapist") language in the conversations.

In general, we find that non-coded language tends to have higher transcription error than coded-language (two-tailed Mann-Whitney U-test, $p < 0.05$ for both WER and semantic distance). Within non-applicable codes, we note that NAT shows higher WER and semantic distance. Since in Table 3 we saw that client language tends to have higher error overall than therapist language, this may indicate that transcription error is correlated to speech content or topic, because NAC covers all client utterances, while NAT is only applied for non-MITI labeled utterances.

When the ASR system is evaluated in terms of transcribing keywords that are relevant to psychotherapy and counseling, results from Table 5 indicate that correctly retrieving keywords is harder for ASR systems than avoiding incorrect insertion of keywords in the transcription, as precision values are concentrated near 1.0, while recall values are more diverse. Table 6 gives an example of how omission errors can change the semantic content of the utterance for LIWC categories such as "DEATH, BODY". In the context of mental health and psychotherapy, these results suggest that aggregate metrics that compare whole ground truth utterances and ASR transcriptions to compute error rate are not granular enough to capture such cases of ASR failure where mistrancriptions of keywords might result in clinicians or counselors missing signs of patient distress or danger.

## 6 The Role of ASR on the Automatic Evaluation of Psychotherapy

Beyond studying the domain-agnostic and domain-relevant error patterns of the automatic transcription, we also study the relationship between the speech transcription step and the later behavior code classification, where ASR transcriptions are fed as input.

### 6.1 Model Performance

To explore whether the use of noisy ASR transcriptions affects the automatic evaluation of psychotherapy, we focus on a behavioral coding task where we seek to label participants' utterances into a set of predefined codes relevant to counseling quality using transcripts that are either manual or automatically generated.

We use the utterance-level annotations provided with the dataset described in Section 3, which consist of ten codes for therapist language plus two additional codes for annotated language from therapists and clients. We thus conduct a multi-label classification task to assign each utterance in the conversation to any of these 12 labels.

Our experiments are performed using a BERT model as our baseline classifier (Devlin et al., 2019) and our evaluated are conducted using 5-fold cross-validation. BERT is a transformer-based model that has been widely used in NLP. We chose this model since pretrained parameters fine-tuned on large natural language corpora are readily available, and also because due to its design the additional context input could easily supplied through the use of separate token type ids. We used the version implemented in (Wolf et al., 2020) with a learning rate of 2e-5. The input to the model is a sequence of token-level embeddings of each utterance in the conversation and the predicted label is assigned using a multilayer perceptron. The experiments are run on a GeForce RTX 2080 Ti.

We first conduct a set of experiments where we train and test multi-class utterance classifiers using either manual or automatic transcripts. In our first experiment, we aim to measure the model accuracy when using high quality training data i.e., manual transcripts for both, testing and training sets. Second, we substitute the train set for its automatically transcribed version and test on a manually transcribed set to evaluate the potential performance loss when training with noisy transcripts. Third, we again train on manual transcripts but this time test on automatic transcripts to evaluate whether a model built with accurate transcripts (i.e., produced by humans) would be effective while testing on transcriptions that are automatically obtained. Fi-

| Category | N | TP | FN | FP | Recall | Precision | Avg Word Len | Std Word Len |
|---|---|---|---|---|---|---|---|---|
| FAMILY | 926 | 827 | 99 | 8 | 89.31 | 99.04 | 5.10 | 1.38 |
| FEEL | 2470 | 2191 | 279 | 47 | 88.70 | 97.90 | 4.12 | 0.43 |
| POSFEEL | 8614 | 7568 | 1046 | 454 | 87.86 | 94.34 | 4.03 | 0.20 |
| HOME | 1550 | 1360 | 190 | 14 | 87.74 | 9.898 | 4.87 | 1.20 |
| LEISURE | 1966 | 1690 | 276 | 21 | 85.96 | 98.77 | 4.92 | 1.29 |
| JOB | 2077 | 1778 | 299 | 21 | 85.60 | 98.83 | 4.98 | 1.38 |
| OPTIM | 791 | 669 | 122 | 13 | 84.58 | 98.09 | 4.53 | 1.39 |
| SELF | 43100 | 36433 | 6667 | 3338 | 84.53 | 91.61 | 1.34 | 0.66 |
| SOCIAL | 54504 | 45866 | 8638 | 3907 | 84.15 | 92.15 | 3.31 | 1.02 |
| ANX | 268 | 224 | 44 | 2 | 83.58 | 99.12 | 6.07 | 3.00 |
| POSEMO | 20712 | 17152 | 3560 | 840 | 82.81 | 95.33 | 3.73 | 1.29 |
| ANGER | 412 | 341 | 71 | 8 | 82.77 | 97.71 | 4.18 | 1.71 |
| AFFECT | 23044 | 18993 | 4051 | 867 | 82.42 | 95.63 | 3.83 | 1.39 |
| BODY | 1721 | 1398 | 323 | 27 | 81.23 | 98.11 | 4.62 | 1.47 |
| PHYSCAL | 4042 | 3224 | 818 | 57 | 79.76 | 98.26 | 4.87 | 1.49 |
| MONEY | 674 | 534 | 140 | 4 | 79.23 | 99.26 | 4.95 | 1.30 |
| EATING | 2063 | 1633 | 430 | 27 | 79.16 | 98.37 | 5.35 | 1.78 |
| NEGEMO | 2130 | 1684 | 446 | 27 | 79.06 | 98.42 | 4.61 | 1.80 |
| SAD | 755 | 587 | 168 | 13 | 77.75 | 97.83 | 4.90 | 1.35 |
| SCHOOL | 492 | 379 | 113 | 2 | 77.03 | 99.48 | 5.04 | 1.40 |
| SLEEP | 212 | 163 | 49 | 2 | 76.89 | 98.79 | 3.99 | 1.01 |
| DOWN | 552 | 415 | 137 | 3 | 75.18 | 99.28 | 3.36 | 0.77 |
| DEATH | 152 | 112 | 40 | 1 | 73.68 | 99.12 | 3.72 | 0.73 |
| FRIENDS | 110 | 81 | 29 | 0 | 73.64 | 100 | 5.72 | 0.83 |
| SEXUAL | 253 | 186 | 67 | 5 | 73.52 | 97.38 | 4.04 | 0.45 |
| RELIG | 234 | 166 | 68 | 2 | 70.94 | 98.81 | 3.30 | 0.65 |

Table 5: Performance on LIWC-identified Keywords

| Category: DEATH, BODY / Error Type: Omission |
|---|
| Manual: And that's losing all the **weight**, and I really felt like I was **dying** |
| ASR: And to Annette loosen all the way. And I really felt like I was there. |
| Category: MONEY / Error Type: Insertion |
| Manual: Oh money to buy the cigarettes, and not to buy medicine Exactly Because it's expensive. |
| ASR: Money to buy cigarettes, but no **money** for the medicine exactly six months ago |

Table 6: Sample ASR errors for LIWC-identified keywords

nally, we evaluate a fully automatic pipeline, where both, train and test sets are obtained using ASR models. Results for these experiments are shown in Table 9.

## 6.2 Performance Trade-off

As results in Table 7 indicate, the choice of transcription method for both training and testing sets has a significant impact on the classification performance. Here, we see that even the model trained on the same manually transcribed training data can have drastically different reported performance, depending on the transcription method of the testing set. On the other hand, we also note that using ASR transcription as training set leads to a large decrease in performance when tested using manual testing data.

Since manual transcription is the most accurate representation of speech data, working with manual transcriptions would be the optimal choice. However, manual transcription can be expensive, especially for situations where a large amount of data has been collected. Thus, in many cases ASR technologies provide a faster and much more affordable transcription method. However, supervised learning with noisy ASR transcripts may result in the model learning spurious correlations, rather than the desired relationship between certain linguistic patterns and the predicted variables. This in turn leads to lower performances as shown in our experiments, where we observe performance losses up to 15%. Furthermore, consider a real case reported by Miner et al. (2020), where the word "depressed" was incorrectly transcribed into "the preston" in a self-harm counseling session. If an emotion detector were to be trained on the automatically transcribed data, the obvious correlation between "depressed" and "sad, blue" emotions will be lost, and replaced with a spurious one.

These considerations raise the question of what would be the best trade-off between the use of manual and automatic transcription methods in the psychotherapy domain.

To answer this question, we conduct a set of ex-

| Train | Test | Acc. | F-score | | | | | | | |
|---|---|---|---|---|---|---|---|---|---|---|
| | | | QUEST | CR | SR | NAT | NAC | SEEK | NGI | PWOP |
| Manual | Manual | 0.6940 | 0.6071 | 0.5334 | 0.0794 | 0.6919 | 0.8758 | 0.3058 | 0.5186 | 0.0048 |
| Automatic | Manual | 0.5520 | 0.4529 | 0.3642 | 0.0010 | 0.2587 | 0.7587 | 0.0815 | 0.3789 | 0.0127 |
| Manual | Automatic | 0.5289 | 0.4135 | 0.2483 | 0.0002 | 0.2263 | 0.7382 | 0.1060 | 0.2915 | 0.0173 |
| Automatic | Automatic | 0.5645 | 0.5268 | 0.3538 | 0.0020 | 0.2688 | 0.7765 | 0.1209 | 0.4341 | 0.0189 |

Table 7: Classification results for behavioral coding in MI sessions. AF, CON, NPWP, AUTO are not reported as their F-scores are zero

| % of Manual Data | Acc. | F-score | | | | | | | |
|---|---|---|---|---|---|---|---|---|---|
| | | QUEST | CR | SR | NAT | NAC | SEEK | NGI | PWOP |
| 0% | 0.5520 | 0.4529 | 0.3642 | 0.0010 | 0.2587 | 0.7587 | 0.0815 | 0.3789 | 0.0127 |
| 20% | 0.6173 | 0.5820 | 0.5053 | 0.0076 | 0.4360 | 0.8132 | 0.1821 | 0.4865 | 0.011 |
| 40% | 0.6734 | 0.5988 | 0.5225 | 0.0241 | 0.6397 | 0.8601 | 0.2981 | 0.4943 | 0.0276 |
| 60% | 0.6827 | 0.5966 | 0.5298 | 0.0336 | 0.6700 | 0.8678 | 0.2314 | 0.4996 | 0.0021 |
| 80% | 0.6914 | 0.6061 | 0.5340 | 0.0810 | 0.6866 | 0.8726 | 0.3073 | 0.5119 | 0.0534 |
| 100% | 0.6940 | 0.6071 | 0.5334 | 0.0794 | 0.6919 | 0.8758 | 0.3058 | 0.5186 | 0.0048 |
| Majority Class Classifier | 0.4321 | 0.0 | 0.0 | 0.0 | 0.6034 | 0.0 | 0.0 | 0.0 | 0.0 |

Table 8: Classification results for behavioral coding for incremental fraction of manual transcripts in training set. The majority class classifier outputs the majority label in the training dataset for each instance

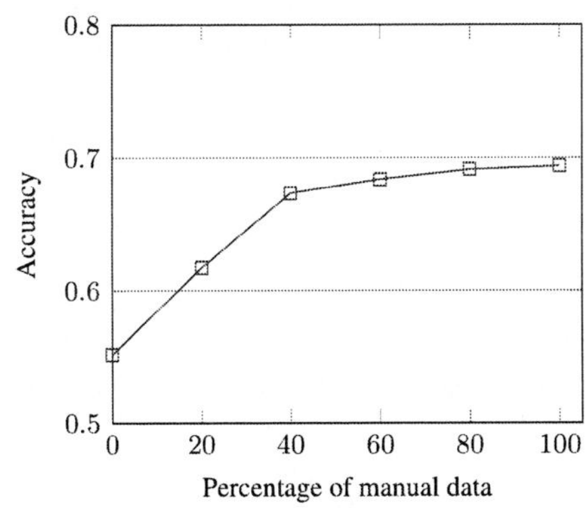


Figure 1: Classification accuracy as the fraction of manually transcribe data increases in the training set

periments where we gradually mix manually and automatically transcribed data during the training phase of the classification model. To ensure that the model is learning fairly, we ensured that each utterance only appears once in the entire dataset, without appearing both in the manual or ASR sets. By progressively adding more manual data in the training set, we emulate practical settings where only a fraction of data can be manually transcribed due to cost or time constrains. More specifically, we start with a full training set using ASR transcription, and increase the percentage of manual data at 20% increments. Note that reported accuracy is measured in a manually transcribed testing set.

As shown in Figure 1, the performance of the trained system does increase as the fraction of manual data increases. However, this is not shown as a linear relationship, as most of the performance gain occurs in the first few additions of the manual data. Although further study is warranted to explain how the small fraction of manual transcription leads to a noticeable increase in performance, this result indicates that even a small amount of manual transcription effort can improve the system performance in a meaningful way, and thus manual transcription is more cost-effective in its early stages than its later stages. For example, in the context of this experiment, practitioners can expect approximately 85% of the performance improvement of full manual transcription at the price of manually transcribing only 40% of the dataset.

### 6.3 Can (noisy) Local Context Help?

ASR error correction is an ongoing research topic in signal processing and natural language processing communities, and several techniques, including post-editing and domain adaptation, have been proposed (Mani et al., 2020). However, in this paper, we explore a simpler strategy based on context augmentation considering the distributional hypothesis in semantic theory, which states that words appearing in the same contexts tend to have similar meaning (Harris, 1954). We thus hypothesize that augmenting the target utterance with local context consisting of neighboring utterances can alleviate the effect of noisy transcription.

To this end, we compare BERT-based classifiers with different amounts of local context in addition to the target utterance (Devlin et al., 2019). The results shown in Table 9 are averaged over the re-

| | Accuracy | Macro F1 |
|---|---|---|
| No Context | 0.5645 | 0.2085 |
| Context = 1 | 0.5762 | 0.2297 |
| Context = 2 | 0.5772 | 0.2290 |

Table 9: Classification results for behavioral coding when using local context

sult of five-fold cross validation. The "No Context" model is given a single utterance as input, and the final label by computing softmax after the final linear layer. For the "Context = $n$" models, $n$ previous and following utterances surrounding the target utterance are also provided to the BERT model, as a concatenation. Note that through the use of separate token type ids, BERT allows practitioners to separately designate a sequence of context tokens, distinct from the target tokens. Overall, models that integrate context information outperform the base model in terms of average accuracy and Macro F1 with small but consistent performance gains, thus suggesting that the system's performance can be improved using this simple strategy as opposed to conducting expensive manual transcription.

## 7 Limitations

Our work has several limitations that should be addressed through future work. First, our study only considers Google's ASR and although this a popular choice there are several other commercial and open source alternatives. Initially, we also explored the use of Amazon Transcribe Medical[3]; however initial experiments did not show much variation with respect to the use of Google ASR. Nonetheless, further analysis is needed to evaluate how well the findings of this work will generalize to other ASR systems. Second, the computed WER and semantic distance are noisy, since the timestamps we used to align manual and automatic transcriptions were obtained through forced alignment. Furthermore, we did not evaluate the speaker diarization performance of the ASR system in identifying speaker's role. Current ASR systems, including Google's speech-to-text, offer the functionality to automatically assign speaker identities to transcribed utterances, and this feature might be useful for automatically assigning speaker roles to each utterance. Finally, we limited our focus to the behavioral coding task.

[3]https://aws.amazon.com/transcribe/medical/

## 8 Conclusion and Lessons Learned

In this work, we conducted an evaluation of automatic speech recognition in the counseling domain using conversations between counselors and clients. To measure the degree of transcription error introduced by the use of an ASR system, we conducted domain-agnostic and domain-relevant evaluations using WER and semantic distance. Our analysis showed that while WER and semantic distance are in the 35 to 40% range when conducting a domain agnostic evaluation, the transcription error is slightly lower when considering transcription segments that are relevant to the domain i.e., utterances identified as important in evaluating the quality of counseling.

Moreover, we examined how the ASR step fits in and impacts the larger pipeline of an NLP system for behavioral coding in psychotherapy by comparing how the use of ASR data in place of manually transcribed data affects the performance of the downstream NLP system. Finally, we empirically showed that augmenting the system input with local context may alleviate the impact of noisy transcription. Given the results and analyses of this work, we conclude with the following lessons we learned in this study, on using ASR for NLP applications in psychotherapy and counseling: (1) Aggregate error measures are not sufficient by themselves, and must be complemented with domain-specific evaluations. (2) ASR error rates and performances differ across speaker roles and demographics as well as utterance content/topics. (3) Even a relatively small amount of manual transcription effort can help counteract noisy ASR and improve performance during the training of NLP models for psychotherapy applications.


## Acknowledgment

This material is based in part upon work supported by the Precision Health initiative at the University of Michigan, by the National Science Foundation (grant #1815291), and by the John Templeton Foundation (grant #61156). Any opinions, findings, and conclusions or recommendations expressed in this material are those of the author and do not necessarily reflect the views of the Precision Health initiative, the National Science Foundation, or John Templeton Foundation.

## References

M. Adda-Decker and L. Lamel. 2005. Do speech recognizers prefer female speakers? In *Interspeech*.

Tim Althoff, Kevin Clark, and Jure Leskovec. 2016. Large-scale analysis of counseling conversations: An application of natural language processing to mental health. *Transactions of the Association for Computational Linguistics*, 4:463–476.

Jacob Devlin, Ming-Wei Chang, Kenton Lee, and Kristina Toutanova. 2019. Bert: Pre-training of deep bidirectional transformers for language understanding.

Nikolaos Flemotomos, Victor R. Martinez, Zhuohao Chen, Karan Singla, Victor Ardulov, Raghuveer Peri, Derek D. Caperton, James Gibson, Michael J. Tanana, Panayiotis Georgiou, Jake Van Epps, Sarah P. Lord, Tad Hirsch, Zac E. Imel, David C. Atkins, and Shrikanth Narayanan. 2021. Automated evaluation of psychotherapy skills using speech and language technologies.

Sharon Goldwater, Dan Jurafsky, and Christopher D. Manning. 2008. Which words are hard to recognize? prosodic, lexical, and disfluency factors that increase ASR error rates. In *Proceedings of ACL-08: HLT*, pages 380–388, Columbus, Ohio. Association for Computational Linguistics.

Google. 2020. Google. cloud speech-to-text

Zellig S. Harris. 1954. Distributional structure. *Word*, 10(2-3):146–162.

Zac E. Imel, M. Steyvers, and David C. Atkins. 2015. Computational psychotherapy research: scaling up the evaluation of patient-provider interactions. *Psychotherapy*, 52 1:19–30.

Suyoun Kim, Abhinav Arora, Duc Le, Ching-Feng Yeh, Christian Fuegen, Ozlem Kalinli, and Michael L. Seltzer. 2021. Semantic distance: A new metric for asr performance analysis towards spoken language understanding.

Xiang Kong, Jeung-Yoon Choi, and Stefanie Shattuck-Hufnagel. 2016. Evaluating automatic speech recognition systems in comparison with human perception results using distinctive feature measures.

Bohan Li, Hao Zhou, Junxian He, Mingxuan Wang, Yiming Yang, and Lei Li. 2020. On the sentence embeddings from pre-trained language models.

Tatiana Likhomanenko, Qiantong Xu, Vineel Pratap, Paden Tomasello, Jacob Kahn, Gilad Avidov, Ronan Collobert, and Gabriel Synnaeve. 2020. Rethinking evaluation in asr: Are our models robust enough?

Anirudh Mani, Shruti Palaskar, Nimshi Venkat Meripo, Sandeep Konam, and Florian Metze. 2020. Asr error correction and domain adaptation using machine translation.

Adam S. Miner, Albert Haque, Jason Alan Fries, S. L. Fleming, D. Wilfley, G. Terence Wilson, A. Milstein, D. Jurafsky, B. Arnow, W. Stewart Agras, Li Fei-Fei, and N. Shah. 2020. Assessing the accuracy of automatic speech recognition for psychotherapy. *NPJ Digital Medicine*, 3.

Robert M Ochshorn and Max Hawkins. gentle forced-aligner.

Youngja Park, Siddharth Patwardhan, K. Visweswariah, and S. C. Gates. 2008. An empirical analysis of word error rate and keyword error rate. In *INTERSPEECH*.

James W. Pennebaker, Martha E. Francis, and Roger J. Booth. 2001. *Linguistic Inquiry and Word Count*. Lawerence Erlbaum Associates, Mahwah, NJ.

Verónica Pérez-Rosas, Rada Mihalcea, Kenneth Resnicow, Satinder Singh, and Lawrence An. 2016. Building a motivational interviewing dataset. In *Proceedings of the Third Workshop on Computational Linguistics and Clinical Psychology*, pages 42–51, San Diego, CA, USA. Association for Computational Linguistics.

Verónica Pérez-Rosas, Rada Mihalcea, Kenneth Resnicow, Satinder Singh, Lawrence An, Kathy J. Goggin, and Delwyn Catley. 2017. Predicting counselor behaviors in motivational interviewing encounters. In *Proceedings of the 15th Conference of the European Chapter of the Association for Computational Linguistics: Volume 1, Long Papers*, pages 1128–1137, Valencia, Spain. Association for Computational Linguistics.

Verónica Pérez-Rosas, Xinyi Wu, Kenneth Resnicow, and Rada Mihalcea. 2019. What makes a good counselor? learning to distinguish between high-quality and low-quality counseling conversations. In *Proceedings of the 57th Annual Meeting of the Association for Computational Linguistics*, pages 926–935, Florence, Italy. Association for Computational Linguistics.

Daniel Povey, Arnab Ghoshal, Gilles Boulianne, Nagendra Goel, Mirko Hannemann, Yanmin Qian, Petr Schwarz, and Georg Stemmer. 2011. The kaldi speech recognition toolkit. In *In IEEE 2011 workshop*.

Nils Reimers and Iryna Gurevych. 2019. Sentence-bert: Sentence embeddings using siamese bert-networks. In *Proceedings of the 2019 Conference on Empirical Methods in Natural Language Processing*. Association for Computational Linguistics.

Siqi Shen, Charles Welch, Rada Mihalcea, and Verónica Pérez-Rosas. 2020. Counseling-style reflection generation using generative pretrained transformers with augmented context. In *Proceedings of the 21th Annual Meeting of the Special Interest Group on Discourse and Dialogue*, pages 10–20, 1st virtual meeting. Association for Computational Linguistics.

Piotr Szymański, Piotr Żelasko, Mikolaj Morzy, Adrian Szymczak, Marzena Żyła-Hoppe, Joanna Banaszczak, Lukasz Augustyniak, Jan Mizgajski, and Yishay Carmiel. 2020. WER we are and WER we think we are. In *Findings of the Association for Computational Linguistics: EMNLP 2020*, pages 3290–3295, Online. Association for Computational Linguistics.

Rachael Tatman. 2017. Gender and dialect bias in YouTube's automatic captions. In *Proceedings of the First ACL Workshop on Ethics in Natural Language Processing*, pages 53–59, Valencia, Spain. Association for Computational Linguistics.

Thomas Wolf, Lysandre Debut, Victor Sanh, Julien Chaumond, Clement Delangue, Anthony Moi, Pierric Cistac, Tim Rault, Remi Louf, Morgan Funtowicz, Joe Davison, Sam Shleifer, Patrick von Platen, Clara Ma, Yacine Jernite, Julien Plu, Canwen Xu, Teven Le Scao, Sylvain Gugger, Mariama Drame, Quentin Lhoest, and Alexander Rush. 2020. Transformers: State-of-the-art natural language processing. In *Proceedings of the 2020 Conference on Empirical Methods in Natural Language Processing: System Demonstrations*, pages 38–45, Online. Association for Computational Linguistics.

B. Xiao, Zac E. Imel, P. Georgiou, David C. Atkins, and Shrikanth S. Narayanan. 2015. "rate my therapist": Automated detection of empathy in drug and alcohol counseling via speech and language processing. *PLoS ONE*, 10.

Justine Zhang and Cristian Danescu-Niculescu-Mizil. 2020a. Balancing objectives in counseling conversations: Advancing forwards or looking backwards. In *Proceedings of the 58th Annual Meeting of the Association for Computational Linguistics*, pages 5276–5289, Online. Association for Computational Linguistics.

Justine Zhang and Cristian Danescu-Niculescu-Mizil. 2020b. Balancing objectives in counseling conversations: Advancing forwards or looking backwards. In *Proceedings of ACL*.

# Qualitative Analysis of Depression Models by Demographics

**Carlos Aguirre**
Johns Hopkins University
caguirr4@jhu.edu

**Mark Dredze**
Johns Hopkins University
mdredze@cs.jhu.edu

## Abstract

Models for identifying depression using social media text exhibit biases towards different gender and racial/ethnic groups. Factors like representation and balance of groups within the dataset are contributory factors, but difference in content and social media use may further explain these biases. We present an analysis of the content of social media posts from different demographic groups. Our analysis shows that there are content differences between depression and control subgroups across demographic groups, and that temporal topics and demographic-specific topics are correlated with downstream depression model error. We discuss the implications of our work on creating future datasets, as well as designing and training models for mental health.

## 1 Introduction

Models of mental health trained on social media data exhibit biases in downstream performance on different gender and racial/ethnic demographic groups (Aguirre et al., 2021). An important factor is that minority groups (People of Color in general) are underrepresented in datasets and thus models under perform compared to majority groups. While size and balance of datasets contribute to the gap in performance, there may be differences in the manner in which depressive behavior is exhibited across demographic groups, creating problems in generalization.

Difference in depression prevalence across demographics have long been known (Brody et al., 2018), although there is no clear explanation for why this is the case (Hasin et al., 2018). On social media, demographic-based mental health analyses have used matched control samples (Dos Reis and Culotta, 2015), which allow for comparison of behaviors across groups (Coppersmith et al., 2014; Amir et al., 2019). These types of analyses have focused on downstream performance of trained models (Aguirre et al., 2021) and how they show differences in depression rates, but there have been no qualitative studies investigating these demographic differences (Chancellor and De Choudhury, 2020; Harrigian et al., 2020b).

Others have used qualitative studies to analyze behaviors and performance of machine learning models in general (Chen et al., 2018). Previous work has analyzed representative sentences (Ettinger, 2020), hashtags (Sykora et al., 2020), performed a thematic analysis by using the Linguistic Inquiry and Word Count dictionary (Wolohan et al., 2018) or trained topic models (Harrigian et al., 2020a; Yazdavar et al., 2017; Mitchell et al., 2015).

We propose a qualitative language analysis to reveal what differences occur, and how these differences can contribute to downstream performance. What language trends characterize depression and how do these vary across demographic groups? We use an analysis method similar to Mueller et al. (2021) but instead of training an Latent Dirichelt Allocation (LDA) (Blei et al., 2003) topic model and performing Point-Wise Mutual inference to obtain topics related to demographics, we train a Partially-Labeled LDA model (Ramage et al., 2011) which allows us to assign labels to demographic groups as well as depression and control groups to obtain label-specific topics to our user groups.

We base our analysis on datasets from previous work using Twitter. We train simple text-based models based on previous work on these datasets (Harrigian et al., 2020a; Aguirre et al., 2021). We use a labeled topic model to characterize what content indicates depression and how this content varies by demographic group.

Our analysis shows variations in content between depression and control subgroups across demographic groups, however, most of these differences are due to non-clinical phenomena e.g. vi-



*Proceedings of the Seventh Workshop on Computational Linguistics and Clinical Psychology*, pages 169–180
June 11, 2021. 

ral content trends such as *TV shows awards*. Further, model error analysis corroborates that temporal trends and nongeneralizable topics of demographic groups are correlated with downstream model error. Our qualitative analysis approach can be utilized to analyze language differences across demographics on other datasets and mental health tasks. We discuss the implications of our work on creating new datasets, as well as designing and training language models for mental health.

## 2 Ethical Considerations

Given the sensitive nature of mental health topics and demographics of individuals, additional precautions (based on *depression diagnoses* (Benton et al., 2017a); *gender identity* (Larson, 2017); *race/ethnicity identity* (Wood-Doughty et al., 2020)) were taken during this study. Data sourced from external research groups was retrieved according to each datasets respective data use policy. For gender labels, due to current limitations on datasets and methods, we consider the *folk perception* of gender, as described in Larson (2017), and for race/ethnicity labels we use the mutually-exclusive non-Hispanic White, non-Hispanic Black,non-Hispanic Asian and Hispanic/Latinx, following Wood-Doughty et al. (2020). We acknowledge that both our gender and racial/ethnic categories do not fully capture many individuals' gender and/or race/ethnicity. Additionally, we acknowledge the limitations of the demographic inference methods employed to obtain the demographic labels that have been raised in multiple previous studies (Mueller et al., 2021; Aguirre et al., 2021). While we carefully consider these issues, we believe the urgency of understanding mental health models (Aguirre et al., 2021) warrants our work and hope that our results provide sufficient evidence to justify further study in this area. This research was deemed exempt from review by our Institutional Review Board (IRB) under 45 CFR ğ 46.104.

## 3 Data

We use two datasets for depression identification on Twitter from previous studies: the *CLPsych 2015 Shared Task* (Coppersmith et al., 2015b), and the multi-disorder *multitask* learning for mental health dataset (Benton et al., 2017b).

**CLPsych.** The dataset contains publicly available tweets of individuals where the diagnosed group was collected by self-report through regular expression matching, e.g. "`I was diagnosed with <disorder>`". Control individuals were approximated by matching inferred age and gender using tools from the World Well-Being Project (Sap et al., 2014) from a pool of random accounts. While the original dataset collected four conditions, we select the depression users (475) and their matched control users resulting in 950 individuals.

| label | topic | tokens | Δ | $E$ |
|---|---|---|---|---|
| | | **Female White** | | |
| Depression | mental health | men mental ppl trans #mentalhealth sex | | 0.936 |
| Female White | Body Neg. | fat weight eating line cross die cut body | | 0.906 |
| Depression | UK language | lovely mum favourite uk mate cos london | | 0.876 |
| Depression | Game/Media | anime luigi art games draw mario character | | 0.919 |
| Depression | One Direction | harry louis zayn direction niall liam | | 0.903 |
| Female White | School | class college weekend homework break | | 0.294 |
| Control | Sports | team football fine state season congrats | | 0.093 |
| Latent | AAVE | nigga gotta yo niggas bout bitches tho | | 0.345 |
| Control | Arabic | libra يا و ما ، لا على الله في من beer | | 0.111 |
| Latent | Rap/Music | yall lmfao smh nah tho drake kanye album | | 0.433 |
| | | **Female PoC** | | |
| Depression | Game/Media | anime luigi art games draw mario character | | 0.944 |
| Female White | Body Neg. | fat weight eating line cross die cut body | | 0.971 |
| depression | mental health | men mental ppl trans #mentalhealth sex | | 0.922 |
| Depression | UK language | lovely mum favourite uk mate cos london | | 0.925 |
| Depression | 5 SOS | #vote5sos luke #kca michael calum ashton | | 0.992 |
| Female White | School | class college weekend homework break | | 0.29 |
| Control | Sports | team football fine state season congrats | | 0.14 |
| Control | Arabic | libra يا و ما ، لا على الله في من beer | | 0.111 |
| | | **Male White** | | |
| Depression | UK language | lovely mum favourite uk mate cos london | | 0.851 |
| Depression | mental health | men mental ppl trans #mentalhealth sex | | 0.827 |
| latent | Politics | police trump president state america | | 0.614 |
| Latent | Media | book movie star film story books episode | | 0.602 |
| Depression | Game/Media | anime luigi art games draw mario character | | 0.729 |
| Female White | School | class college weekend homework break | | 0.205 |
| Latent | Relationship | text boyfriend care relationship not_want | | 0.347 |
| Control | Sports | team football fine state season congrats | | 0.116 |
| Latent | Spanish | que la el en es te un mi se lo por los | | 0.135 |
| Control | Arabic | libra يا و ما ، لا على الله في من beer | | 0.111 |
| | | **Male PoC** | | |
| Depression | Game/Media | anime luigi art games draw mario character | | 0.845 |
| depression | One Direction | harry louis zayn direction niall liam | | 0.997 |
| Depression | 5 SOS | #vote5sos luke #kca michael calum ashton | | 0.996 |
| Female White | Body Neg. | fat weight eating line cross die cut body | | 0.902 |
| Female PoC | Pop Culture | jacob jack vine dm fans #fifthharmony | | 0.999 |
| Female White | School | class college weekend homework break | | 0.207 |
| Control | Sports | team football fine state season congrats | | 0.051 |
| Control | Arabic | libra يا و ما ، لا على الله في من beer | | 0.111 |

Table 1: Top and bottom 5 topics, as measured by the change of prevalence between depression and control group Δ, per demographic group on *Multitask* dataset. Only showing topics were Δ is statistically significant with bootstrapping (iterations = 1000, CI = 0.95).

**Multitask.** This dataset combines subsets of several datasets (Coppersmith et al., 2015a,b,c). All methods used the same collection process: self-report through regular expression matching, and control individuals by matching inferred age and gender with the same tool. Additionally, the complete public history of tweets is collected for each individual as opposed to the latest 3000 tweets on *CLPsych* resulting in a bigger dataset. We select the depression users (1400) and their matched control users resulting in 2800 individuals.

While both dataset collection methods are nearly identical, the time period in which the tweets were collected, and the number of tweets and individuals are different for each dataset, likely leading to different types of depression indicators. Note that while there is an overlap between **Multitask** and **CLPsych** of 110 individuals, it is a small percentage of both datasets ($\sim 4\%$ and $\sim 10\%$ respectively).

## 4 Methodology

**Demographic Labels.** While both datasets utilized gender and age inferences to match control and disorder groups at collection time, these models are now out-dated and labels for race/ethnicity were not made available. We obtain new race/ethnicity and gender labels from the work of Aguirre et al. (2021). Demographic statistics for both datasets are available in Appendix A. Since the race/ethnicity minority groups are extremely underrepresented in the datasets, we combine them to create a Person of Color (PoC) group.

**Mental Health Models.** We create mental health models for these datasets based on recent work (Harrigian et al., 2020a; Aguirre et al., 2021). Following standard pre-processing procedures, we filter numeric values, username mentions, retweets and urls from raw tweet text. For model features, we considered TF-IDF vector representations, mean-pooled 200 dimensional Twitter GloVe embeddings (Pennington et al., 2014), Linguistic Inquiry Word Count (LIWC) representations (Pennebaker et al., 2007), and features based on topic distributions learned via LDA. We train $\ell_2$-regularized logistic regression models on both datasets and follow hyper-parameter tuning procedures from Harrigian et al. (2020a); Aguirre et al. (2021).

### 4.1 Topic Model Analysis

**Model.** We use a topic model analysis to identify topic distribution differences between demographic groups. We train on each dataset (separately) a Partially Labeled LDA model (Ramage et al., 2011), which incorporates per-label latent topics into an LDA model. We assign both *depression* and *demographic* labels to individuals, with $K = 5$ topics per label and 20 latent topics not associated with any labels for a total of 50 topics, following the number of topics from previous work. Intuitively, this has the effect of *credit attribution* – associating words to either *depression*, *demographic* groups or latent to dataset topics for each individual.

**Metrics.** To measure topic prevalence between groups, we use the enrichment ($E$) metric from Marlin et al. (2012); Ghassemi et al. (2014):

$$E(c') = \frac{\sum_i^d \mathbb{1}(c_i = c')y_i \cdot q_{ic}}{\sum_i^d \mathbb{1}(c_i = c')q_{ic}}$$

The metric $E$ has the effect of highlighting topics regardless of topic importance within the group. In order to preserve topic importance, we take the non-normalized average difference in $E$ between control and depression groups ($\Delta$). For each document $i$, and corresponding label $y_i$, topic $c_i$, and topic probability $q_{ic}$:

$$\Delta(c') = \frac{1}{n}\sum_i^n \mathbb{1}(y_i = 1)\mathbb{1}(c_i = c')q_{ic} - \mathbb{1}(y_i = 0)\mathbb{1}(c_i = c')q_{ic}$$

Where negative values are topics most aligned with control group and positive values are aligned with depression.

Finally, To measure error rate attributions to topics, we use the topic error rate $\hat{E}$ metric from Chen et al. (2018):

$$\hat{E}(c') = \frac{\sum_i^d \mathbb{1}(y_i \neq \hat{y})\mathbb{1}(c_i = c')q_{ic}}{\sum_i^d \mathbb{1}(c_i = c')q_{ic}}$$

**Data Processing.** In addition to removing numeric values, username mentions, retweets and urls, we also remove English stopwords, pronouns[1] and emojis in order to create more coherent topics for our annotators. Removing stopwords and pronouns has the potential to erase depression signals as previous studies have found signals on pronoun usage, and also suppress voices and languages that do not fit certain norms. A full list of stopwords and pronouns is provided in Appendix B. We excluded topics from our results that did not have any coherent semantic groupings as annotated by one of the authors and 2 volunteers by looking at the top 15 most probable words per topic, obtaining a fair multi-annotator agreement Fleiss' Kappa $\kappa = 0.332$. After, topics were the majority of annotators selected as coherent where labeled by one of the authors.

[1]English stopwords and pronouns were obtained from *NLTK* tool (Bird et al., 2009)

| | label | topic | tokens | Female | | Male | |
|---|---|---|---|---|---|---|---|
| | | | | White | PoC | White | PoC |
| **Multitask** | Female-PoC | Pop Culture | jacob jack vine dm fans ugly #theyretheone meet | 0.172 | 0.002 | 0.137 | **0.999** |
| | depression | One Direction | harry louis zayn direction niall liam | 0.087 | 0.024 | 0.155 | **0.951** |
| | Male-White | Video Games | tap games gta stream gg pc xbox glitch #gamergate | 0.156 | 0.001 | 0.341 | **0.988** |
| | Female-White | Justin Bieber | justin retweet bieber ily tour babe meet #mtvstars | 0.143 | 0.039 | 0.164 | **0.766** |
| | depression | Game/Media | anime luigi art games draw mario character | 0.140 | 0.136 | 0.307 | **0.776** |
| **CLPsych** | Female-White | Justin Bieber | bieber #emazing #mtvstars beliebers | 0.141 | 0.709 | **0.717** | 0.017 |
| | Female-White | One Direction | direction niall liam louis zayn leo #mtvhottest fandom | 0.315 | **0.837** | 0.106 | 0.563 |
| | Female-White | People's Choice | demi miley austin lovato #peopleschoice vote album | 0.231 | 0.154 | **0.698** | 0.001 |
| | control | Beauty | wedding #love #fashion #nails #beauty #hair #beautiful | **0.592** | 0.051 | 0.004 | 0.008 |
| | Female-PoC | AAVE | n**gas smh yo gone somebody everybody mad ima | 0.247 | **0.577** | 0.151 | 0.097 |

Table 2: Top 5 topics as measured by topic error rate $\hat{E}$. Higher value represents higher prevalence on individuals that mental health models misclassified.

## 5 Analysis

The topic model identified label-specific topics for depression and control. Appendix C shows the topics for both datasets as well as the top 10 most probable words per topic. Some *depression* topics are reasonable e.g. `mental-health` (in both datasets) and `social media stats` (may be related to internet statistics and popularity). Similarly, *control* topics like `sports` and `beauty` are active, positive and self-caring topics that are reasonable for being representative of our control group. However, some topics in both *depression* and *control* groups are not clearly tied to the groups e.g. for depression group, topics like `One Direction` and `5 Seconds of Summer`. These might be topics introduced by temporal phenomena impeding model generalization (Harrigian et al., 2020a), rather than representative topics for those labels.

### 5.1 Content Differences

We characterize the difference in content between depression and control groups for each demographic. Table 1 shows the top and bottom 5 most prevalent topics with respect to the depression subgroups per demographic category as measured by $\Delta$ on the *Multitask* dataset where only the topics with statistically significant $\Delta$ are shown, as computed by bootstrapping with 1000 iterations with a CI of 95%.

Some *depression* topics (`One Direction` and `5 Seconds of Summer`) are not representative across demographics, while reasonable topics e.g. `mental-health` are representative of depression across demographic groups. Additionally, the topics most prevalent in the control subgroups (`School` and `Sports`) are the same across all demographics and represent qualities that are not related to depression, showing the robustness of these indicators and the well-formed nature of the control group in the dataset.

The `Body Negative` topic, attributed to the *Female-White* label, is very prevalent on depression subgroups for both female groups but is not prevalent on male subgroups, suggesting that there are differences on depression language online between gender groups.

For Male PoC individuals, the `mental-health` topic for depression is not prevalent in the depression subgroup while `One Direction` is prevalent. Given that the Male PoC group has the fewest users in the dataset, this suggests that its depression subgroup is not a representative group of individuals for depression yielding spurious topics, confirming prior work on dataset size being a factor on difference in performance across demographics (Aguirre et al., 2021).

Further, topics representing non-English language (`Arabic` and `Spanish`) or minority accent (AAVE) are more prevalent in control subgroups of demographics where those are not expected e.g. `Arabic` and `AAVE` on *Female-White* group. Perhaps this is evidence of demographic label noise, further exacerbating the need of obtaining self-reported demographic labels on mental health datasets for more concrete analysis.

### 5.2 Depression Model Errors

We analyze the predictions of our depression models to identify content differences between demographics that are correlated to models errors. Table 2 shows the top 5 topics that are most prevalent on individuals that were wrongly classified by the models on each dataset. Expanding results from Section 5.1, we find that topics that are not representative across demographics e.g. `One Direction`, are correlated with downstream errors in classification of mental health models. This suggests that topics that are prevalent of depression subgroups and are not related to depression are misleading the model.

Additionally, the majority of topics most prevalent on model errors (e.g. `Justin Bieber`, `One Direction` and `People's Choice`) apart from not being related to mental health and not representative across demographics, are influenced by temporal phenomena e.g. short term events such People's Choice Awards, that stem from the time period in which the dataset was collected. Such topics are not generalizable, corroborating evidence from prior work on challenges in model generalizations of temporal themes (Harrigian et al., 2020a).

Further, some topics prevalent on model errors are the effect of dataset balance. For example, the topic `AAVE` is a *Female-PoC* labeled topic, but it also is very prevalent on model errors, suggesting that there are very few examples of AAVE in the dataset and the mental health model is oversensitive to this language. On the other hand, the topic `beauty`, labeled as *control*, is over-represented in the dataset. This suggest that datasets should be balanced based on demographics, following prior work (Aguirre et al., 2021).

## 6 Conclusion

We performed a qualitative analysis to find content differences related to mental health across demographic groups. We showed that there are content differences between depression and control subgroups, while most of these differences are due to non-clinical phenomena e.g. temporal topics. Additionally, we find that dataset size might be a factor in these content differences. Furthermore, model error analysis corroborates that temporal topics and demographic-specific topics are correlated with downstream model error.

Our findings support prior work on the importance of methods that seek to generalize temporal topics (Harrigian et al., 2020a). We also find supporting evidence of the importance of dataset size as well as dataset balance in order to generalize to minority groups (Aguirre et al., 2021). Though in our analysis we only consider one mental health disorder (*depression*), our methodology was able to generalize across two datasets. This suggests that it is a valid method for qualitative analysis on finding content differences in other mental health datasets. Additionally, while we were limited in our demographic labels by current demographic models and dataset sizes, we showed that our approach is valid across two demographic axes and could be expanded to include other demographic axes (such as age and economic status), and include genders and racial/ethnic groups outside of the ones considered in this work. We hope our work warrants further studies of mental health language differences across more diverse demographic groups yielding more inclusive datasets and research.

## Acknowledgements

We thank the anonymous reviewers for their helpful feedback. Also, we thank Keith Harrigian and Rachel Wicks for volunteering time and effort for topic annotations, and Elizabeth Salesky for helpful feedback.

## References

Carlos Aguirre, Keith Harrigian, and Mark Dredze. 2021. Gender and racial fairness in depression research using social media. In *Proceedings of the 16th conference of the European Chapter of the Association for Computational Linguistics (EACL)*.

Silvio Amir, Mark Dredze, and John W. Ayers. 2019. Mental health surveillance over social media with digital cohorts. In *Proceedings of the Sixth Workshop on Computational Linguistics and Clinical Psychology*, pages 114–120, Minneapolis, Minnesota. Association for Computational Linguistics.

Adrian Benton, Glen Coppersmith, and Mark Dredze. 2017a. Ethical research protocols for social media health research. In *Proceedings of the First ACL Workshop on Ethics in Natural Language Processing*, pages 94–102.

Adrian Benton, Margaret Mitchell, and Dirk Hovy. 2017b. Multitask learning for mental health conditions with limited social media data. In *Conference of the European Chapter of the Association for Computational Linguistics (EACL)*. Association for Computational Linguistics.

S Bird, E Loper, and E Klein. 2009. Natural language processing with python oreilly media inc.

David M Blei, Andrew Y Ng, and Michael I Jordan. 2003. Latent dirichlet allocation. *Journal of machine Learning research*, 3(Jan):993–1022.

Debra Brody, Laura Pratt, and Jeffery Hughes. 2018. Prevalence of depression among adults aged 20 and over: United states, 2013-2016. *NCHS data brief*, pages 1–8.

Stevie Chancellor and Munmun De Choudhury. 2020. Methods in predictive techniques for mental health status on social media: a critical review. *NPJ digital medicine*, 3(1):1–11.

Irene Chen, Fredrik D Johansson, and David Sontag. 2018. Why is my classifier discriminatory? *arXiv preprint arXiv:1805.12002*.

Glen Coppersmith, Mark Dredze, and Craig Harman. 2014. Quantifying mental health signals in twitter. In *Proceedings of the workshop on computational linguistics and clinical psychology: From linguistic signal to clinical reality*, pages 51–60.

Glen Coppersmith, Mark Dredze, Craig Harman, and Kristy Hollingshead. 2015a. From ADHD to SAD: Analyzing the language of mental health on Twitter through self-reported diagnoses. In *Proceedings of the 2nd Workshop on Computational Linguistics and Clinical Psychology: From Linguistic Signal to Clinical Reality*, pages 1–10, Denver, Colorado. Association for Computational Linguistics.

Glen Coppersmith, Mark Dredze, Craig Harman, Kristy Hollingshead, and Margaret Mitchell. 2015b. CLPsych 2015 shared task: Depression and PTSD on Twitter. In *Proceedings of the 2nd Workshop on Computational Linguistics and Clinical Psychology: From Linguistic Signal to Clinical Reality*, pages 31–39.

Glen Coppersmith, Ryan Leary, Eric Whyne, and Tony Wood. 2015c. Quantifying suicidal ideation via language usage on social media. In *Joint Statistics Meetings Proceedings, Statistical Computing Section, JSM*, volume 110.

Virgile Landeiro Dos Reis and Aron Culotta. 2015. Using matched samples to estimate the effects of exercise on mental health via twitter. In *Proceedings of the AAAI Conference on Artificial Intelligence*, volume 29.

Allyson Ettinger. 2020. What BERT is not: Lessons from a new suite of psycholinguistic diagnostics for language models. *Transactions of the Association for Computational Linguistics*, 8:34–48.

Marzyeh Ghassemi, Tristan Naumann, Finale Doshi-Velez, Nicole Brimmer, Rohit Joshi, Anna Rumshisky, and Peter Szolovits. 2014. Unfolding physiological state: Mortality modelling in intensive care units. In *Proceedings of the 20th ACM SIGKDD international conference on Knowledge discovery and data mining*, pages 75–84.

Keith Harrigian, Carlos Aguirre, and Mark Dredze. 2020a. Do models of mental health based on social media data generalize? In *Findings of the Association for Computational Linguistics: EMNLP 2020*, pages 3774–3788, Online. Association for Computational Linguistics.

Keith Harrigian, Carlos Aguirre, and Mark Dredze. 2020b. On the state of social media data for mental health research. *arXiv preprint arXiv:2011.05233*.

Deborah S. Hasin, Aaron L. Sarvet, Jacquelyn L. Meyers, Tulshi D. Saha, W. June Ruan, Malka Stohl, and Bridget F. Grant. 2018. Epidemiology of Adult DSM-5 Major Depressive Disorder and Its Specifiers in the United States. *JAMA Psychiatry*, 75(4):336–346.

Brian Larson. 2017. Gender as a variable in natural-language processing: Ethical considerations. In *Proceedings of the First ACL Workshop on Ethics in Natural Language Processing*, pages 1–11, Valencia, Spain. Association for Computational Linguistics.

Benjamin M Marlin, David C Kale, Robinder G Khemani, and Randall C Wetzel. 2012. Unsupervised pattern discovery in electronic health care data using probabilistic clustering models. In *Proceedings of the 2nd ACM SIGHIT international health informatics symposium*, pages 389–398.

Margaret Mitchell, Kristy Hollingshead, and Glen Coppersmith. 2015. Quantifying the language of schizophrenia in social media. In *Proceedings of the 2nd workshop on Computational linguistics and clinical psychology: From linguistic signal to clinical reality*, pages 11–20.

Aaron Mueller, Zach Wood-Doughty, Silvio Amir, Mark Dredze, and Alicia L Nobles. 2021. Demographic representation and collective storytelling in the me too twitter hashtag activism movement. *Proceedings of the ACM on Human-Computer Interaction, CSCW*.

James W Pennebaker, Roger J Booth, and Martha E Francis. 2007. Operators manual: Linguistic inquiry and word count: LIWC2007. *Austin, Texas: LIWC. net http://homepage. psy. utexas. edu/HomePage/Faculty/Pennebaker/Reprints/LIWC2007_OperatorManual. pdf (accessed 1 October 2013)*.

Jeffrey Pennington, Richard Socher, and Christopher D Manning. 2014. Glove: Global vectors for word representation. In *Proceedings of the 2014 conference on empirical methods in natural language processing (EMNLP)*, pages 1532–1543.

Daniel Ramage, Christopher D Manning, and Susan Dumais. 2011. Partially labeled topic models for interpretable text mining. In *Proceedings of the 17th ACM SIGKDD international conference on Knowledge discovery and data mining*, pages 457–465.

Maarten Sap, Gregory Park, Johannes Eichstaedt, Margaret Kern, David Stillwell, Michal Kosinski, Lyle Ungar, and H Andrew Schwartz. 2014. Developing age and gender predictive lexica over social media. In *Proceedings of the 2014 Conference on Empirical Methods in Natural Language Processing (EMNLP)*, pages 1146–1151.

Martin Sykora, Suzanne Elayan, and Thomas W Jackson. 2020. A qualitative analysis of sarcasm, irony and related #hashtags on twitter. *Big Data & Society*, 7(2):2053951720972735.

JT Wolohan, Misato Hiraga, Atreyee Mukherjee, Zeeshan Ali Sayyed, and Matthew Millard. 2018. Detecting linguistic traces of depression in topic-restricted text: Attending to self-stigmatized depression with NLP. In *Proceedings of the First International Workshop on Language Cognition and Computational Models*, pages 11–21.

Zach Wood-Doughty, Paiheng Xu, Xiao Liu, and Mark Dredze. 2020. Using noisy self-reports to predict twitter user demographics. *arXiv preprint arXiv:2005.00635*.

Amir Hossein Yazdavar, Hussein S Al-Olimat, Monireh Ebrahimi, Goonmeet Bajaj, Tanvi Banerjee, Krishnaprasad Thirunarayan, Jyotishman Pathak, and Amit Sheth. 2017. Semi-supervised approach to monitoring clinical depressive symptoms in social media. In *Proceedings of the 2017 IEEE/ACM International Conference on Advances in Social Networks Analysis and Mining 2017*, pages 1191–1198.

# A Dataset Demographics

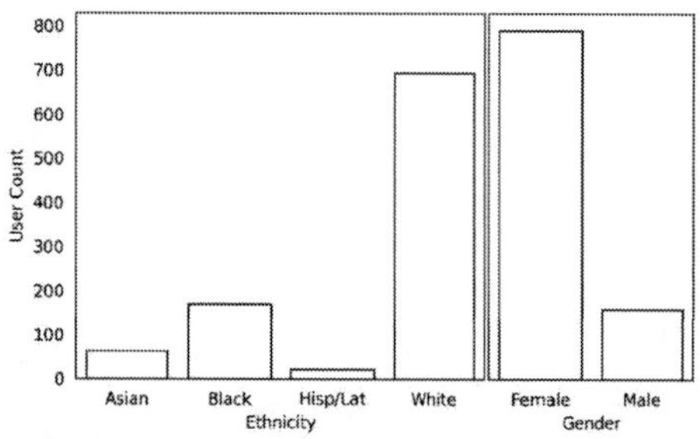


Figure 1: User count by gender and racial/ethnic demographic groups on **CLPsych** dataset.

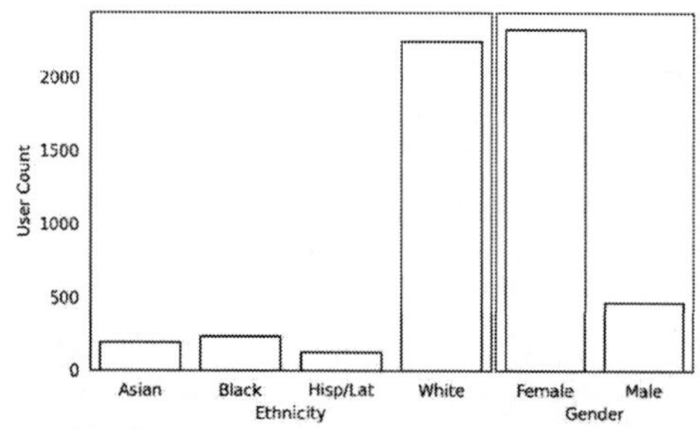


Figure 2: User count by gender and racial/ethnic demographic groups on **Multitask** dataset.

## B Partially Labeled LDA

We need a procedure to identify topic distribution differences between demographic groups. Prior work have accomplished this by training an LDA topic model and either using pointwise mutual inference (PMI) (Mueller et al., 2021) or an enrichment metric $E$ (Marlin et al., 2012; Ghassemi et al., 2014) to measure how distinctive a given topic is of a given demographic group.

Instead, we train[2], on both datasets separately, a Partially Labeled LDA model (Ramage et al., 2011), which incorporates per-label latent topics to an LDA model. Unlike LDA, each document $d$ can only use the topics associated with the set of labels $L_d$ assigned to $d$, where each label $l \in L_d$ is assigned some number of topics $K$. The model computes the joint likelihood of observed words $w$, observed labels $l$ and topic assignments $z$, given available labels $\Lambda$, and document-topic $\alpha$, topic-word $\eta$ priors from a Dirichtlet distribution $P(w, l, z|\Lambda, \alpha, \eta)$. We assign both *depression* and *demographic* labels to individuals, with $K = 5$ topics per label and 20 latent topics not associated with any labels for a total of 50 topics. Intuitively, this has the effect of *credit attribution* – associating words to either *depression*, *demographic* groups or latent to dataset topics for each individual.

To measure topic difference between groups (**RQ1**) we use the enrichment ($E$) metric from Marlin et al. (2012); Ghassemi et al. (2014):

$$E(c') = \frac{\sum_i^d \mathbb{1}(c_i = c')y_i \cdot q_{ic}}{\sum_i^d \mathbb{1}(c_i = c')q_{ic}}$$

To measure error rate (**RQ2**) we use the topic error rate $\hat{E}$ metric from Chen et al. (2018):

$$\hat{E}(c') = \frac{\sum_i^d \mathbb{1}(y_i \neq \hat{y})\mathbb{1}(c_i = c')q_{ic}}{\sum_i^d \mathbb{1}(c_i = c')q_{ic}}$$

Additionally, in order to preserve topic importance, we take the non-normalized average difference in $E$ between control and depression groups ($\Delta$). For each document $i$, and corresponding label $y_i$, topic $c_i$, and topic probability $q_{ic}$:

$$\Delta(c') = \frac{1}{n}\sum_i^n \mathbb{1}(y_i = 1)\mathbb{1}(c_i = c')q_{ic} - \mathbb{1}(y_i = 0)\mathbb{1}(c_i = c')q_{ic}$$

Where negative values are topics most aligned with control group and positive values are aligned with depression. In addition to filtering numeric values, username mentions, retweets and urls, we also filter stopwords, pronouns and emojis to obtain more coherent topics. We excluded topics from our results that did not have any coherent semantic groupings as annotated by one of the authors by looking at top 10 most probable words per topic.

[2]Model implementation based on Tomotopy python library `https://bab2min.github.io/tomotopy/v0.10.2/en/` which provides Gibbs-sampling based implementations of multiple *LDA models.

| i | me | my | myself | we | our | ours | ourselves | you |
|---|---|---|---|---|---|---|---|---|
| you've | you'll | you'd | your | yours | yourself | yourselves | he | him |
| himself | she | she's | her | hers | herself | it | it's | its |
| they | them | their | theirs | themselves | what | which | who | whom |
| that | that'll | these | those | am | is | are | was | were |
| been | being | have | has | had | having | do | does | did |
| a | an | the | and | but | if | or | because | as |
| while | of | at | by | for | with | about | against | between |
| through | during | before | after | above | below | to | from | up |
| in | out | on | off | over | under | again | further | then |
| here | there | when | where | why | how | all | any | both |
| few | more | most | other | some | such | no | nor | not |
| own | same | so | than | too | very | s | t | can |
| just | don | don't | should | should've | now | d | ll | m |
| re | ve | y | ain | aren | aren't | couldn | couldn't | didn |
| doesn | doesn't | hadn | hadn't | hasn | hasn't | haven | haven't | isn |
| ma | mightn | mightn't | mustn | mustn't | needn | needn't | shan | shan't |
| shouldn't | wasn | wasn't | weren | weren't | won | won't | wouldn | wouldn't |

Table 3: English stopwords from NLTK (Bird et al., 2009)

| he | she | they | i | him | her | we | me | it |
|---|---|---|---|---|---|---|---|---|
| us | them | myself | ourselves | yourself | yourselves | himself | itself | herself |
| my | our | ours | your | yours | their | its | mine | theirs |

Table 4: English pronouns from NLTK (Bird et al., 2009)

## C PLDA Topics

| label | Topic | words | weight |
|---|---|---|---|
| depression | Mental Health | men mental ppl trans #mentalhealth sex health woman racist depression illness racism rape gender | 0.0166 |
| | UK Language | xx lovely bit xxx mum favourite uk mate cos london ffs australia brilliant bloody | 0.0140 |
| | One Direction | harry louis zayn direction niall liam в stats unfollowers followers #emabiggestfans1d и на не fandom | 0.0107 |
| | 5 Seconds of Summer | #vote5sos luke #kca michael calum ashton seconds clifford hood summer hemmings #mtvstars #5sosfam | 0.0058 |
| control | Arabic | libra الله في من enter #nyc لا على aries و ما ، check يا beer | 0.0069 |
| | Portuguese | #android que e é discovered location não j eu lyn pra london um com streets | 0.0021 |
| | Sports | team football fine state season nails posted touch college congrats basketball proud coach | 0.0225 |
| Female White | Body Negative | fat weight eating line cross die cut body anymore skinny loves kill pain | 0.0154 |
| | School | class summer college weekend car homework dad break friday semester dog netflix room hour | 0.0563 |
| | Justin Beiber | justin retweet bieber dm #callmecam ily tour gain babe meet #mtvstars jacob pls proud | 0.0136 |
| | TV Shows | proud #thewalkingdead season episode #supernatural strong dead #love :d saved sam | 0.0061 |
| Female PoC | Dating | se dating singles z je surveys polls za yahoo politics health si po ṕro | 0.0003 |
| | Spanish | la en el que con por un los es gracias para del las se una | 0.0019 |
| | Pop culture | jacob jack vine dm fans ugly #fifthharmony #theyretheone meet sebastian af indirect #shawnformmva | 0.0032 |
| Male White | German | ich die und daily der das eyes hazel #supergirl ist nicht es top stories zu | 0.0006 |
| | Cities | #albuquerque #tpp israel vote u.s. obama #tcot #newmexico support war #faceofmlb transport | 0.0021 |
| | Canada/Music | #nowplaying team season #winnipeg load #spotify #canada football ask band final #music #indie player | 0.0068 |
| | Video Games | full added tap games liked beer menu gta stream gg pc xbox glitch #gamergate xd | 0.0025 |
| Male PoC | Social Media | followers #retweet goodmorning fast #teamfollowback retweets mentions #follow2befollowed followed | 0.0010 |
| | Video Games | #gamergate #notyourshield http anti- games gg sjws htt h sjw gamers anti harassment ht wu | 0.0003 |
| | AAVE | bro smh yall gotta bruh tho vine im team lebron season fam nba its dont | 0.0015 |
| latent | Politics | police trump president state america obama law country killed news vote gun government american rights | 0.0226 |
| | Zodiac | cancer others although current seems leo seem capricorn energy mind surgery gemini | 0.0193 |
| | AAVE | nigga gotta yo niggas bout bitches tho lil af bruh cuz n hoes bro dude | 0.0419 |
| | Media | book movie star film story books episode series reading writing art write post blog | 0.0232 |
| | Social Events | check party friday album top tickets weekend adam posted tour meet fans congrats vote | 0.0382 |
| | Music | yall lmfao smh nah tho drake kanye mad men bae gotta saying album boo wtf | 0.0304 |
| | People | kids child woman mother lady sister season married brother movie dad daughter sex | 0.0348 |
| | Spanish | que la el en es te un mi se lo por los con las para | 0.0087 |
| | Relationship | tired text anymore boyfriend care relationship honestly mood kinda forever babe leave sick | 0.1121 |

Table 5: Label, topic title, top words and topic importance of *Multitask* dataset.

| label | Topic | words | weight |
|---|---|---|---|
| depression | Mental Health | #mentalhealth mental depression link submitted comment health asked anxiety disorder illness | 0.0063 |
| | Social Stats | stats loves unfollowers photoset followed happen daily follower unfollower | 0.0049 |
| | Pop culture | pls luke michael ilysm hemmings babe ashton penguin zayn pizza clifford calum niall | 0.0045 |
| control | Peoples Choice | katy perry #peopleschoice others skinny share roar fiber sticker unlocked glee darren #musicvideo | 0.0031 |
| | Beauty | wedding #love #fashion #cute #nails #me #beauty #hair #beautiful #taurus #instagood | 0.0035 |
| | Life | movies virgo college favorite win shopping seeing classes study puppies studying loved | 0.0062 |
| Female White | Peoples Choice | demi miley austin lovato #peopleschoice vote album tour sign cyrus selena xoxo miley's | 0.0080 |
| | Justin Bieber | bieber #emazing de que #mtvstars el la beliebers #mtvhottest en justin's reason #kca foto te | 0.0070 |
| | One Direction | direction niall liam louis zayn leo #mtvhottest fandom luke fans album xx story babe styles | 0.0136 |
| | Internet Abv. | ily omfg crying aw dm meet retweet bye idk picture quote literally ill cry bby | 0.0228 |
| Female PoC | AAVE | n**gas smh yo gone somebody everybody mad ima hoes swear nobody af lil yall | 0.0190 |
| | Spanish | mi :p que de la te tu ke en r el se luv b h | 0.0042 |
| Male White | Nascar | bob fans mate race #bbq #nascar palace bbq win top racing league grilling fame | 0.0037 |
| | Work | č dundee salary hiking camping scotland engineer manager angus trail jobs sales | 0.0016 |
| Male PoC | AAVE | niggas bout wit yo smh aint bro gone yu everybody yea lil hoes bitches tryna | 0.0043 |
| | Music | outta line #soundcloud #new #retweet feat #rt download mixtape prod essay ft | 0.0037 |
| latent | Relationships | care anymore hurt smile relationship alone reason enough fall not_want change thinking feelings | 0.1058 |
| | AAVE | lmfao fuckin thats ill yo bitches dude lil high bout smoke wtf cuz hit black | 0.0422 |
| | Life | awesome needs r friday vote lady w pic chris retweet la halloween h movie lead | 0.0229 |
| | Politics | welcome west america state history star obama bill v national second john question david government | 0.0255 |
| | Sports | team season win games football goal fans vs congrats playing run beat final ball fan | 0.0287 |
| | Social Stats | followed stats unfollowers bro follower moment question yea hahahaha bus awkward teacher thats la unfollower | 0.0287 |
| | Relationships | babe idk ugly rn bae boyfriend wtf mad bored honestly annoying w k tbh kiss | 0.0467 |

Table 6: Label, topic title, top words and topic importance of *CLPsych* dataset.

# Safeguarding against spurious AI-based predictions: The case of automated verbal memory assessment

**Chelsea Chandler** and **Peter W. Foltz**
Department of Computer Science &
Institute of Cognitive Science,
University of Colorado Boulder
chelsea.chandler@colorado.edu
peter.foltz@colorado.edu

**Alex S. Cohen**
Department of Psychology &
Center for Computation & Technology,
Louisiana State University
acohen@lsu.edu

**Terje B. Holmlund**
Department of Clinical Medicine,
University of Tromsø -
The Arctic University of Norway
terje.holmlund@uit.no

**Brita Elvevåg**
Department of Clinical Medicine,
University of Tromsø -
The Arctic University of Norway &
Norwegian Centre for eHealth Research
brita@elvevaag.net

## Abstract

A growing amount of psychiatric research incorporates machine learning and natural language processing methods, however findings have yet to be translated into actual clinical decision support systems. Many of these studies are based on relatively small datasets in homogeneous populations, which has the associated risk that the models may not perform adequately on new data in real clinical practice. The nature of serious mental illness is that it is hard to define, hard to capture, and requires frequent monitoring, which leads to imperfect data where attribute and class noise are common. With the goal of an effective AI-mediated clinical decision support system, there must be computational safeguards placed on the models used in order to avoid spurious predictions and thus allow humans to review data in the settings where models are unstable or bound not to generalize. This paper describes two approaches to implementing safeguards: (1) the determination of cases in which models are unstable by means of attribute and class based outlier detection and (2) finding the extent to which models show inductive bias. These safeguards are illustrated in the automated scoring of a story recall task via natural language processing methods. With the integration of human-in-the-loop machine learning in the clinical implementation process, incorporating safeguards such as these into the models will offer patients increased protection from spurious predictions.

## 1 Introduction

Artificial intelligence (AI)-based systems that incorporate language and behavioral data hold promise of increasing sensitivity, equity, and access in the assessment and treatment of mental illness through the use of remote and continuous monitoring via clinical decision support systems. This is due to the fact that the pattern and content of language, as well as additional measures of behavior, such as timing and neuropsychological task scores, provide rich information that can be traced back to an individuals' overall mental state.

In order to demonstrate clinical translational value there are numerous risks and factors that are necessary to consider. First, it is important to collect data from large samples of the population across differing ages, cultures, genders, clinical conditions, and stages of disorder. Second, it is critical to create models that are explainable, transparent, and generalizable (Chandler et al., 2020b) in order to nurture trust from both patients and clinicians. And finally - the area that this paper will address - it is necessary to add safeguards to models such that they are capable of flagging cases that show attribute noise (i.e., abnormalities in feature values) or class noise (i.e., erroneous or missing class labels), and of determining the extent to which models will generalize to unseen data. These safeguards will enable a human-in-the-loop system where humans are required to review data abnormalities.

AI is used in a wide range of applications within mental health, notably within clinical research settings where data are used to aid in understanding the nature of diagnoses and to improve diagnostic accuracy (for reviews see Shatte et al., 2019; Su et al., 2020; Thieme et al., 2020), as well as in making complex and potentially lifesaving de-



*Proceedings of the Seventh Workshop on Computational Linguistics and Clinical Psychology*, pages 181–191
June 11, 2021. 

cisions (e.g., in suicidology - for review see Cox et al., 2020). Acoustic measurements of speech have been analyzed in automated applications for detecting Mild Cognitive Impairment and dementia (Roark et al., 2011; König et al., 2015), as well as serious mental illness (Cohen et al., 2019) and depression (McGinnis et al., 2019).

In the domain of techniques that specifically leverage natural language processing (NLP), there are a growing number of reports of using these methods on social media data, notably to data mine publicly shared written reports of mood on platforms such as Twitter and Reddit (Zirikly et al., 2019; Peng et al., 2019; Wu et al., 2012). There is also a growing interest in using such methods on electronic medical records to assist in the extraction of diagnostic information or to enhance understanding of medical conditions (Ryu et al., 2016; Wang et al., 2012; Metzger et al., 2017). A broad range of NLP metrics such as incoherence and tangentiality have been used to automatically assess the clinical state of patients with schizophrenia (Elvevåg et al., 2007) and predict the risk of psychosis onset (Bedi et al., 2015; Rosenstein et al., 2015; Corcoran et al., 2018). Deep language models and NLP feature-based models have also been shown to differentiate the language of healthy controls from those diagnosed with Mild Cognitive Impairment or dementia (Orimaye et al., 2018; Eyigoz et al., 2020).

There is clear evidence that the clinical data used in AI-based research applications hold predictive power in detection and diagnosis, prognosis, support and treatment, and as a second opinion measurement for illness severity, but it is unclear about the degree to which these models will be stable on new data. Many psychiatric studies that harness AI tend to do so on relatively small datasets (i.e., 10-100 participants) in fairly homogeneous populations (e.g., the WEIRD (Western, Educated, Industrialized, Rich, and Democratic) phenomenon - Henrich et al., 2010; and the predominance of male participants in psychiatric research studies - Longenecker et al., 2010). These shortcomings may lead to insufficient accuracy on unseen data retrieved from different experimental settings (e.g., in a lab vs. remote; prompted free speech vs. natural; as a component of a larger testing battery vs. on its own), populations (e.g., southern vs. northern; different English speaking countries; monolingual vs. multilingual participants), and clinical states (e.g., hallucinating vs. not hallucinating). One must keep in mind that in small datasets, spurious features may not be generalizable to a larger population, especially if they are not of any apparent clinical relevance (Chandler et al., 2020b; Whelan and Garavan, 2014). While these research experiments are noteworthy, they must be re-evaluated on larger and more diverse sets of participants to test for robustness and generalizability.

Incorrect or ill-advised decisions and predictions in psychiatry can be dangerous and life altering for patients, and the difficulty in decision making is further confounded by the very short time frame in which changes in mental state occur and the associated clinical decisions must be made. Thus, we must build systems that have the ability to instantaneously flag data abnormalities - both in the research phase and when translated into real clinical use - and pass these cases on for human review. Furthermore, rather than selecting a preferred machine learning model based on metrics such as accuracy, sensitivity, or correlation as is common in AI and NLP applications, we must seek to understand the underlying mechanisms and the context in which they will be used (Ethayarajh and Jurafsky, 2020; Hand, 2006).

Researchers in machine learning have proposed assessing models with stability metrics which define ways to quantify and compare the stability of results rather than simply focusing on the aforementioned metrics (Turney, 1995; Lange et al., 2002). Specifically, Zhu and Wu (2004) differentiated data-based noise and outliers into class noise and attribute noise, and advocated for analyzing their effects on machine learning models separately. Uncertainty estimation, as well as in- and out-of-distribution error detection has been critically important in the use of AI in a wide range of applications such as self driving cars (Mohseni et al., 2020), general medicine (Kompa et al., 2021), education (Foltz et al., 2013), and in many other domains.

In this paper we illustrate an example of NLP and machine learning methods applied to the automated scoring of a story recall task, a core component of psychiatric neuropsychological assessments. We focus on two approaches to safeguarding such a model: 1) the detection of attribute and class noise that can affect the predictions of a model and 2) the evaluation of the extent to which the model may or may not generalize to unseen data. We first applied methods to determine where noise exists with an outlier detection algorithm and data visualization. For the issue of model generalizability, we studied the effect of dataset size on the results, and we illustrate how such results change as we randomly remove portions of our training data. Additionally, we show the results of this particular story recall model applied to a new collection of data. We advocate that these computational safeguards, which have major implications in regard to their use in human-in-the-loop clinical support systems, must be placed on each machine learning model that is developed to automate or assist in clinical assessments.

## 2 Experimental overview

### 2.1 The *d*MSE

The data in the present work were collected from a mobile phone application (the *delta* Mental Status Examination, henceforth called *d*MSE) designed to assess patient state via various neuropsychological assessments, with many relying on patient language (Chandler et al., 2020a; Cohen et al., 2019; Holmlund et al., 2019; Holmlund et al., 2020). A total of 12 behavioral assessment tasks were employed to specifically assess the language, cognition, motor skill, and mental state of patients - areas where assessment is critical in those with serious mental illness - and integrated into the *d*MSE smart device application. The behavioral assessment tasks were similar to standardly employed neuropsychological tests (for an overview of neuropsychological testing, see Lezak et al., 2012), but adapted such that they could be remotely and frequently self-administered with variations of each task presented over time (Chandler et al., 2020a; Holmlund et al., 2019). As an automated measurement tool that can be used remotely, frequently and self-administered, this approach has the potential to enable greater access to mental health services. It permits patients to be monitored longitudinally outside of clinical institutions and can alert clinicians to critical changes in mental states, thereby providing greater availability to assistance, regardless of age, gender, ethnicity, location, or socioeconomic status.

The data comprised N = 25 patients and N = 79 presumed healthy undergraduate students from Louisiana State University who all provided informed written consent. These participants completed N = 118 and N = 226 sessions (i.e., one completion of the full battery of tasks in a single use of the application) with the *d*MSE, with an average of 4.72 (stdev = 1.14) and 2.90 (stdev = 0.90) per person, respectively. The patients were severely mentally ill outpatients on the psychosis spectrum. Two-thirds of the patients met the criteria for schizophrenia (N = 16), and the remaining met the criteria for major depressive disorder (N = 8) and bipolar disorder (N = 1). This study was approved by the Louisiana State University Institutional Review Board (#3618) and participants provided their informed written consent before participation. The application was designed specifically for use in remote settings, such as rural Louisiana and Northern Norway, where access to in-person clinical support can be quite difficult.

### 2.2 The story recall regression model

The machine learning model we use to illustrate safeguarding techniques automatically scored a variant of the immediate and delayed Logical Memory story recall task (of the Wechsler Memory test; Wechsler, 1997) that was employed in the *d*MSE. The story recall task is critical in neuropsychological assessment as memory function is of core interest in the evaluation of many neurodevelopmental, neurodegenerative and neuropsychiatric conditions, as well as in brain injuries (Baddeley and Wilson, 2002). Further, it is of enormous interest in mental illness research because of its value as a critical endophenotype (Cirillo and Seidman, 2003), as well as the fact that the process of recollecting has similarities to what is required by patients when their medical history is taken.

In our version of this task, a participant listens to a short story of on average 74 words (min = 62, max = 87) and then is asked to retell it both immediately and after a delay of 30 minutes in as much detail as possible, thus following the same format as the traditional Wechsler version. Stories were either narrative or instructional. The narrative stories contain two characters, a setting, an action that caused a problem, and a resolution. The instructional passages described how to accomplish some sort of goal, such as how to assemble a skateboard or how to clean a fish bowl. This *d*MSE story recall task was developed such that there could be many different versions capable of being scored with automated NLP methods (e.g., Chandler et al., 2021, Holmlund et al., 2020) rather than traditional rubric-based methods.

Three trained human raters with clinical experience assigned scores to the recall transcriptions based on the quality and amount of details (e.g., characters, events, dates, descriptors, feelings) recalled. The rubric was on a scale from 1 to 6, with 1 indicating no details were recalled, and 6 indicating all major and almost all minor details were recalled. Each participant completed one immediate narrative recall, one immediate instructional recall, and one delayed narrative recall per session. After the removal of responses with no words, the dataset contained N = 846 samples (N = 285 immediate narrative, N = 285 immediate instructional, and N = 276 delayed narrative).

A ridge regression model was created to predict the rating a trained professional would assign to a story recall. The model was trained on (1) the number of word types (i.e., unique words) in the recall, (2) the number of common word types between the original story and the recall, and (3) the BERTScore (Zhang et al., 2020) between the original story and the retell (the model was created in the same manner as that of Chandler et al., 2019 besides a change in the last feature from the word mover's distance to BERTScore). BERTScore is a similarity metric that was created to produce a score of how close a machine generated translation is to the gold standard(s) of some piece of text. Specifically, it creates a matrix of BERT (Devlin et al., 2019) cosine distances between words in one text to words in another. Alignment between words in both texts is produced greedily with the maximum cosine distance for each word in one text to another in the reference. All distances are averaged and inverse document frequency weightings are optionally incorporated.

The ridge regression model was trained and tested using 10-fold cross-validation and controlled such that sessions from the same participant did not occur simultaneously in both the train and test sets. The rating prediction model resulted in an average Pearson r correlation with human ratings of r = 0.91. These results indicate that we can automatically derive a range of semantic and surface level features from spoken recalls, and that these features can be harnessed to accurately predict the ratings of expert humans.

## 3 Effects of attribute and class noise

We begin our analysis of computational safeguards by discussing the determination of attribute and class noise in the context of model stability. Model stability analysis allows us to establish how unusual variations in input data will affect the output of the model. Put simply, we wish to find where in the feature space models may be the most unstable. We illustrate an approach that will allow researchers to detect attribute and class noise in data that could be due to construct-irrelevance or errors in assumptions.

| Attribute 1: Number word types | Attribute 2: Number common word types | Attribute 3: BERTScore | Class rating |
|---|---|---|---|
| 4 | 2 | <u>0.91</u> | 1 |
| 3 | <u>'x'</u> | 0.70 | 1 |
| 36 | 25 | 0.93 | 6 |
| 4 | 3 | 0.71 | ***6*** |

Table 1: Hypothetical subset of story recall data showing attribute noise (underlined) and class noise (bold and italicized). First, 0.91 in the first row constitutes potential attribute noise as the average BERTScore for examples with a rating of 1 is 0.80 (stdev = 0.05), and furthermore the average BERTScore for examples with 4 word types and 2 common words is 0.79 (stdev = 0.04) and 0.80 (stdev = 0.05), respectively. Thus, it is far out of the expected distribution. Second, 'x' in the second row constitutes attribute noise because this attribute expects numbers and there is a string in its place. Thus, it is erroneous. The class label of 6 in the last row constitutes class noise as the distribution of the feature values resembles a much lower recall score.

Specifically, attribute noise is where values of individual attributes do not make sense; whether they are erroneous or missing. Class noise is where a label does not make sense given the distribution of the features for other data with the same label; whether it is mislabeled or contradictory. In order to make the notions of attribute noise and class noise concrete, see Table 1 for a hypothetical distribution of the story recall data with an emphasis on what could potentially constitute both types of noise. In this section, we explore instability that could be due to outliers in training data, disagreement between features, or incorrect assumptions of the data.

Our first outlier analysis was based on research-stage settings where we have access to both attribute values and class labels. While this exact approach may not always be feasible in the eventual clinical application stage (since there are not always ground truth class labels available), the approach itself can nonetheless be harnessed in

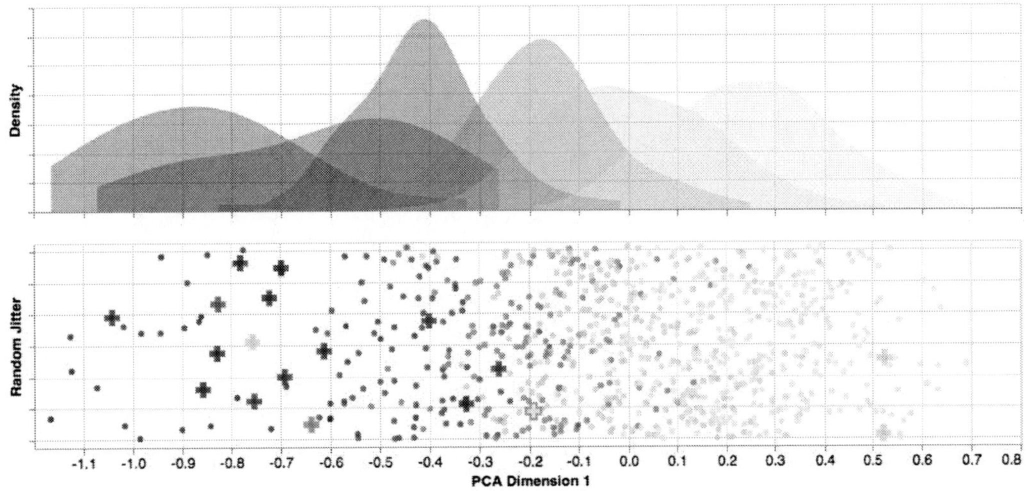


Figure 1: Distribution of the first dimension of Principal Component Analysis (PCA) of the 3 features of the story recall data separated by rating. The darker colored peak on the left represents the lowest rating (1 point) which increases by one point per peak to the lighter colored peak on the right hand side (6 points). Outliers found with the Isolation Forest algorithm are shown with a cross and the color of the cross represents the human rating given to that example.

the same manner but with the omission of ratings, classes, or labels. Here, we discovered outliers using the Isolation Forest algorithm (Liu et al., 2018). Most outlier detection algorithms first find the normal region of data and subsequently define anything outside of this defined region to be an outlier. The Isolation Forest algorithm, on the other hand, discovers minority data points that have attribute values that differ from those of the usual instances. Specifically, the algorithm isolates examples by selecting an attribute at random and then selecting a random split value between the maximum and minimum values of the selected feature. Anomalous examples will have shorter paths from the root to the leaves in their isolation trees than the normal examples since they need fewer partitions to be isolated. This algorithm is well-suited for high dimensional datasets and has proven to be an effective way of detecting outliers and anomalies (Ding and Fei, 2013). Furthermore, it works especially well for behavioral data as "normal" regions tend to be more variable than in other domains.

The current outlier analysis was specifically based on the number of types (i.e., unique words), the number of common types between the original story and the recall, the BERTScore between the original story and the recall, and the human rating given to the recall. Figure 1 shows the results of applying the Isolation Forest algorithm to the story recall data. It is shown that 18 outliers were detected. Such instances would be flagged for human review, where researchers can determine if attribute or class noise is present and either fix the erroneous values or exclude them from the modeling in the case that the examples are entirely invalid. When the approach is used in clinical settings to flag attribute noise, clinicians can review the raw data and make determinations for themselves rather than relying on a machine prediction.

Out of the 18 examples flagged by the Isolation Forest algorithm, 9 were determined to be invalid responses (i.e., participants stating that they simply do not remember or responses that are insufficient for data analysis) and 9 were valid responses with either sparse amounts of language or large amounts of language but poor performance. The average absolute error on the outliers was 1.34 (stdev = 0.80); the valid response outliers generated a higher absolute error (average = 1.63, stdev = 0.91) than the invalid response outliers (average = 1.05, stdev = 0.63). The performance of the model on outlier data is far lower than the models overall performance.

As the contamination threshold of the Isolation Forest algorithm is increased (i.e., the criteria for an outlier is relaxed), additional responses are chosen that mirror the behavior of these 18. This is a parameter that would need to be tuned such that

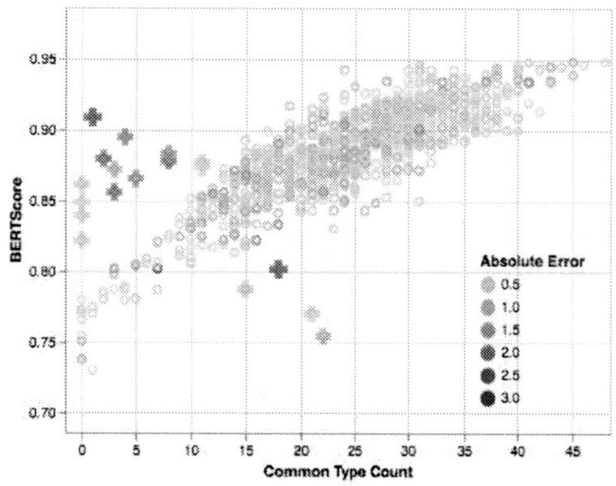


Figure 2: Scatter plot depicting the relationship between the number of common word types and the BERTScore of each example. The color represents the absolute error between the model rating and the human rating in each instance. Cross symbols indicate attribute noise (with two specific examples colored red and detailed at a high level in the text).

all true outliers are detected yet it does not extend into the normal data range. Furthermore, this parameter will need to be learned by investigating the true distribution of these phenomena and will depend on the application. Interestingly, the model performance did not change with the removal of these outliers. As approximately 2% of the data was flagged in this experiment, the model behaved indifferently to their exclusion. The exclusion of extremes (which help the performance of the model) combined with noise (which harm the performance of the model) potentially balanced out the effects of both. This Isolation Forest analysis can be performed on the same data without ratings in the eventual clinical stage to find attribute noise and extremes. We also present an analysis of features alone that can be done in any stage of the modeling process.

A basic noise detection approach that can be used at any stage of the modeling process is to simply find the examples with low attribute agreement (assuming that the attributes are collinear). Figure 2 depicts the distribution of two of the most predictive features of the story recall rating prediction model (the number of common word types and the BERTScore between the original story and the recall). There is a steady agreement between the two features, with some outliers (marked with crosses) outside of the diagonal where the features do not agree. The color of the circles represent how far off the model rating was from the human rating. Two examples with exceptionally high error (~2.5-3.0) are identified in red. The bottom-most red example is a response with a mixture of correct and incorrect (random) details, as well as incoherent language. The top-most red example is a response with a high BERTScore even though only a recitation of the title of the story was spoken. This high disagreement between features in turn uncovered a faulty feature score potentially due to flawed weighting parameters in the BERTScore model. We have shown that examples located off of the diagonal in plots such as these should be passed on for human evaluation as disagreement in two objective collinear attributes of story recall may raise concern.

Finding these outliers is critical because if a model has not been exposed to certain combinations of features or labels in its training set, then we cannot assume that it will produce accurate predictions in such settings. Outliers are important to detect both in the research stage in order to update or exclude certain examples from affecting the model in a negative manner and in the clinical setting so that spurious decisions are not made on abnormal data.

## 4 Effects of model generalizability

As previously stated, one of the most critical safeguards to spurious AI-based predictions is using large, diverse, and representative data (Cirillo et al., 2020), but this is not always possible. When using human behavioral data in machine learning algorithms, researchers inadvertently make the assumption that there is one canonical representation of specific groups of humans (i.e., those with serious mental illness), but this is simply not true. Those with psychiatric disorders exhibit extremely diverse symptoms and behaviors. Human behavior displays patterns indicative of a chaotic system (Paulus and Braff, 2003; Guess and Sailor, 1993), which holds true for behavior within one person as well as behavior within a group. To approach the topic of generalizable data, we first explored whether choosing different subsets within a training dataset would affect the output of the resulting model and whether there are spurious results when using smaller subsets.

The story recall regression model was trained on N = 846 samples, a large size relative to clinical experiments in the mental health domain. We used stratified sampling to create smaller subsets of the data that retain the proportions of each rating

| Percent of data (N) | Average model rating correlation (stdev) | Average BERTScore correlation (stdev) | Average common types correlation (stdev) |
|---|---|---|---|
| 100% (846) | 0.91 | 0.86 | 0.82 |
| 75% (634) | 0.91 (0.01) | 0.86 (0.01) | 0.82 (0.01) |
| 50% (423) | 0.90 (0.01) | 0.82 (0.01) | 0.79 (0.01) |
| 25% (212) | 0.88 (0.02) | 0.81 (0.03) | 0.79 (0.02) |

Table 2: The change in the average and standard deviation (stdev) of the correlations between (1) the human rating and the model rating, (2) the human rating and BERTScore, and (3) the human rating and common word types as smaller subsets of the data are randomly chosen in a stratified manner for training and testing. The first column displays the percent of data and the number of data points used in each data reduction setting.

and tested how the model behaved on these smaller subsets. Table 2 depicts the changing accuracy of the model and correlations of features to human ratings when these smaller subsets of the data were used for training and testing. We found the average correlation over a 10-fold cross-validation of the sampled subsets controlled such that sessions from the same participant did not occur simultaneously in both the training and testing sets. So as to show the low effect on the randomness involved in sampling smaller subsets, we report these metrics after 10 random re-samplings. It is shown that this regression model is stable when smaller subsets of the training data are used. Had the model shown significant drops in accuracy when restricting the dataset size, it could be concluded that the model was unstable or had overfit the training data.

Since experiments based on subsets of data retrieved from the same experimental population and setting do not necessarily show the true extent of model generalizability, we also performed transfer tests of the story recall model. Specifically, a second dataset was collected from inpatients at a substance abuse program in Louisiana (N = 99), most of whom suffered from co-occurring mood, psychotic, anxiety and personality spectrum disorders, as well as an additional collection from presumed healthy undergraduates at Louisiana State University (N = 124). Together, the inpatients and the presumed healthy undergraduates completed N = 1254 story recalls. A ridge regression model with the same NLP features as previously reported was trained on the initial dataset and tested on the new dataset, as well as vice versa. The first experiment resulted in a Pearson r correlation of 0.86 and the reverse an r of 0.84. Here, we conclude that the story recall regression model will generalize to differing clinical populations as well as illness severities. The same may not hold true for differing cultural populations as language differences may prove to be a confounding variable in transferring such a model. We thus advocate testing models on each new population prior to implementation.

Neuropsychological task scoring is a much more objective application area than other modeling applications in this field in which less is known and gold standard labels are often disagreed upon (e.g., disease detection, mental state tracking, and so on). Thus, generalizability is much more critical to test in these other applications and will potentially not yield such robust conclusions. Nonetheless, the understanding of when a model will yield accurate output and when it will not is an extremely important endeavor. Finding representative data is of the utmost importance in machine learning. In some cases, such as the story recall regression model, it is best to get as much data from as many people as possible. In other cases, especially when dealing with extreme diversity between individuals or subsets of individuals, it may be best to only use data that behaves in a similar fashion to the example currently being tested.

## 5 Discussion

Mental health is extremely dynamic as it can change on the scale of seconds, minutes, hours, or days, and language offers an objective and potentially unobtrusive way to assay such changes. Mental state in some conditions can change quickly with fatal consequences (e.g., suicide attempts) and more frequent monitoring of language and behavioral data, combined with machine learning methods, has the potential to offer clinicians unprecedented support in tracking patient state. Language can be harnessed for many applications as it offers a quantitative conceptualization of a person's underlying thought processes and mental health. Tracking such phenomena is extremely important yet increasingly complex, and as such there is a need for greater reliance on model outputs in this field.

In experiments involving NLP methods, it is common to deal with high dimensionality from features like word embeddings, parser outputs, and so on, which makes interpretation and understanding of models difficult. Features often go beyond normal distributions and as such there tends to be high variability in data distributions. Thus, it is especially important to create methods and tools that allow us to better understand the feature space and determine whether attributes or classes may violate assumptions.

An eventual goal of this line of work is to have a human-in-the-loop system where models analyze streams of high dimensional patient data and produce predictions of mental state and well-being. In the research stage of this implementation, real data must be analyzed to determine what normal distributions of attributes and classes appear to be. Aberrant instances of patient data can be flagged and reviewed by researchers to either update or exclude from models. Researchers must also test their models' generalizability by collecting additional samples or performing validation techniques to verify performance on unseen data. This process will allow for models to be based on the most accurate and representative data.

In the eventual clinical decision support system implementation, models must be realized such that attribute outliers are not predicted on, but rather the raw data is passed to a clinician to make a judgment. If the outlier is due to faulty feature values, clinicians can update these values or they can create their own labels and update the system such that future similar cases would not necessarily need to be verified by a human. In such a situation, there is a "best of both worlds" where models can execute the tasks that they are best at (high dimensional data analysis) and humans can execute the tasks that they are best at (handling anomalies and interpreting patient data).

For NLP and machine learning methods to be adopted in current research experiments as well as in eventual clinical practice, they require critical peer evaluation. What is needed is transparency in terms of data collection, validation, reproducibility, and clinical agreement in the association of language features to underlying illness. This paper showcases how essential it is that clinicians are involved in all stages of development. As such, it is a large step towards bringing more ethics and transparency into AI-based studies in mental health. Ethics review boards must demand this type of transparency and fairness in the creation of models so that systems that harness machine learning can be implemented in real clinical practice with low risk. Some discussion of this path forward has been brought to light by Friesen et al. (2021) who reported on IRBs as a means of ethics oversight in health research that harnesses AI.

## 6 Conclusion

This paper illustrates the importance of understanding the assumptions and distributions that underlie training data and the algorithms used, as well as the need to flag data that have characteristics that violate these assumptions. Not only is this knowledge important, but so too is having the tools to do this. We found model instabilities in a story recall regression model with the use of outlier detection algorithms and error analyses with respect to varying input. We advocate that approaches such as these be incorporated into machine learning and NLP-based clinical research and implementation. With the complexities inherent to models based on many features, high numbers of parameters, highly variable human behavioral data, and extremely high (and potentially fatal) stakes for mistakes, it is critical to establish methods beyond model designer intuition in assuring robustness and that predictions cannot be made on out of range data or data that lies in areas of instability. It should now be obvious that high predictive power on a relatively small dataset does not entail clinical relevance or generalizability, and that it is essential to use larger data sets, have more data collection outside of controlled settings, incorporate modeling safeguards, and use human-in-the-loop methodologies at all steps of the process.

## Acknowledgements

Parts of this project were funded by grant 231395 from the Research Council of Norway awarded to Brita Elvevåg.

## References

Alan Baddeley and Barbara A. Wilson. 2002. Prose recall and amnesia: implications for the structure of working memory. *Neuropsychologia*, 40:1737–1743.

Gillinder Bedi, Facundo Carrillo, Guillermo A. Cecchi, Diego Fernández Slezak, Mariano Sigman, Natália B. Mota, Sidarta Ribeiro, Daniel C. Javitt, Mauro Copelli, and Cheryl M. Corcoran. 2015. Automated analysis of free speech predicts psychosis onset in high-risk youths. *npj Schizophrenia*, 1.

Chelsea Chandler, Peter W. Foltz, Jian Cheng, Jared C. Bernstein, Elizabeth P. Rosenfeld, Alex S. Cohen, Terje B. Holmlund, and Brita Elvevåg. 2019. Overcoming the bottleneck in traditional assessments of verbal memory: Modeling human ratings and classifying clinical group membership. In *Proceedings of the Sixth Workshop on Computational Linguistics and Clinical Psychology*, pages 137–147.

Chelsea Chandler, Peter W. Foltz, Alex S. Cohen, Terje B. Holmlund, Jian Cheng, Jared C. Bernstein, Elizabeth P. Rosenfeld, and Brita Elvevåg. 2020a. Machine learning for ambulatory applications of neuropsychological testing. *Intelligence-Based Medicine*, 1–2.

Chelsea Chandler, Peter W. Foltz, and Brita Elvevåg. 2020b. Using machine learning in psychiatry: The need to establish a framework that nurtures trustworthiness. *Schizophrenia Bulletin*, 46:11–14.

Chelsea Chandler, Terje B. Holmlund, Peter W. Foltz, Alex S. Cohen, and Brita Elvevåg. 2021. Extending the usefulness of the verbal memory test: The promise of machine learning. *Psychiatry Research*, 297.

Davide Cirillo, Silvina Catuara-Solarz, Czuee Morey, Emre Guney, Laia Subirats, Simona Mellino, Annalisa Gigante, Alfonso Valencia, María José Rementeria, Antonella Santuccione Chadha, and Nikolaos Mavridis. 2020. Sex and gender differences and biases in artificial intelligence for biomedicine and healthcare. *npj Digital Medicine*, 3.

Michael A. Cirillo and Larry J. Seidman. 2003. Verbal declarative memory dysfunction in schizophrenia: from clinical assessment to genetics and brain mechanisms. *Neuropsychol. Rev*, 13:43–77.

Alex S. Cohen, Taylor L Fedechko, Elana K. Schwartz, Thanh P. Le, Peter W. Foltz, Jared Bernstein, Jian Cheng, Terje B. Holmlund, and Brita Elvevåg. 2019. Ambulatory vocal acoustics, temporal dynamics and serious mental illness. *Journal of Abnormal Psychology*, 128:97–105.

Cheryl M. Corcoran, Facundo Carrillo, Diego Fernández-Slezak, Gillinder Bedi, Casimir Klim, Daniel C. Javitt, Carrie E. Bearden, and Guillermo A. Cecchi. 2018. Prediction of psychosis across protocols and risk cohorts using automated language analysis. *World Psychiatry*, 17:67–75.

Christopher R. Cox, Emma H. Moscardini, Alex S. Cohen, and Raymond P. Tucker. 2020. Machine learning for suicidology: A practical review of exploratory and hypothesis-driven approaches. *Clin Psychol Rev*, 82.

Jacob Devlin, Ming-Wei Chang, Kenton Lee, and Kristina Toutanova. 2019. BERT: Pre-training of deep bidirectional transformers for language understanding. In *Proceedings of the 2019 Conference of the North American Chapter of the Association for Computational Linguistics: Human Language Technologies, Volume 1 (Long and Short Papers)*, pages 4171–4186, Minneapolis, Minnesota. Association for Computational Linguistics.

Zhiguo Ding and Minrui Fei. 2013. An anomaly detection approach based on isolation forest algorithm for streaming data using sliding window. In *PIFAC Proceedings Volumes*, volume 46, pages 12–17.

Brita Elvevåg, Peter W. Foltz, Daniel R. Weinberger, and Terry E. Goldberg. 2007. Quantifying incoherence in speech: an automated methodology and novel application to schizophrenia. *Schizophrenia Research*, 93:304–316.

Kawin Ethayarajh and Dan Jurafsky. 2020. Utility is in the eye of the user: A critique of nlp leaderboards. In *Proceedings of the 2020 Conference on Empirical Methods in Natural Language Processing*, pages 4846–4853.

Elif Eyigoz, Sachin Mathur, Guillermo Cecchi Mar Santamaria, and Melissa Naylor. 2020. Linguistic markers predict onset of alzheimer's disease. *EClinicalMedicine*, 28:304–316.

Peter W. Foltz, Mark Rosenstein, and Karen E. Lochbaum. 2013. Improving performance of automated scoring through detection of outliers and understanding model instabilities. In *Presented at the National Council on Measurement in Education Conference*, San Francisco, CA.

Phoebe Friesen, Rachel Douglas-Jones, Mason Marks, Robin Pierce, Katherine Fletcher, Abhishek Mishra, Jessica Lorimer, Carissa Véliz, Nina Hallowell, Mackenzie Graham, Mei Sum Chan, Huw Davies, and Taj Sallamuddin. 2021. Governing ai-driven health research: Are irbs up to the task? *Ethics Hum Res*, 43:35–42.

Doug Guess and Wayne Sailor. 1993. Chaos theory and the study of human behavior: Implications for special education and developmental disabilities. *The Journal of Special Education*, 27:16–34.

David J. Hand. 2006. Classifier technology and the illusion of progress. *Statistical Science*, 21:1–14.

Joseph Henrich, Steven J. Heine, and Ara Norenzayan. 2010. The weirdest people in the world. *Behav Brain Sci*, 33:61–83.

Terje B. Holmlund, Chelsea Chandler, Peter W. Foltz, Alex S. Cohen, Jian Cheng, Jared C. Bernstein, Elizabeth P. Rosenfeld, and Brita Elvevåg. 2020. Applying speech technologies to assess verbal memory in patients with serious mental illness. *npj Digital Medicine*, 3.

Terje B. Holmlund, Peter W. Foltz, Alex S. Cohen, Håvard D. Johansen, Randi Sigurdsen, Pål Fugelli, Dagfinn Bergsager, Jian Cheng, Jared Bernstein, Elizabeth Rosenfeld, and Brita Elvevåg. 2019. Moving psychological assessment out of the controlled laboratory setting and into the hands of the individual: Practical challenges. *Psychological Assessment*, 31:292–303.

Benjamin Kompa, Jasper Snoek, , and Andrew L. Beam. 2021. Second opinion needed: communicating uncertainty in medical machine learning. *npj Digital Medicine*, 4.

Alexandra König, Aharon Satt, Alexander Sorin, Ron Hoory, Orith Toledo-Ronen, Alexandre Derreumaux, Valeria Manera, Frans Verhey, Pauline Aalten, Phillipe H. Robert, and Renaud David. 2015. Automatic speech analysis for the assessment of patients with predementia and alzheimer's disease. *Alzheimer's Dementia: Diagnosis, Assessment Disease Monitoring*, 1:112–124.

Tilman Lange, Mikio L. Braun, Volker Roth, and Joachim M. Buhmann. 2002. Stability-based model selection. In *Proceedings of the 15th International Conference on Neural Information Processing Systems*, pages 633—-642.

Muriel D. Lezak, Diane B. Howieson, Erin D. Bigler, and Daniel Tranel. 2012. *Neuropsychological assessment (5th Ed.)*. Oxford University Press.

Fei T. Liu, Kai M. Ting, and Zhi-Hua Zhou. 2018. Isolation forest. In *Proceedings of the Eighth IEEE International Conference on Data Mining*, pages 413–422, Pisa, Italy.

Julia Longenecker, Jamie Genderson, Dwight Dickinson, James Malley, Brita Elvevåg, Daniel R. Weinberger, and James Gold. 2010. Where have all the women gone?: participant gender in epidemiological and non-epidemiological research of schizophrenia. *Schizophrenia Research*, 119:240–245.

Ellen W. McGinnis, Steven P. Anderau, Jessica Hruschak, Reed D. Gurchiek, Nestor L. Lopez-Duran, Kate Fitzgerald, Katherine L. Rosenblum, Maria Muzik, and Ryan S. McGinnis. 2019. Giving voice to vulnerable children: Machine learning analysis of speech detects anxiety and depression in early childhood. *IEEE Journal of Biomedical and Health Informatics*, 23:2294–2301.

Marie-Hélène Metzger, Nastassia Tvardik, Quentin Gicquel, Côme Bouvry, Emmanuel Poulet, and Véronique Potinet-Pagliaroli. 2017. Use of emergency department electronic medical records for automated epidemiological surveillance of suicide attempts: a french pilot study. *International Journal of Methods in Psychiatric Research*, 26:e1522.

Sina Mohseni, Mandar Pitale, Vasu Singh, and Zhangyang Wang. 2020. Practical solutions for machine learning safety in autonomous vehicles. In *The AAAI Workshop on Artificial Intelligence Safety (Safe AI)*.

Sylvester Olubolu Orimaye, Jojo Sze-Meng Wong, and Chee Piau Wong. 2018. Deep language space neural network for classifying mild cognitive impairment and alzheimer-type dementia. *PLoS ONE*, 13:e0205636.

Martin T. Paulus and David L. Braff. 2003. Chaos and schizophrenia: does the method fit the madness? *Neuroscience Perspectives*, 53:3–11.

Zhichao Peng, Qinghua Hu, and Jianwu Dang. 2019. Multi-kernel svm based depression recognition using social media data. *International Journal of Machine Learning and Cybernetics*, 10:43–57.

Brian Roark, Margaret Mitchell, John-Paul Hosom, Kristy Hollingshead, and Jeffrey Kaye. 2011. Spoken language derived measures for detecting mild cognitive impairment. *IEEE Transactions on Audio, Speech, and Language Processing*, 19:2081—2090.

Mark Rosenstein, Peter W. Foltz, Lynn E. DeLisi, and Brita Elvevåg. 2015. Language as a biomarker in those at high-risk for psychosis. *Schizophrenia Research*, 165:249—250.

Euijung Ryu, Alanna M. Chamberlain, Richard S. Pendegraft, Tanya M. Petterson, William V. Bobo, , and Jyotishman Pathak. 2016. Quantifying the impact of chronic conditions on a diagnosis of major depressive disorder in adults: a cohort study using linked electronic medical records. *BMC Psychiatry*, 16.

Adrian B. R. Shatte, Delyse M. Hutchinson, and Samantha J. Teague. 2019. Machine learning in mental health: A scoping review of methods and applications. *Psychological Medicine*, 49:1426–1448.

Chang Su, Zhenxing Xu, Jyotishman Pathak, and Fei Wang. 2020. Deep learning in mental health outcome research: a scoping review. *Transl Psychiatry*, 10.

Anja Thieme, Danielle Belgrave, and Gavin Doherty. 2020. Machine learning in mental health: A systematic review of the hci literature to support the development of effective and implementable ml systems. *ACM Transactions on Computer-Human Interaction*, 27:Article 34.

Peter Turney. 1995. Technical note: Bias and the quantification of stability. *Machine Learning*, 20:23–33.

Zhuoran Wang, Anoop D. Shah, A. Rosemary Tate, Spiros Denaxas, John Shawe-Taylor, and Harry Hemingway. 2012. Extracting diagnoses and investigation results from unstructured text in electronic health records by semi-supervised machine learning. *PLoS One*, 7:e30412.

David Wechsler. 1997. *Wechsler Memory Scale - Third Edition, WMS-III: Administration and scoring manual*. The Psychological Corporation.

Robert Whelan and Hugh Garavan. 2014. When optimism hurts: inflated predictions in psychiatric neuroimaging. *Biol Psychiatry*, 75:746–748.

Jheng-Long Wu, Liang-Chih Yu, and Pei-Chann Chang. 2012. Detecting causality from online psychiatric texts using inter-sentential language patterns. *BMC Medical Informatics and Decision Making*, 12.

Tianyi Zhang, Varsha Kishore, Felix Wu, Kilian Q. Weinberger, and Yoav Artzi. 2020. Bertscore: Evaluating text generation with bert. In *International Conference on Learning Representations*.

Xingquan Zhu and Xindong Wu. 2004. Class noise vs. attribute noise: A quantitative study. *Artificial Intelligence Review*, 22:177–210.

Ayah Zirikly, Philip Resnik, Özlem Uzuner, and Kristy Hollingshead. 2019. Clpsych 2019 shared task: Predicting the degree of suicide risk in reddit posts. In *Proceedings of the Sixth Workshop on Computational Linguistics and Clinical Psychology*, pages 24–33.

# Towards the Development of Speech-Based Measures of Stress Response in Individuals

**Archna Bhatia** and **Toshiya Miyatsu** and **Peter Pirolli**
Institute for Human and Machine Cognition
15 SE Osceola Ave
Ocala, FL 34471

## Abstract

Psychological and physiological stress in the environment can induce a different stress response in different individuals. Given the causal relationship between stress, mental health, and psychopathologies, as well as its impact on individuals' executive functioning and performance, identifying the extent of stress response in individuals can be useful for providing targeted support to those who are in need. In this paper, we identify and validate features in speech that can be used as indicators of stress response in individuals to develop speech-based measures of stress response. We evaluate effectiveness of two types of tasks used for collecting speech samples in developing stress response measures, namely Read Speech Task and Open-Ended Question Task. Participants completed these tasks, along with the verbal fluency task (an established measure of executive functioning) before and after clinically validated stress induction to see if the changes in the speech-based features are associated with the stress-induced decline in executive functioning. Further, we supplement our analyses with an extensive, external assessment of the individuals' stress tolerance in the real life to validate the usefulness of the speech-based measures in predicting meaningful outcomes outside of the experimental setting.

## 1 Introduction

Various psychological and physiological stress conditions, e.g., an approaching deadline, an interview not going well, a combat situation, or extreme temperatures, can have an impact on an individual due to various (maladaptive) physiological and mental processes (Yaribeygi et al., 2017; Sapolsky, 1996). Long-term exposure to stress can play a significant role in the formation and exacerbation of mental disorders, such as anxiety disorders, depression, and schizophrenia (Gomes and Grace, 2017; Yang et al., 2015; Tafet and Nemeroff, 2015; Esch et al., 2002). Stress in one's environment may result in a relatively immediate (whether short-term or long-lasting) effect on performance. For example, a condescending interviewer may lead an interviewee to not be able to respond at all or a sudden combat situation may lead an individual to make more errors. However, different individuals respond differently to the same stress conditions depending on their mental and physiological constitution, experiences, training and preparedness, among other factors. Identifying the degree of stress response in individuals would be helpful in reducing stress' impact on their health and performance both at the individual and at the community level. For example, an automatic measure of stress response can be used by an individual for self monitoring and deciding to use a management strategy of daily stress reduction exercises when needed. Similarly, community members can be supported through targeted allocation of mental health resources. An automatic measure of stress response can provide additional information about individuals' response to stress to the clinicians treating them so that appropriate and timely therapeutic support can be provided.

Previously, self-report inventories of stressors and their symptoms have been used to measure individuals' stress response (e.g., Bland et al., 2012; Tatar et al., 2018; Rushall, 1990). However, these inventories are limited in their scope (e.g., sports, school) and utilities. Further, self-report inventories have inherent problems. For example, individuals may not be fully aware of the effect stressors have on them, or they may not answer questions truthfully. Therefore, development of more objective yet accessible measures of stress response are needed.

An extensive body of research has shown the impact of stress on speech, e.g., Jackson et al. (2016); Jena and Singh (2016); Schuller et al. (2014); Giddens et al. (2013); Lierde et al. (2009); He et al. (2008); Dietrich et al. (2008); Hansen and Patil



*Proceedings of the Seventh Workshop on Computational Linguistics and Clinical Psychology*, pages 192–203
June 11, 2021. 

(2007); Fernandez and Picard (2003); Brenner and Shipp (1988); Brenner et al. (1983) have associated certain changes in speech with exposure to stress. For example, Brenner and Shipp (1988) reported an increase in fundamental frequency, amplitude and speech rate in extreme levels of stress. They also reported changes in the energy distribution, frequency jitter and amplitude shimmer in stressed speech. Features, such as Mel-frequency Cepstrum Coefficients, have been found to be affected by emotional states including anxiety/stress to be useful for identifying or classifying these emotional states, e.g., see Vaikole et al. (2020); Dhole and Kale (2020); Tomba et al. (2018); Hansen and Patil (2007). The primary goal of the current project is to extend this work by identifying and validating a set of speech-based features that can be used as individual difference measures of stress response.

To achieve this goal, we use two continuous speech sample collection methods, namely a Read Speech Task and an Open-ended Question Task. Read speech provides much cleaner data than spontaneous speech and hence can be very useful for modeling speech related phenomena. It has been extensively used in speech processing studies, for example, see Pernkopf et al. (2009); Nakamura et al. (2008); Pruthi and Espy-Wilson (2007, 2004); Garofolo et al. (1993). Open-ended Question Task, on the other hand, provides more naturalistic data, which can complement the information available in read speech. The notion that they may provide different types of information is confirmed by works such as Schuppler (2017) which discussed the need for developing methods for different speaking styles instead of just focusing on read speech.

For the purpose of developing measures for stress response, we consider `stress response` and `stress tolerance` to be two facets of the same phenomenon, where stress response refers to how an individual responds to or is affected by stress, whereas stress tolerance refers to how tolerant an individual is to stress (i.e., how well they can still perform tasks under stress). We focus on investigating speech and identifying relevant acoustic features to develop a speech-based measure of stress response. We expect such a measure to also be informative about an individual's stress tolerance. To this end, we establish the relationship of speech features with a complex, ecologically valid stress tolerance measure based on trained judges' stress tolerance assessment of individuals as described in Section 2.3.

The rest of the paper is organized as follows: In Section 2, we describe our data and data collection procedures. In Section 3, we discuss our methodology to extract acoustic features from the speech produced by individuals in stress conditions and to prepare the performance assessment measures for evaluating the usefulness of the extracted features. In Section 4, we present our findings based on the analysis of extracted features from speech in terms of their relationship with stress conditions as well as with cognitive performance and real-life stress tolerance measures. In Section 5, we discuss our results and implications for development of speech-based measures for stress response in individuals. In addition to the features identified by our investigation to be informative of an individual's stress response, we also provide recommendations with respect to the types of tasks used to collect speech samples to extract these features. In Section 6, we conclude and briefly discuss future work. This is followed by Section 7 where we discuss a few use cases and ethical considerations for this work.

## 2 Data

The data used for our investigation were collected from 13 male participants. They were recruited from among the candidates going through a week-long selection assessment process at a US military unit who had provided consent prior to the selection week and remained on-site until the end of the selection week. They were provided a description of the study before obtaining their consent. The data collection was conducted the day after the selection week was over. All protocols for data collection were approved by the Institutional Review Board at the appropriate branch of the US military (where data had to be collected) as well as the Institutional Review Board at the Florida Institute for Human and Machine Cognition (where the research activity had to take place) prior to data collection.

Two types of data were collected: speech samples and selection assessment data. The speech samples were recorded in a lab setting in two stress conditions, namely `Neutral` and `Stress`. Section 2.1 provides details about stress induction and Section 2.2 provides more details about speech data collection. In addition to the speech samples, participants' scores that were assigned during the selection week by trained and experienced US mil-

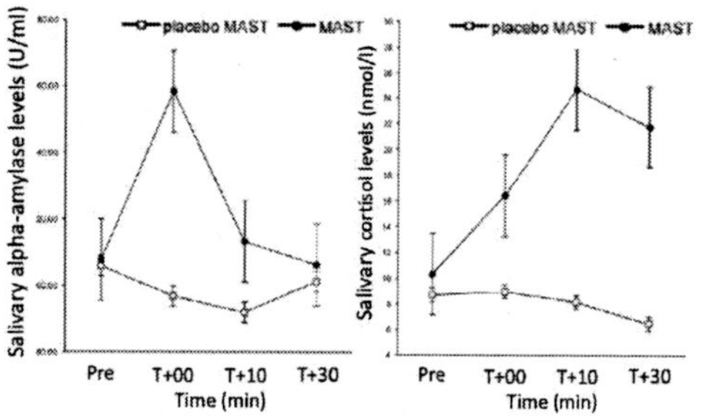

Figure 1: A demonstration of salivary gluco-corticol stress response induced by MAST (Smeets, et al., 2012)

itary assessment personnel were obtained to augment our analyses. More details about the selection assessment data are provided in Section 2.3.

### 2.1 Stress Induction

In order to collect speech samples from participants in the two stress conditions, `Neutral` and `Stress`, a version of the Maastricht Acute Stress Test (MAST) (Smeets et al., 2012) was used for stress induction. MAST is a clinically certified stress induction which reliably elicits gluco-corticol stress response that could be measured through an increase in established biomarkers of stress, such as alpha-amylase and cortisol, see Figure 1. MAST combines two established methods of stress induction: social stress through the Trier Social Stress Test and physiological stress through the Cold Pressor Test. In our version, participants were instructed to sit in front of a video recorder and look straight at it because their facial expression would be analyzed later. Then, they completed several rounds of hand immersion trials (HITs) and Mental Arithmetic (MA) trials. In the HITs, they were asked to submerge their hand for a set duration (see Figure 2) in a container filled with ice water which was kept at 4°C. In the MA trials, they were asked to count backwards by 17 starting from 2043, and whenever they made a mistake or took more than three seconds to say the next number, the experimenter gave them negative feedback and asked them to start over. The participants also performed a Verbal Fluency Task (see Section 2.2) during stress induction. Figure 2 shows the task sequence during stress induction.

### 2.2 Speech Data

Speech samples were collected from the participants through three tasks: Read Speech Task, Open-ended Question Task and Verbal Fluency Task. Participants completed these tasks under both the `Neutral` condition and the `Stress` condition. Read Speech Task and Open-ended Question Task were used to extract acoustic features from the two types of continuous speech samples, the read speech samples and the naturalistic speech samples respectively. Verbal Fluency Task, on the other hand, was used as an assessment for cognitive performance under the two stress conditions.

**Read Speech Task.** Participants read out loud a 243 words passage about the psychological construct 'grit' in both stress conditions. The passage was borrowed from an online blog on grit (Doyle, 2020).[1] It was modified to include all phonemes in American English to enable a rich set of analyses (including at the phonemic level).

**Open-Ended Question Task.** Participants were asked to speak for two minutes in response to four open-ended questions each to obtain their naturalistic speech samples in the two stress conditions. The questions focused on stimulation seeking and suppression of emotions in the `Neutral` condition and on response to distress and reappraisal of negative emotions in the `Stress` condition. The topics of these questions were derived from two well-established works regarding stress response, namely defensive reactivity (Kramer et al., 2012) and emotion regulation (Gross, 2014). Specifically, these questions sought information from participants about how they reacted to stressful situations, e.g., "Recall and describe the most recent event in which you were stressed about something. How quickly/slowly did you recover from it?".

**Verbal Fluency Task.** Verbal Fluency Task, a well-established measure of executive functioning (Shao et al., 2014), was used to assess participants' cognitive performance in the `Neutral` and `Stress` conditions. Participants were asked to say out loud as many words as they could remember in 1 minute that belonged to a given category. 'Body parts', 'fruits', 'words starting with A', and 'words starting with F' were used in the `Neutral` condition and 'animals' and 'words starting with C' were used in the `Stress` condition).

[1] `https://www.aceable.com/blog/aceable-essay-on-grit/`

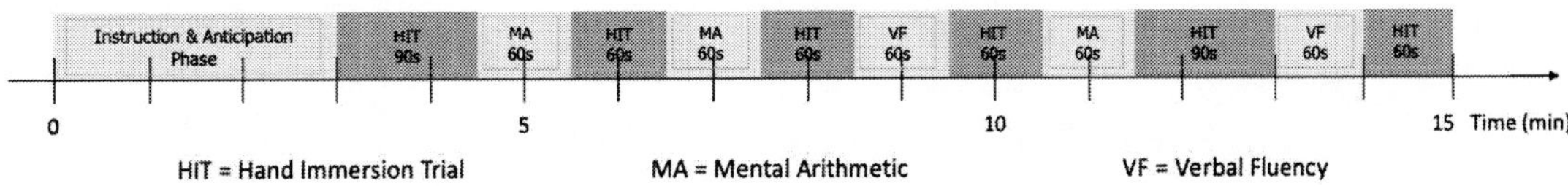


Figure 2: A schematic illustration of the task sequence within the version of MAST employed in the current study

### 2.3 Selection Assessment Data

In addition to the participants' responses to Verbal Fluency Task used for assessing their cognitive performance, we also obtained selection assessment data from the host military organization. Specifically, these data were scores assigned by trained assessment personnel based on their observations of attributes demonstrated by the participants in a given task, such as teamwork, leadership, and stress tolerance. The data consisted of 61 scores from 17 tasks performed by the participants during the week-long selection assessment.

## 3 Methodology

### 3.1 Feature Extraction from Speech Samples

In order to identify indicators of stress response in the speech signal, we extracted a number of acoustic features from the speech samples collected from the participants under the two stress conditions. We used Librosa (McFee et al., 2020, 2015), a Python library for audio and music analysis, to extract the acoustic features from the speech samples. Both time domain and frequency domain features were extracted. The time domain features included Amplitude Envelope, Root Mean Square Energy and Zero Crossing Rate. The frequency domain features included Magnitude Spectrum, Short-time Fourier transform Spectrum, and 13 Mel-frequency Cepstrum Coefficients (MFCCs) as well as their first and second derivatives. Many of these time domain and frequency domain features have been found to be affected by stress (Hansen and Patil, 2007; Fernandez and Picard, 2003). Features were extracted from speech using overlapping frames with frame size of 1024 and hop size of 512 waveform samples. The mean of the feature values extracted for each frame in a speech sample was used as the extracted feature from the speech sample for the analyses discussed in Section 4. Additionally, duration of the signal was also taken as a feature from Read Speech Task because the rate of speech has previously been found to be affected by stress (Wikibooks, 2018; Brenner and Shipp, 1988; Brenner et al., 1983). Thus, a total of 45 features from Read Speech Task and 44 features from Open-ended Question Task were extracted from the speech samples of participants corresponding to each of the stress conditions.

### 3.2 Performance Score Based on Verbal Fluency Task

The speech data collected from Verbal Fluency Task were transcribed using speech_recognition, a Python client for the Google Speech-to-Text API (Google, 2019). The generated transcriptions were manually checked for any errors by one of the authors of this paper. Although the transcription had a high accuracy (> 90%), manual checking of errors and manual counting of the total number of words recalled by participants in the transcriptions was necessary because the transcription often recognized two components of a multiword item as two words (e.g., 'dragon' and 'fruits' for 'dragon fruits', 'polar' and 'bear' for 'polar bear').[2] The transcriptions were then used to compute the performance metric 'Word Recall', the total number of words recalled by the participant in one minute.

Given the different base rates for different categories (i.e., the semantic categories of 'body parts', 'fruits' and 'animals', and the phonetic categories of 'words starting with A', 'words starting with F' and 'words starting with C'), we first normalized (Z-score) the performance metric 'Word Recall' for each of the categories. We, then, computed the mean of the normalized word recall for the four `Neutral` categories (viz. 'body parts', 'fruits', 'words starting with A', and 'words starting with F') and for the two `Stress` categories (viz. 'animals' and 'words starting with C'), and subtracted the resulting score in the `Neutral` condition from the score in the `Stress` condition for each participant. Thus obtained score represented the change in cognitive performance due to stress (in *SD*) relative to other participants.

[2]Inspired by the relevance of the error rates in mental health contexts, since speech-to-text systems' performance may also decline as the speech is affected under stress, in our future investigations, we may examine the error rate also as a feature for an individual's stress response.

### 3.3 Stress Tolerance Score Based on Selection Assessment Data

Of the 61 selection assessment scores assigned by the host agency's experienced personnel based on participant attributes demonstrated in 17 tasks performed during selection week (see Section 2.3), seven were on stress tolerance. Given that these scores came from a diverse set of tasks (e.g., a team building exercise) we computed a correlation (Pearson's Correlation) among all of them to see if these scores converged to measure the same construct (i.e., stress tolerance). These seven stress tolerance scores showed reasonable convergence, $r(7) = .41$. Thus, an average of these stress tolerance scores was taken as a complex, ecologically valid measure of the participants' stress tolerance.

## 4 Analysis and Findings

One can expect that the features in speech indicative of an individual's stress response can also be taken to indicate that the corresponding speech was produced in a `Stress` condition. Hence, one strategy to identify potentially relevant features associated with stress response from among the 45 extracted features (44 in case of Open-ended Question Task) is to find the ones that can distinguish between the `Neutral` and the `Stress` conditions. Hence, we performed a paired sample t-test on our data where the extracted features in the two stress conditions were taken as the two sets of observations pre- and post- stress induction.[3] We found that a number of speech features, extracted from Read Speech Task and Open-ended Question Task, showed statistically significant difference between the means in the observations corresponding to the two stress conditions, as shown in Table 1.

We found that duration and a time domain feature Zero Crossing Rate extracted from Read Speech Task showed significant difference between the two stress conditions with $p < .005$. [4] The rest of the features that showed significant difference with varying $p$-values ($p < .01$ or $p < .05$) were all associated with the frequency domain features MFCCs or their derivatives.

| Read Speech | Open-ended Question |
|---|---|
| Duration*** | - |
| Zero Crossing Rate*** | Zero Crossing Rate* |
| MFCC2** | MFCC2*** |
| Delta MFCC4* | - |
| Delta Delta MFCC4* | - |
| - | Delta Delta MFCC12* |
| - | Delta Delta MFCC11* |
| - | Delta Delta MFCC2* |
| - | Delta Delta MFCC10* |

Table 1: Speech features that show statistically significant difference between the `Neutral` and `Stress` conditions for the two speech tasks, Read Speech Task and Open-ended Question Task. *** indicates $p < .005$, ** indicates $p < .01$, * indicates $p < .05$

Some features showed significant difference when extracted from either of the two speech tasks. For example, the time domain feature Zero Crossing Rate showed significant difference when extracted from the Open-ended Question Task speech samples as well, although with comparatively less significance than when extracted from the Read Speech Task speech samples. In contrast, the frequency domain feature MFCC2 showed more significance when extracted from the Open-ended Question Task samples than when extracted from the Read Speech Task samples.

The frequency domain features Delta MFCC4 (first derivative of MFCC4) and Delta Delta MFCC4 (second derivative of MFCC4) showed significant difference between the means for observations in the two stress conditions when they were extracted from Read Speech Task, but not when they were extracted from Open-ended Question Task. Similarly, second derivatives of MFCC12, MFCC11, MFCC2 and MFCC10 showed significant difference in the two stress conditions when they were extracted from Open-ended Question Task but not when they were extracted from Read Speech Task.

While more features extracted from the speech samples collected for Open-ended Question Task showed significant difference in the means between the two stress conditions, the features extracted from the Read Speech Task speech samples showed a higher significant difference between the two conditions, in general.

We then correlated the extracted features with Verbal Fluency Task-based scores representing the change in cognitive performance due to stress to

[3]We plan to collect more data to increase sample size to confirm our findings and increase reliability of the results. With the larger dataset in the future, for a more robust set of significant features, Bonferroni or similar corrections may be applied to the multiple comparisons.

[4]In this paper, we have specified a $p$ value of $< .005$ to indicate high significance but these denote the cases where the $p$ values are very close to though slightly greater than $< .001$.

provide a measurement of stress response in participants (performance score), as described in Section 3.2. Similarly, we correlated these acoustic features with the selection assessment-based scores that provided a measurement of stress tolerance in participants (stress tolerance score), as described in Section 2.3. Table 2 presents the correlations (Pearson's Correlation Coefficients) of the acoustic features with the performance score (left) and the stress tolerance score (right) for the two speech tasks. We explored these correlations to test two hypotheses as follows. First, some of the acoustic features show significant correlations with stress response/stress tolerance. Second, these features overlap with features identified to be differential between the two stress conditions based on inferential statistics (e.g., the paired sample t-test above).

In regards to the hypothesis about the relationship between acoustic features and performance scores/stress tolerance scores, we found that there were a number of acoustic features in both speech tasks that showed high to moderate correlations with performance scores as well as with stress tolerance scores as shown in Table 2. Many of these were found to be highly significant with $p$-values $< .005$ as in the case of acoustic feature Delta Delta MFCC7 for correlations with performance scores for Open-ended Question Task, and acoustic features magnitude spectrum for correlations with stress tolerance scores for Read Speech Task and MFCC11 for correlations with stress tolerance scores for Open-ended Question Task. Additionally, there were other features with which also correlations were significant with $p$-values of $< .01$ or $< .05$. For example, Delta Delta MFCC1 showed strong correlations with performance scores for Open-ended Question Task with $p$-values $< .01$. Similarly, the duration feature, representing rate of speech, was found to be moderately correlated with performance scores for Read Speech Task with $p$-values $< .05$. Delta Delta MFCC13 was moderately correlated with stress tolerance in Read Speech Task with $p$-values $< .05$, and so were features MFCC1, MFCC8 and MFCC9 in Open-ended Question Task. The full set of correlations corresponding to all extracted features are provided in Appendix A.

In regards to the second hypothesis about the highly differential features for stress conditions to also be correlated with the performance and the stress tolerance scores, we found that there is some overlap between the two sets of features. However, not all features that significantly distinguished the stress conditions were also strongly/moderately correlated with the performance and the stress tolerance scores for the two speech tasks. Duration extracted from Read Speech Task was found to be significantly differential for stress conditions and it showed moderate correlation with the performance score in Read Speech Task which was also significant. However, it was not found to be correlated with stress tolerance (with $r(13) = .184$ for Read Speech Task and .313 for Open-ended Question Task). Zero Crossing Rate, on the other hand, was found to be significantly differential for the stress conditions for both Read Speech Task and Open-ended Question Task, but it did not show significant correlations with the performance and the stress tolerance scores for either of the tasks. Similarly, MFCC2, while significantly differential for stress conditions for both the tasks, showed low correlations with the performance and the stress tolerance scores. Delta MFCC4 and Delta Delta MFCC4 were significantly differential for Read Speech Task but showed low correlations with both the scores for both the tasks. Delta Delta MFCC12, Delta Delta MFCC11, Delta Delta MFCC2 and Delta Delta MFCC10 were significantly differential for Open-ended Question Task, but showed low correlations with both the scores for both the tasks.

While exploring the above two hypotheses, we found a subset of acoustic features (e.g., duration, magnitude spectrum, some of the MFCCs or their derivatives) that showed strong to moderate correlations with the performance score (stress response) or the stress tolerance score, answering our first question setforth in Section 1. Next, we explored the effectiveness of speech tasks in indicating an individual's stress response to answer the second question setforth in Section 1.

Based on the correlations in Table 2, we found that Open-ended Question Task resulted in more acoustic features than Read Speech Task that showed strong correlations with performance scores (Delta Delta MFCC7 and Delta Delta MFCC1) as well as stress tolerance scores (MFCC11, MFCC1, MFCC8 and MFCC9) that were significant. However, in Read Speech Task also, we found a few acoustic features that were not identified by Open-ended Question Task but still showed strong correlations with the stress tolerance scores (magnitude spectrum and Delta Delta

| Feature | Performance Score | | Stress Tolerance Score | |
|---|---|---|---|---|
| | RST | OQT | RST | OQT |
| **Duration related features:** | | | | |
| Duration (Speech Rate) | -.628* | - | | - |
| **Time domain features:** | | | | |
| Zero Crossing Rate | .546 | | | |
| **Frequency domain features:** | | | | |
| Magnitude Spectrum | | | -.753*** | |
| Short-time Fourier Transform Spectrum | -.502 | | | .515 |
| MFCC1 | | | | .617* |
| MFCC7 | | .524 | | |
| MFCC8 | | | | -.583* |
| MFCC9 | | | | -.554* |
| MFCC11 | | .516 | | -.735*** |
| Delta Delta MFCC1 | | -.710** | | |
| Delta Delta MFCC7 | | .759*** | | |
| Delta Delta MFCC8 | | .519 | | |
| Delta Delta MFCC10 | | | | .534 |
| Delta Delta MFCC13 | | | .589* | |

Table 2: Moderate to high correlations (> .5) between the difference scores (`Stress - Neutral`) from among the 45 features extracted from Read Speech Task (RST) and from Open-ended Question Task (OQT) and the stress-induced change in Verbal Fluency Task performance (left) and the stress tolerance score from the selection assessment data (right). The duration feature was dropped for OQT since the task duration itself was set to two minutes. *** indicates $p < .005$, ** indicates $p < .01$, * indicates $p < .05$

MFCC13) that were significant. Other features were also identified in these tasks that showed moderate correlations with the performance scores (e.g., Zero Crossing Rate and Short-time Fourier Transform spectrum in Read Speech Task, and duration, MFCC7, Delta Delta MFCC8 and MFCC11 in Open-ended Question Task) as well as with the stress tolerance scores (e.g., Delta Delta MFCC10 and Short-time Fourier Transform Spectrum in Open-ended Question Task) but they were not found to be significant. Thus, we found that the two tasks provided complementary information in terms of features that were strongly correlated with stress response/stress tolerance with significance.

## 5 Towards Building a Speech-Based Measure of Stress Response/Tolerance

In Section 4, we identified a number of acoustic features that strongly/moderately correlated with Verbal Fluency Task-based performance scores and external assessment-based stress tolerance scores. We found that duration (speech rate) and frequency domain features, such as magnitude spectrum and some of the MFCCs or their derivatives, showed strong to moderate correlations with stress response/stress tolerance that were significant. Time domain features, on the other hand, did not show significant correlations with either stress response or stress tolerance. Although this finding needs to be confirmed with larger data, this may be taken to indicate the effectiveness of duration and frequency domain features over time domain features in a speech-based measure for stress response/tolerance. It should be noted that some of these duration and frequency domain features, e.g., speech rate, MFCCs and their derivatives have also been found to be indicative of stress in prior works on stress detection from the speech signal, e.g., Vaikole et al. (2020); Dhole and Kale (2020); Tomba et al. (2018); Brenner and Shipp (1988), to name a few. Also, opportunities to validate stress response or any predictive features against rich, ecologically valid datasets like the Selection Assessment Data used in our experiments are rare, and it is encouraging to see significant correlations between some of these features and the stress tolerance score based on the Selection Assessment Data (Table 2).

We compared the two speech collection tasks, Read Speech Task and Open-ended Question Task, for their effectiveness in providing useful information in speech features for stress response/tolerance. The findings showed that these two tasks provided complementary information in that a different set of features correlated with stress response/tolerance when samples were collected through the two tasks. This makes sense given the fact that Read Speech Task provides cleaner speech samples whereas Open-ended Question Task provides more naturalistic speech. This suggests that a speech-based measure for stress response/tolerance would benefit from using both these tasks for data collection. However, if only one task needs be used in order to minimize time, effort, and other resources expended in data collection, Open-ended Question Task should be used since Open-ended Question Task samples led to a larger number of correlated acoustic features which showed strong correlations with high significance values.

Another finding from Section 4 was that the features helpful in distinguishing between the stress conditions may not necessarily be the features that indicate stress response/tolerance. This may reflect that certain aspects of an individual's speech, while being affected by environmental stressors, may not involve the same mechanism as performance decline. However, there may be other aspects of speech that are involved in this way and hence correlated with stress response/tolerance. This may suggest that all speech features are not equal when identifying their relationship with stress response (performance decline) or stress tolerance. Environmental stress' effect on an individual's speech does not imply an effect on their cognitive performance to the same extent or in the same way. This finding supports a recommendation that while speech features employed for stress detection can be potential candidates for stress response/tolerance prediction, they do not provide an exhaustive set of features useful for such predictions and hence further features should be explored while developing measures for stress response/tolerance.

The flipside of the above finding is that a number of features that did not show significant difference between the `Neutral` and `Stress` conditions nevertheless showed a good correlation with the stress-induced change in executive functioning (i.e., Verbal Fluency Task) and with the external assessment of stress tolerance. To provide further support for the recommendation made above, this finding may suggest that subtle differences that do not reach significance in distinguishing between the `Neutral` and `Stress` conditions may still be useful as an indicator for stress response. Alternatively, this mismatch between the features that appear more frequently in the stress conditions and the features that predict stress response/tolerance elsewhere could be due to the difference in the type of stress between our stress induction method and the stress experienced by the participants during the selection week (for the Selection Assessment Data) or the small sample size. Hence, for development of a reliable speech-based measure for stress response/tolerance, further exploration would be useful to test the mismatch hypothesis above, and with larger datasets.

## 6 Conclusion and Future Work

The main contribution of this paper is to present a proof of concept identifying potential language markers for stress response which we plan to extend and refine in the future with more focused and larger trials. Specifically, we have explored the usefulness of specific acoustic features extracted from speech produced in two stress conditions for their relationship with stress response and stress tolerance. We identified duration and a number of frequency domain features that are significantly correlated with stress response or tolerance. In the future, we plan to extract further acoustic features to test a more extensive list of features as potential candidates for involvement in the development of the speech-based measure.

We also tested the effectiveness of two continuous speech sample collection tasks, Read Speech Task and Open-ended Question Task, in developing speech-based measures for stress response and tolerance. We found that both tasks provided complementary information in the speech features and hence it can be beneficial to use both these tasks for data collection. However, if one of the tasks needs to be selected, Open-ended Question Task would provide more information that has consequences for stress response/tolerance.

Since this study involved only 13 participants, in order to confirm the current results and develop a more reliable and robust measure of stress response/tolerance, we plan to extend it further by increasing the sample size. The participants in the current study belong to a very particular population, young males aspiring to serve in a military unit. One can reasonably assume that this special population tends to have high stress tolerance. Testing general population with the current experimental design might reveal greater range of stress response scores. Hence, training on the general population may lead to greater predictive validity.

Additionally, in this study, we focused on correlations as a measure of the relationship between extracted acoustic features and stress response/tolerance. In our future work, we plan to use the identified acoustic features to develop and train machine learning models to predict the stress response/tolerance scores. Additionally, Open-ended Question Task's capability of extracting semantics-based features, such as sentiment and topic, will also be used to develop robust models for stress response and stress tolerance predictions.

## 7 Possible Use Cases and Ethical Considerations

The set of features identified in the current paper has a few possible use cases. First, as stress plays a central role in the development of psychopathology, it is possible that those who show greater response to stress are more likely to develop psychopathology. An assessment based on these features may be used to identify those who are at a greater risk to develop psychopathology later in life. These features could be used to build a real-time sensor to detect signs of stress response. Such a sensor could automatically identify when people are experiencing a heightened sense of stress and help appropriate parties to reach out for mitigation.

**Privacy.** Building such real-time sensors that can detect stress or stress response in individuals based on the speech/language they produce, however, calls for ethical considerations. For example, as the smart home assistance devices (Amazon's Alexa or Google home) become increasingly common in households, the companies that operate the devices can easily collect the speech data and detect stress. This information could be used for commercial and other purposes (e.g., showing an advertisement for vacation or spa upon detecting stress) without users' consent. In general, the previous discussions of privacy concerns regarding speech data have focused on the identifying information in the speech signal and the semantic content of the speech. The possible use case of the current study could result in information about one's emotional state also to be collected by third parties without consent.

To mitigate some of these undesired consequences, free public programs could be planned that educate users of technology what capabilities modern technologies have, how they can be used both for the good and for sabotage depending on who controls it, what privacy preferences are available to the users and how they can choose these preferences in a more informed manner to be able to benefit from the capabilities without being sabotaged. Developers of such technologies also have a responsibility to ensure easy access and ease of selection of preferences related to users' privacy.

**Equality.** In order to develop a fully automated measure for stress response expanding on the approach illustrated in this paper, one needs to rely on the speech-to-text systems' output to compute an individual's performance score. However, automatic speech recognition systems may perform differently for different social groups (Koenecke et al., 2020) or populations with different mental health conditions (Miner et al., 2020), for example. Hence, an automatic measure for stress response that uses speech-to-text transcriptions may not work as well for certain populations as it would for others. This could result in unintended biases against individuals belonging to certain social/clinical groups through misdiagnoses or misclassifications. To overcome such unintended results, one needs to account for the differences in performance of the components used to develop the automated measure of stress response for different populations. Involving individuals from different populations while developing such automated systems, e.g., by training systems on data obtained from them, can be helpful.

**Study Ethics Statement.** Approval of the oversight military Institutional Review Board (IRB) was obtained prior to starting the study. Informed consent was obtained from all study participants. The IRB protocol was followed without exception during performance of this research. The findings from the collected data are reported in aggregated forms and no identifying information is released.

## Acknowledgements

This work was supported by DARPA/AFRL Contract FA8650-19-C-7944 within the DARPA Measuring Biological Aptitude (MBA) program. All statements of fact, opinion or conclusions contained herein are those of the authors and should not be construed as representing the official views or policies of DARPA, AFRL or the U.S. Government. Authors thank Vanessa Oviedo, Shelby Greene, and Joel Schooler for their assistance with data collection, and the anonymous reviewers for their feedback. Authors also thank colleagues Timothy Broderick, Arash Mahyari, Ian Perera, Kurtis Gruters, Ursula Schwuttke and Vandana Puri for discussions and/or feedback. This document was approved by DARPA for Public Release, Distribution Unlimited.

## References

Helen W Bland, Bridget F. Melton, Paul Welle, and Lauren Bigham. 2012. Stress Tolerance: New Challenges for Millennial College Students. *College Student Journal*, 46(2):362–375.

Malcolm Brenner and Thomas Shipp. 1988. Voice Stress Analysis. NASA. Langley Research Center, Mental-State Estimation 1987.

Malcolm Brenner, Thomas Shipp, E.T. Doherty, and P. Morrissey. 1983. Voice Measures of Psychological Stress: Laboratory and Field Data. In Ingo R. Titze and Ronald C. Scherer, editors, *Vocal Fold Physiology, Biomechanics, Acoustics, and Phonatory Control*, pages 239–248. The Denver Center for the Performing Arts, Denver.

N.P. Dhole and S.N. Kale. 2020. Stress Detection in Speech Signal Using Machine Learning and AI. In Debabala Swain, Prasant Kumar Pattnaik, and Pradeep K. Gupta, editors, *Machine Learning and Information Processing. Advances in Intelligent Systems and Computing*, volume 1101. Springer, Singapore.

Maria Dietrich, Katherine Verdolini Abbott, Jackie Gartner-Schmidt, and Clark A. Rosen. 2008. The Frequency of Perceived Stress, Anxiety, and Depression in Patients with Common Pathologies Affecting Voice. *Journal of Voice*, 22(4):472–488.

Krista Doyle. 2020. Aceable Field Notes: An Essay on Grit.

Tobias Esch, George B. Stefano, Gregory L. Fricchione, and Herbert Benson. 2002. The Role of Stress in Neurodegenerative Diseases and Mental Disorders. *Neuroendocrinology Letters*, 23(3):199–208.

Raul Fernandez and Rosalind W. Picard. 2003. Modeling Drivers' Speech under Stress. *Speech Communication*, 40(1-2):145–159.

John S. Garofolo, Lori F. Lamel, William M. Fisher, Jonathan G. Fiscus, and Denise S. Pallett. 1993. DARPA TIMIT Acoustic-Phonetic Continous Speech Corpus CD-ROM. NIST Speech Disc 1-1.1. NASA STI/Recon Technical Report N.

Cheryl L. Giddens, Kirk W. Barron, Jennifer Byrd-Cravenand Keith F. Clark, and A. Scott Winter. 2013. Vocal Indices of Stress: A Review. *Journal of Voice*, 27(3):390.E21–390.E29.

Felipe V. Gomes and Anthony A. Grace. 2017. Adolescent stress as a driving factor for schizophrenia development—a basic science perspective. *Schizophrenia Bulletin*, 43(3):486–489.

Google. 2019. Python Client for Cloud Speech API. Google Cloud Client Libraries for google-cloud-speech.

James J. Gross. 2014. Emotion Regulation: Conceptual and Empirical Foundations. In James J. Gross, editor, *Handbook of Emotion Regulation*, pages 3–20. The Guilford Press.

John H.L. Hansen and Sanjay A. Patil. 2007. Speech under Stress: Analysis, Modeling and Recognition. In Müller C., editor, *Speaker Classification I. Lecture Notes in Computer Science*, volume 4343, pages 108–137. Springer, Berlin, Heidelberg.

Ling He, Margaret Lech, Sheeraz Memon, and Nicholas Allen. 2008. Recognition of Stress in Speech using Wavelet Analysis and Teager Energy Operator. In *Proceedings of INTERSPEECH*, pages 605–608, Brisbane, Australia.

Eric S. Jackson, Mark Tiede, Deryk Beal, and D.H. Whalen. 2016. The Impact of Social-Cognitive Stress on Speech Variability, Determinism, and Stability in Adults Who Do and Do Not Stutter. *Journal of Speech, Language, and Hearing Research*, 59(6):1295–1314.

Bhagyalaxmi Jena and Sudhansu Sekhar Singh. 2016. Psychological Stress Speech Analysis: A Review. *International Journal of Engineering Research & Technology*, 4(28):1–4.

Allison Koenecke, Andrew Nam, Emily Lake, Joe Nudell, Minnie Quartey, Zion Mengesha, Connor Toups, John R. Rickford, Dan Jurafsky, and Sharad Goel. 2020. Racial Disparities in Automated Speech Recognition. In *Proceedings of the National Academy of Sciences*, volume 117, pages 7684–7689.

Mark D. Kramer, Christopher Patrick, and Robert F. Krueger. 2012. Delineating Physiologic Defensive Reactivity in the Domain of Self-Report: Phenotypic and Etiologic Structure of Dispositional Fear. *Psychological Medicine*, 42(6):1305–1320.

Kristiane Van Lierde, S. Van Heule, S. De Ley, E. Mertens, and S. Claeys. 2009. Effect of Psychological Stress on Female Vocal Quality: A Multiparameter Approach. *Folia Phoniatr Logop*, 61(2):105–111.

Brian McFee, Vincent Lostanlen, Alexandros Metsai, Matt McVicar, Stefan Balke, Carl Thomé, Colin Raffel, Frank Zalkow, Ayoub Malek, Dana,

Kyungyun Lee, Oriol Nieto, Jack Mason, Dan Ellis, Eric Battenberg, Scott Seyfarth, Ryuichi Yamamoto, Keunwoo Choi, viktorandreevichmorozov, Josh Moore, Rachel Bittner, Shunsuke Hidaka, Ziyao Wei, nullmightybofo, Darío Hereñú, Fabian-Robert Stöter, Pius Friesch, Adam Weiss, Matt Vollrath, and Taewoon Kim. 2020. librosa/librosa: 0.8.0 (version 0.8.0). Librosa 0.8.0.

Brian McFee, Colin Raffel, Dawen Liang, Daniel PW Ellis, Matt McVicar, Eric Battenberg, and Oriol Nieto. 2015. librosa: Audio and Music Signal Analysis in Python. In *Proceedings of the 14th Python in Science Conference*, pages 18–24.

Adam S. Miner, Albert Haque, Jason A. Fries, Scott L. Fleming, Denise E. Wilfley, G. Terence Wilson, Arnold Milstein, Dan Jurafsky, Bruce A. Arnow, W. Stewart Agras, Li Fei-Fei, and Nigam H. Shah. 2020. Assessing the Accuracy of Automatic Speech Recognition for Psychotherapy. *NPJ Digital Medicine*, 3(1):1–8.

Masanobu Nakamura, Koji Iwano, and Sadaoki Furui. 2008. Differences between Acoustic Characteristics of Spontaneous and Read Speech and their Effects on Speech Recognition Performance. *Computer Speech Language*, 22(2):171–184.

Franz Pernkopf, Tuan Van Pham, and Jeff A. Bilmes. 2009. Broad Phonetic Classification using Discriminative Bayesian Networks. *Speech Communication*, 51(2):151–166.

Tarun Pruthi and Carol Y. Espy-Wilson. 2004. Acoustic Parameters for Automatic Detection of Nasal Manner. *Speech Communication*, 43(3):225–239.

Tarun Pruthi and Carol Y. Espy-Wilson. 2007. Acoustic Parameters for the Automatic Detection of Vowel Nasalization. In *INTERSPEECH*, pages 1925–1928, Antwerp, Belgium.

Brent S. Rushall. 1990. A Tool for Measuring Stress Tolerance in Elite Athletes. *Journal of Applied Sport Psychology*, 2(1):51–66.

Robert M. Sapolsky. 1996. Why Stress is Bad for Your Brain. *Science*, 273(5276):749–750.

Björn Schuller, Stefan Steidl, Anton Batliner, Julien Epps, Florian Eyben, Fabien Ringeval, Erik Marchi, and Sebastian Schnieder. 2014. The INTERSPEECH 2014 Computational Paralinguistics Challenge: Cognitive Physical Load. In *Proceedings of the Annual Conference of the International Speech Communication Association (INTERSPEECH)*.

Barbara Schuppler. 2017. Rethinking classification results based on read speech, or: why improvements do not always transfer to other speaking styles. *International Journal of Speech Technology*, 20:699–713.

Zeshu Shao, Esther Janse, Karina Visser, and Antje S. Meyer. 2014. What do verbal fluency tasks measure? Predictors of verbal fluency performance in older adults. *Frontiers in Psychology*, 5:772.

Tom Smeets, Sandra Cornelisse, Conny W.E.M. Quaedflieg, Thomas Meyer, Marko Jelicic, and Harald Merckelbach. 2012. Introducing the Maastricht Acute Stress Test (MAST): A quick and non-invasive approach to elicit robust autonomic and glucocorticoid stress responses. *Psychoneuroendocrinology*, 37(12):1998–2008.

Gustavo E. Tafet and Charles M. Nemeroff. 2015. The Links Between Stress and Depression: Psychoneuroendocrinological, Genetic, and Environmental Interactions. *The Journal of Neuropsychiatry and Clinical Neurosciences*, 28(2):77–88.

Arkun Tatar, Gaye Saltukoğlu, and Ercan Özmen. 2018. Development of a Self Report Stress Scale Using Item Response Theory-I: Item Selection, Formation of Factor Structure and Examination of Its Psychometric Properties. *Archives of Neuropsychiatry*, 55(2):161–170.

Kevin Tomba, Joel Dumoulin, Elena Mugellini, Omar Abou Khaled, and Salah Hawila. 2018. Stress Detection Through Speech Analysis. In *Proceedings of the 15th International Joint Conference on e-Business and Telecommunications (ICETE)*, volume 1, pages 394–398.

S. Vaikole, S. Mulajkar, A. More, P. Jayaswal, and S. Dhas. 2020. Stress Detection through Speech Analysis using Machine Learning. *International Journal of Creative Research Thoughts (IJCRT)*, 8(5).

Wikibooks. 2018. Speech-Language Pathology/Stuttering/Stress-Related Changes.

Longfei Yang, Yinghao Zhao, Yicun Wang, Lei Liu, Xingyi Zhang, Bingjin Li, and Ranji Cui. 2015. The Effects of Psychological Stress on Depression. *Current neuropharmacology*, 13(4):494–504.

Habib Yaribeygi, Yunes Panahi, Hedayat Sahraei, Thomas P. Johnston, and Amirhossein Sahebkar. 2017. The Impact of Stress on Body Function: A Review. *EXCLI Journal*, 16:1057–1072.

# A Full Set of Correlations of Difference Scores for Extracted Acoustic Features with the Performance Score and the Stress Tolerance Score

| Feature | Performance Score | | Stress Tolerance Score | |
|---|---|---|---|---|
| | RST | OQT | RST | OQT |
| **Duration related features:** | | | | |
| Duration (Speech Rate) | **-.628*** | - | .184 | - |
| **Time domain features:** | | | | |
| Amplitude Envelope | [illegible] | [illegible] | [illegible] | [illegible] |
| Root Mean Square Energy | -.381 | -.410 | -.055 | .377 |
| Zero Crossing Rate | **.546** | -.115 | -.470 | .359 |
| **Frequency domain features:** | | | | |
| Magnitude Spectrum | .050 | .337 | **-.753***** | -.492 |
| Short-time Fourier Transform Spectrum | **-.502** | -.271 | -.140 | **.515** |
| MFCC1 | -.282 | -.341 | -.247 | **.617*** |
| MFCC2 | .017 | -.297 | -.217 | .405 |
| MFCC3 | .019 | .138 | .030 | -.300 |
| MFCC4 | -.464 | .134 | .223 | .202 |
| MFCC5 | -.447 | -.207 | .099 | -.094 |
| MFCC6 | .167 | -.245 | .293 | -.125 |
| MFCC7 | -.155 | **.524** | -.050 | -.486 |
| MFCC8 | -.272 | .432 | -.451 | **-.583*** |
| MFCC9 | -.423 | .048 | .132 | **-.554*** |
| MFCC10 | .153 | -.343 | -.009 | -.206 |
| MFCC11 | -.209 | **.516** | -.037 | **-.735***** |
| MFCC12 | .128 | .288 | .104 | -.387 |
| MFCC13 | .411 | -.250 | -.167 | .363 |
| Delta MFCC1 | .063 | -.279 | -.052 | .360 |
| Delta MFCC2 | .061 | .124 | -.156 | .028 |
| Delta MFCC3 | -.037 | .125 | -.240 | -.028 |
| Delta MFCC4 | -.349 | -.191 | -.015 | .090 |
| Delta MFCC5 | -.055 | .221 | -.308 | .064 |
| Delta MFCC6 | -.268 | .358 | .016 | -.151 |
| Delta MFCC7 | -.401 | .439 | .051 | -.187 |
| Delta MFCC8 | -.159 | .086 | .025 | -.205 |
| Delta MFCC9 | -.149 | -.280 | -.169 | .014 |
| Delta MFCC10 | .469 | -.261 | -.413 | .046 |
| Delta MFCC11 | .329 | .129 | -.474 | -.177 |
| Delta MFCC12 | -.181 | -.350 | -.480 | .203 |
| Delta MFCC13 | -.073 | -.197 | -.406 | .068 |
| Delta Delta MFCC1 | -.183 | **-.710**** | .129 | .322 |
| Delta Delta MFCC2 | -.293 | -.122 | -.057 | -.118 |
| Delta Delta MFCC3 | -.003 | .323 | -.060 | -.129 |
| Delta Delta MFCC4 | .187 | .090 | -.320 | -.112 |
| Delta Delta MFCC5 | -.413 | -.069 | -.115 | 0.177 |
| Delta Delta MFCC6 | .256 | .531 | .059 | -.199 |
| Delta Delta MFCC7 | .094 | **.759***** | .038 | -.090 |
| Delta Delta MFCC8 | -.163 | **.519** | -.133 | -.137 |
| Delta Delta MFCC9 | -.037 | -.153 | -.225 | -.020 |
| Delta Delta MFCC10 | .353 | -.459 | .200 | **.534** |
| Delta Delta MFCC11 | .301 | -.258 | .097 | .186 |
| Delta Delta MFCC12 | -.008 | -.094 | .293 | -.015 |
| Delta Delta MFCC13 | -.456 | .350 | **.589*** | -.155 |

Table 3: Correlations between the difference scores (`Stress - Neutral`) of the 45 features extracted from Read Speech Task (RST) and Open-ended Question Task (OQT) and the stress-induced change in Verbal Fluency Task performance (left) and the stress tolerance score from the selection assessment data (right). The duration feature was dropped for OQT since the task duration itself was set to two minutes. Moderate to high correlations (> .5) are indicated with the bold font. *** indicates $p < .005$,.** indicates $p < .01$, * indicates $p < .05$

# Towards Low-Resource Real-Time Assessment of Empathy in Counselling

**Zixiu Wu**
Philips Research & University of Cagliari
zixiu.wu@philips.com

**Rim Helaoui**
Philips Research
rim.helaoui@philips.com

**Diego Reforgiato Recupero** and **Daniele Riboni**
University of Cagliari
{diego.reforgiato, riboni}@unica.it

## Abstract

Gauging therapist empathy in counselling is an important component of understanding counselling quality. While session-level empathy assessment based on machine learning has been investigated extensively, it relies on relatively large amounts of well-annotated dialogue data, and real-time evaluation has been overlooked in the past. In this paper, we focus on the task of low-resource utterance-level binary empathy assessment. We train deep learning models on heuristically constructed empathy vs. non-empathy contrast in general conversations, and apply the models directly to therapeutic dialogues, assuming correlation between empathy manifested in those two domains. We show that such training yields poor performance in general, probe its causes, and examine the actual effect of learning from empathy contrast in general conversation.

## 1 Introduction

As a pillar of psychotherapy, empathy is crucial to effective counselling, owing to its importance in building counsellor[1]-client rapport (Elliott et al., 2011) that can enable more effective interventions and better outcomes (McCambridge et al., 2011; Gaume et al., 2009). In particular, "listening with empathy" is considered a guiding principle (Rollnick et al., 2008) for motivational interviewing (Miller and Rollnick, 2012) (MI), a psychotherapeutic approach widely adopted to elicit positive behaviour change by evoking motivation from clients. Gauging counsellor-side empathy is, therefore, essential to assessing MI integrity (Moyers et al., 2016).

Empathy assessment for MI has conventionally been conducted manually by trained annotators, which requires extensive annotator training and transcript review. Since such a time-consuming and costly setup is difficult to scale up, recent years have seen attempts of automating the process with machine learning, including transcript-based (Xiao et al., 2012; Gibson et al., 2015, 2016), speech-based (Xiao et al., 2014, 2015), and multimodal (Xiao et al., 2016b) methods. Those works are, however, limited in that 1) therapist empathy is only assessed at session-level rather than utterance-level; 2) classical machine learning with heuristic feature engineering is used, while recent deep-learning frameworks have not been utilised for this purpose; 3) the machine-learning-based approaches all assume access to privately-owned sizeable corpora of therapeutic dialogues with empathy annotation at session level, but in reality such well-annotated data are often very limited, even more so at utterance level; and 4) the link between empathy manifested in general conversation and in MI counselling remains unexplored.

In this work, we make the first attempt (to the best of our knowledge) at addressing those limitations while probing the correlation between empathy manifestations in different domains. Specifically, we employ pre-trained language models such as BERT (Devlin et al., 2019) for text-based binary classification of utterance-level therapist empathy, optionally taking the conversation context as input. We consider any counsellor utterance to be empathetic if it shows empathy, and non-empathetic if it does not (ranging from neutral to apathetic). Our models have no access to counselling conversations during their training and validation, as we experiment with learning from contrast of empathy vs. non-empathy in out-of-domain (OOD) training data. To that end, we leverage publicly available datasets of general conversations with heuristic empathy labels (Rashkin et al., 2019; Zhong et al., 2020) for OOD training, investigating the connections between general-conversational empathy and therapeutic empathy, as illustrated in Figure 1.

To benchmark the models, we manually anno-

[1]We use "counsellor" and "therapist" interchangeably in this work.



*Proceedings of the Seventh Workshop on Computational Linguistics and Clinical Psychology*, pages 204–216
June 11, 2021. ©2021 Association for Computational Linguistics

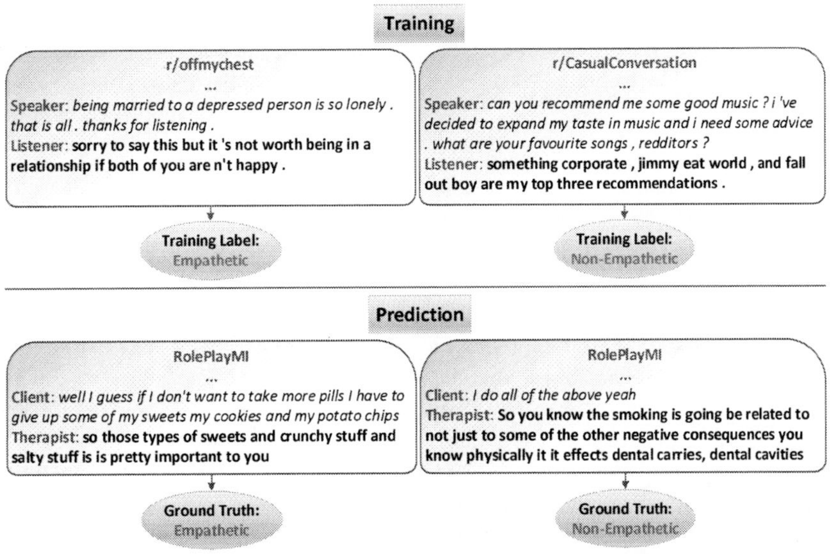


Figure 1: Training a binary empathy classifier on heuristically constructed empathetic vs. non-empathetic utterances in general conversations (i.e. out-of-domain w.r.t. MI), and then testing it on MI conversations. In this case, the empathy contrast for training is *r/OffMyChest* vs. *r/CasualConversation*. The classifier can take only the listener/therapist utterance (**bold**) as input or additionally use the preceding speaker/client utterance (*italic*).

tated utterance-level empathy for a subset of transcribed high- vs. low-quality counselling demonstrations (Pérez-Rosas et al., 2019) that are publicly available. We also build unsupervised baselines for the task by **a)** formulating binary empathy classification as natural language inference (NLI), as proposed by Yin et al. (2019), and **b)** tackling the surrogate task of client-counsellor agreement via NLI, under the assumption that an empathetic reply from the counsellor tends to show accordance with the client utterance in the preceding turn.

Our experiments show that models trained on OOD empathy contrast are not sufficiently accurate predictors of MI empathy/non-empathy, even though the benefit of such training can be observed when compared to training on OOD data without empathy contrast. Upon probing, we argue that more fine-grained (e.g. sentence-level) empathy annotation and prediction could yield better results.

## 2 Related Work

### 2.1 Machine-Learning-Based Approaches to Empathy Analysis for MI

Prior work has approached assessment of empathy in MI delivery via speech and linguistic features.

Among text-based methods, Xiao et al. (2012) proposed one of the earliest approaches for utterance-level empathy classification using an n-gram language model. Psycholinguistic norm features are used in addition to other linguistic features in the work of (Gibson et al., 2015). More recently, Gibson et al. (2016) utilised long short-term memory networks (LSTMs) (Hochreiter and Schmidhuber, 1997) to generate turn-level behavioural acts that are further processed by a deep neural network to predict session-level empathy.

Speech features have also been examined. Xiao et al. (2014) investigated features such as jitter and shimmer from speech signals, Xiao et al. (2015) studied speech rate entrainment, while Pérez-Rosas et al. (2017) used an array of acoustic and linguistic features to train their multimodal models.

There are also a number of recent studies on data-driven MI behaviour coding based on text (Cao et al., 2019; Tanana et al., 2016; Xiao et al., 2016a; Gibson et al., 2018), speech (Singla et al., 2020), and both (Chen et al., 2019; Flemotomos et al., 2021) , but they are less relevant to this work due to their lack of explicit empathy modelling.

Different from the research listed above, this work addresses utterance-level empathy classification instead of session-level assessment, similar to Wu et al. (2020) which proposes utterance-level prediction of whether the therapist needs to show empathy given the context.

### 2.2 Data-Driven Text-Based Research on Empathy in General Conversation

Recent years have witnessed a boom of research on data-driven analysis and application of empathy in general conversations.

In terms of empathy analysis for open-domain conversations, Zhou et al. (2021) addressed scoring empathy grounded in specific situations, Welivita and Pu (2020) created a taxonomy of empathetic response intents in social dialogues, while Guda et al. (2021) proposed to take user demographic information into account for empathy prediction.

As therapeutic conversation data is scarce, recent works on empathy analysis have also turned to peer-support dialogues from online communities. Zhou and Jurgens (2020) analysed Reddit[2] conversations for the relationships between condolence, distress and empathy, Hosseini and Caragea (2021) studied empathy seeking and providing with dialogues from a cancer survivor network, and Sharma et al. (2020) proposed an empathy framework of reaction-interpretation-exploration for conversations from mental-health-related online forums.

While early general empathetic chatbots (Zhou and Wang, 2018; Lubis et al., 2018) were mostly based on recurrent neural networks and produced emotion-conditioned output, their more recent counterparts are predominantly based on pre-trained language models and leverage emotions in various ways, including emotion detection as an auxiliary objective (Lin et al., 2020), emotion-based mixture-of-experts decoding (Lin et al., 2019), and rewarding response candidates likely to induce positive user emotion (Shin et al., 2020).

## 3 Data

We leverage[3] two types of data: general conversations and transcripts of MI demonstration videos.

We define an utterance as everything said by an interlocutor in their turn in a 2-person conversation, which is the most widely used definition of utterance in the literature of deep-learning-based conversational intelligence. This differs from some utterance definitions in psychotherapy. For example, an "utterance" in this work is identical to a "volley" as defined in the motivational interviewing skill code (MISC) (Miller et al., 2003), while an "utterance" in MISC is "a complete thought" that "ends either when one thought is completed or a new thought begins with the same speaker, or by an utterance from the other speaker".

### 3.1 General Conversations

Our general conversation data is from two datasets: **P**ersona-based **E**mpathetic **C**onversation (*PEC*) (Zhong et al., 2020) and *EmpatheticDialogues* (*ED*) (Rashkin et al., 2019). Their statistics are listed in Table 1. For each 2-interlocutor dialogue, we consider the initiator of the conversation as the **speaker** and the other as the **listener**.

*PEC* consists of general conversations crawled from 3 subreddits: *r/Happy*[4] (*r/H*), *r/OffMyChest*[5] (*r/OMC*), and *r/CasualConversation*[6] (*r/CC*). Reddit users exchange happy experiences and thoughts in *r/H*, share emotional stories that cannot be told easily in *r/OMC*, and simply talk casually in *r/CC*. Since the original *PEC* dataset includes conversations between more than two participants and some conversations are actually subsets of other conversations (e.g. a 2-turn conversation that in effect constitutes the first 2 turns of a 4-turn conversation), we retain only the non-subset conversations that are between 2 interlocutors, in order to align with the counsellor-client nature of therapeutic conversations, and the filtered *PEC* contains around 56% of the conversations in the original one.

*EmpatheticDialogues* (abbreviated as *ED*) is comprised of 23.1K general conversations from MTurker pairs. The speaker of each dialogue was first given an emotion label (e.g. "Afraid"), then described a situation where they had felt the emotion before (e.g. "I've been hearing noises around the house at night"), and finally initiated the conversation about this situation with a listener.

#### 3.1.1 Empathy vs. Non-Empathy

We divide the general conversation data into 2 parts: empathetic-listener conversations and non-empathetic-listener ones. Specifically, we assign "empathetic" labels to all the listener utterances of the dialogues in *r/H*, *r/OMC* and *ED*, and "non-empathetic" to the counterparts in *r/CC*.

For *PEC*, the heuristic empathy labelling is based on the annotator ratings from the original paper that suggest comments (i.e. listener

[2]Reddit (`https://www.reddit.com/`) is an online platform comprised of subforums (known as **subreddits**), each with a specific topic for Reddit users to discuss.

[3]Identifiable information (e.g. names, dates) was replaced with placeholders prior to the experiments.

[4]`https://www.reddit.com/r/happy/`

[5]`https://www.reddit.com/r/offmychest/`

[6]`https://www.reddit.com/r/CasualConversation`

| | *r/Happy* ‡ | | | *r/OffMyChest* ‡ | | | *r/CasualConversation* ¶ | | | *EmpatheticDialogues* ‡ | | |
|---|---|---|---|---|---|---|---|---|---|---|---|---|
| Split | train | valid | test | train | valid | test | train | valid | test | train | valid | test |
| #Conv | 113.9K | 13.9K | 16.0K | 94.0K | 12.1K | 11.7K | 530.2K | 67.5K | 66.9K | 17.8K | 2.8K | 2.5K |

Table 1: Statistics of *PEC* (*r/Happy*, *r/OffMyChest*, and *r/CasualConversation*) & *EmpatheticDialogues*. For *PEC*, we utilise 2-interlocutor conversations only. #Conv: number of conversations in the data split. We consider *r/Happy*, *r/OffMyChest* and *EmpatheticDialogues* to consist of mostly empathetic (‡) listener utterances and *r/CasualConversation* to be comprised of predominantly non-empathetic (¶) ones. Note that the statistics of *PEC* are about the filtered dataset as described in Section 3.1. See Table 4 for more details.

utterances) in *r/H* and *r/OMC* are significantly more empathetic than those in *r/CC*, and the inter-annotator agreement on this as measured by Fleiss' kappa (Fleiss, 1971) was "substantial". For *ED*, the empathy labelling is intuitive as the authors explicitly instructed the "listeners" to respond empathetically during the data collection.

We note that our heuristic labelling for *PEC* and *ED* is based on the corpus-level labels given by the creators of the datasets, thus it may not be completely accurate at utterance or sentence level. We nevertheless utilise the heuristic labels for our experiments and leave more fine-grained annotation for future work.

### 3.2 Motivational Interviewing

Our counselling conversations are from Pérez-Rosas et al. (2019), who collected the first and only (to the best of our knowledge) publicly available dataset of MI conversations. The dialogues are the transcripts of 152 demonstrations of high-quality (MI adherent) and another 101 of low-quality (MI non-adherent) counselling from video-sharing platforms such as YouTube and Vimeo. The original transcripts were obtained with the automatic captioning tool of YouTube, so the conversations have minor transcription errors and are mostly without punctuation. We refer to this dataset as ROLEPLAYMI, and list its statistics in Table 2.

#### 3.2.1 Manual Empathy Annotation

We select a subset of ROLEPLAYMI to manually annotate utterance-level empathy to build a benchmark dataset for our models. The annotation guideline follows the definition of high empathy in MISC: *Counsellors high on the empathy scale show an active interest in making sure they understand what the client is saying, including the client's perceptions, situation, meaning, and feelings.* We ask the annotators to consider an utterance that shows MISC-defined *high empathy* as **empathetic**, otherwise as **non-empathetic**. Thus, non-empathy in this context can range from neutrality to apathy.

| | **ROLEPLAYMI** | | **ANNO** | |
|---|---|---|---|---|
| MI Quality | High | Low | High | Low |
| #Conv | 152 | 101 | 7 | 14 |
| #T-u | 3928 | 1534 | 217 | 214 |
| %(emp.T-u) | n/a | n/a | 38.7% | 2.3% |
| %(¬Q.T-u) | n/a | n/a | 71.9% | 73.8% |
| $p$(emp \| ¬Q, T-u) | n/a | n/a | **0.50** | **0.03** |
| $p$(emp \| Q, T-u) | n/a | n/a | 0.10 | 0.00 |

Table 2: Statistics of ROLEPLAYMI and ANNO. #Conv: number of conversations in the subset. "T-u" is short for "Therapist Utterance(s)". #T-u: number of therapist utterances in the subset. %(emp.T-u): percentage of empathetic therapist utterances. %(¬Q.T-u): percentage of non-question therapist utterances. $p$(emp | ¬Q, T-u): probability of a non-question therapist utterance being empathetic. $p$(emp | Q, T-u): probability of a question therapist utterance being empathetic. See Table 5 for more details.

We choose 7 transcripts (217 counsellor utterances in total) from the high-quality subset with negligible transcription errors, and 14 transcripts (214 counsellor utterances in total) from the low-quality one. The 431 selected utterances are presented to 2 human annotators for binary utterance-level empathy annotation. One annotator is a senior researcher that has received formal MI training in the past, and the other is a PhD student that has read in depth about MI (incl. Rollnick et al. (2008)). Their annotations show an inter-annotator agreement of 0.71 measured by Cohen's kappa (Cohen, 1968), indicating "substantial agreement". Finally, the annotators discussed their results and resolved the differences. The annotated MI conversations are denoted as ANNO in the rest of the paper.

As Table 2 shows, 38.7% of the therapist utterances in the high-quality subset are empathetic (i.e. 61.3% non-empathetic), while the number for the low-quality subset is 2.3% for empathetic (i.e. 97.7% non-empathetic), suggesting a marked difference between the empathy levels in high- and low-quality counselling.

We note that our empathy annotation is at utterance-level on the punctuation-free MI transcripts, which means an utterance is marked as empathetic as long as a part of the utterance is so, even though the remainder might not be. More fine-grained annotation would be possible with punctuated utterances, which we leave for future work.

### 3.2.2 Question & Empathy

Empirically, we observe that questions in MI do not show empathy in general, which is intuitive since the purpose of questions is to gather more information. Indeed, we notice that the vast majority of the examples of open and closed questions provided by MISC are not empathetic.

Therefore, we additionally conduct binary annotation for each therapist utterance in ANNO as to whether the utterance is (predominantly) a question, by marking an utterance as a question utterance if more than half of the tokens in an utterance constitute at least one open or closed question as defined by MISC. For instance, "it's good to see you up and about how are you feeling after your last little hospitalization" is considered a question utterance, since "how are you feeling after your last little hospitalization" is an open question and makes up more than half of the utterance. We denote the non-question subset of ANNO as $\neg$Q.ANNO.

The relationship between empathy and question found in ANNO confirms our observation: a non-question therapist utterance from high-quality counselling is substantially more likely (0.50) to be empathetic than one from low-quality counselling (0.03), while the same does not hold for question therapist utterances: 0.10 for high-quality and 0.00 for low-quality, which indicates that therapist questions are overall very unlikely to be empathetic.

## 3.3 General-Conversation Empathy vs. Therapeutic Empathy

Comparing ROLEPLAYMI with *PEC & ED*, we noticed a pronounced difference between empathy in general conversation and therapy: an MI-adherent therapist tends to express empathy through non-questions (as shown in Table 2), e.g. "The blood sugars have increased some, so you're concerned that things are not as good as they were last time that we talked". Conversely, participants in general conversations often show empathy via questions, e.g. "Oh no! That's scary! What do you think it is?". Thus, analysing sentence-level empathy (instead of utterance-level) could better separate the empathetic and non-empathetic parts, and more overlap between general-conversation empathy and therapeutic empathy may be found in the non-question sentences. This was not possible in our experiments as ROLEPLAYMI is not punctuated, thus we leave it for future work.

We note that another domain difference is that ROLEPLAYMI consists of transcripts of spoken dialogues whereas *PEC* and *ED* contain "written" chat conversations. The difference is smoothed by the high-quality transcription of the ROLEPLAYMI videos and we therefore do not use specific techniques to address the difference, but we plan to investigate this factor further in future work.

# 4 Binary Empathy Classification

In this section, we first define the task of binary empathy classification, then lay out the out-of-domain empathy contrast strategy behind our supervised models for the task, and finally describe our unsupervised baselines driven by NLI.

## 4.1 Task Definition

We denote $D^{MI} = \{(u_i^C,\ u_i^T,\ e_i)\},\ i = 1,\cdots,N$ as a collection of {*(client utterance, therapist utterance, empathy label)*} tuples, where $u_i^T$ is the therapist reply to the client utterance $u_i^C$, $e_i \in \{\texttt{emp}, \neg\texttt{emp}\}$ denotes if $u_i^T$ shows empathy, and $N$ is the number of such tuples in the dataset. Our task can be formulated as follows: given $u_i^T$ and optionally $u_i^C$ for more context, predict the correct empathy label $e_i$ of $u_i^T$. We use ANNO as $D^{MI}$.

## 4.2 Supervised Learning: Using Out-of-Domain Empathy Contrast

Since our manually annotated subset of ROLEPLAYMI is too small to be a proper training set, we resort to learning from out-of-domain (i.e. non-MI) (OOD) empathy contrast. Specifically, as described in Section 3.1.1 and Figure 1, we utilise all listener utterances in *r/H*, *r/OMC* and *ED* as positive (empathetic) examples and their counterparts in *r/CC* as negative (non-empathetic) examples, as we aim to leverage parallels between general-conversation empathy and psychotherapeutic empathy.

We build 3 empathy vs. non-empathy contrast[7] pairs from general conversations: (*r/H* vs. *r/CC*); (*r/OMC* vs. *r/CC*); (*ED* vs. *r/CC*). For each pair, we sample an equal number of examples from the empathetic (positive) and non-empathetic (negative) subsets to construct a contrast dataset

[7]We use "empathy vs. non-empathy contrast" and "empathy contrast" interchangeably.

| | | |
|---|---|---|
| P[a] | Client: Everyone's getting on me about my drinking. \| Therapist: Kind of like a bunch of crows pecking at you. | Relationship |
| H[b] | The therapist is empathetic towards the patient | Entailment |
| | The client wants to smoke more. | Neutral |
| | The therapist is not listening to the client. | Contradiction |
| [a]P, Premise. | | |
| [b]H, Hypothesis. | | |

Table 3: Natural Language Inference, example utterances from Miller et al. (2003)

$D^{Gen} = \{(u_j^S, u_j^L, e_j)\}$, where in each sample the empathy label $e_j \in \{\texttt{emp}, \neg\texttt{emp}\}$ denotes whether the listener response $u_j^L$ is empathetic towards its preceding speaker utterance $u_j^S$. Our sampling ensures that the 2 classes (i.e. emp & ¬emp) in each pair during training are balanced.

For each contrast pair, we train a 1-utterance general-conversation empathy classifier $cls_{(1)}$ to predict $e_j$ given $u_j^L$, as well as a 2-utterance counterpart $cls_{(2)}$ to predict $e_j$ given $(u_j^S, u_j^L)$. Finally, we apply the trained $cls_{(1)}$ and $cls_{(2)}$ directly on $D^{MI}$, using $u_i^C$ as $u_j^S$ and $u_i^T$ as $u_j^L$.

### 4.3 Unsupervised Baseline: Text Classification as Natural Language Inference

Natural language inference (NLI) is the task of determining if a **hypothesis** is true (*entailment*), false (*contradiction*), or undetermined (*neutral*) given a **premise**[8] (Table 3). Following Yin et al. (2019) where NLI models prove effective as ready-made zero-shot sequence classifiers, we formulate our empathy classification task as an NLI problem.

Assuming only $u_i^T$ is available, we use it as the premise, and define the 1-utterance empathy hypothesis $h_{(1)}$ as "This text is empathetic.". We then utilise an off-the-shelf NLI model $M$ as an unsupervised 1-utterance empathy classifier $nli_{(1)}^E$ to directly predict a label from {*entailment, contradiction, neutral*} given $(u_i^T, h_{(1)})$. We consider $u_i^T$ to be classified as an empathetic utterance only if the predicted label is *entailment*.

We also investigate a client-therapist exchange scenario where both $u_i^C$ and $u_i^T$ are provided. The premise $p_i$ is then formatted as "Client: $u_i^C$ | Therapist: $u_i^T$", and we define the 2-utterance hypothesis as $h_{(2)}$ = "The Therapist is empathetic towards the Client.". We use the same $M$ as an unsupervised 2-utterance empathy classifier $nli_{(2)}^E$ given the input $(p_i, h_{(2)})$. Again, only *entailment* is deemed equivalent to categorising $u_i^T$ as empathetic.

[8]Definition of NLI: `https://paperswithcode.com/task/natural-language-inference`

### 4.4 Unsupervised Baseline: Client-Therapist Agreement as Natural Language Inference

It is our observation from MISC as well as ROLEPLAYMI that an empathetic therapist tends to acknowledge the difficulties and feelings of clients, and hence we experiment with NLI-style modelling for client-therapist agreement.

Specifically, we use $M$ as an unsupervised 2-utterance agreement classifier $nli_{C\rightarrow T}^A$ to measure the agreement between $u_i^C$ and $u_i^T$, using the former as the premise and the latter as the hypothesis. We only interpret an *entailment* prediction from $M$ as the therapist agreeing with the client and hence the therapist empathising with the client.

## 5 Experiments

### 5.1 Implementation

For OOD empathy contrast (Section 4.2), we keep the original train/dev/test splits of *PEC* and *ED*. Since the two datasets in each contrast pair can be vastly different in their sizes (e.g. *ED* has only 17.8K training examples whereas *r/CC* has 530.2K), we always sample the positive and negative subsets so that their sizes are identical to that of *ED*, the smallest dataset, which ensures **a)** the two classes are balanced in each pair, and **b)** different $cls$ models are trained with equal amounts of data and their performances are hence comparable.

To minimise the bias in training data caused by such sampling, we train the classifier of each contrast pair 5 times, each time with its own randomly sampled data. Note that this leads to 5 different groups of class-balanced {train, dev, set} datasets for each pair.

We leverage pre-trained language models for all our experiments. BERT (Devlin et al., 2019) is the backbone of our OOD empathy contrast models and its BERT-BASE-UNCASED variant is chosen. We add a fully connected layer atop the classification token ([CLS]) position of the language model to implement a binary classifier, and train the entire model end-to-end on the empathy contrast pairs. For the backbone $M$ of the unsupervised zero-shot baselines, we use the BART-LARGE variant of BART (Lewis et al., 2020) that has been fine-tuned on MultiNLI (Williams et al., 2018). For more details, see Section B.

To measure model performance on ANNO, we choose Matthews correlation coefficient (MCC) since it is robust to class imbalance, taking into account that only 38.7% of the ANNO examples from the high-quality subset are marked as empathetic and the number is only 2.3% for low-quality. We also use MCC to measure test set performance to increase comparability.

### 5.2 Results

We examine the performances achieved on ANNO by the models introduced in Section 4 , namely the blue bars in the "OOD$_{(1)}$ w/ Contrast" (1-utterance models trained on OOD empathy contrast, i.e. $cls_{(1)}$), "OOD$_{(2)}$ w/ Contrast" (2-utterance models trained on OOD empathy contrast, i.e. $cls_{(2)}$), and "Baselines" subplots of Figure 2. The value of each blue bar indicates the mean MCC of the 5 models from the corresponding pair, and we use the error bar to simply represent +/- one standard deviation from the mean, in order to illustrate the variation among the scores of the 5 models.

Also, we show in Figure 3 the performances of the OOD models on their respective test sets. In the test set of each of the 5 models from a $(D_+, D_-)$ OOD pair, we have $N_T$ random samples from $D_+$ and another $N_T$ from $D_-$, where $N_T$ is the size of the original test set of *ED*, in line with our sampling method for the OOD training sets. The mean (bar value) - standard deviation (error bar) representation follows that of Figure 2. By comparing the scores of the 5 models from an OOD setup on their own test sets and on ANNO, it becomes clear how the domain shift from general conversation to MI affects the performance of those models.

We first observe that while each test set in the OOD setups is different as we address class imbalance with random sampling, it is still obvious that the OOD models achieve considerably better scores on their test sets but experience significant drops on ANNO. In particular, *ED* vs. *r/CC* (2) reaches over 0.9 MCC on average on its test sets but only around 0.10 on ANNO. This stops any of the OOD empathy contrast models from being a reliable indicator of therapeutic empathy.

There is also considerable variation in the scores on ANNO (but not on the test sets) of the OOD models from the same empathy contrast pair. For instance, while *r/OMC* vs. *r/CC* (2) reaches 0.17 MCC on average, the standard deviation is 0.03. Further, we find that among the 5 models of the *r/OMC* vs. *r/CC* (2) pair, the MCC can be as high as 0.21 and as low as 0.11 despite that **a)** the 5 models only differ in the randomness of their training data sampling, **b)** the models have negligible variation in their test set performances (Figure 3). This pattern is present in all the OOD models, revealing their brittleness w.r.t. MI empathy classification.

As for the choice between 1-utterance and 2-utterance, the effects are mixed. Specifically, *r/H* vs. *r/CC* and *ED* vs. *r/CC* both have decreased performances on ANNO going from 1-utterance to 2-utterance, while *r/OMC* vs. *r/CC* benefits from this transition. In fact, in terms of the average score, *r/OMC* vs. *r/CC* (2) is the best setup. This could be because a client talks more about negative experiences in a therapy session, not unlike how the typical speaker shares emotional stores in *r/OMC*. In contrast, the speakers in *r/H* are more likely to tell positive experiences, which could explain the performance drop resulting from including the speaker utterance in *r/H* vs. *r/CC* (2).

The unsupervised zero-shot baselines do not fare better in general. $nli^E_{(1)}$ and $nli^E_{(2)}$ score around 0.05 and 0.02, respectively, both below most of the mean scores achieved by the OOD empathy contrast models. This can be attributed to the fact that knowledge gained from NLI tasks are not sufficient for reasoning about complex concepts such as empathy. $nli^A_{C\to T}$, on the other hand, shows better results and outperforms half of the OOD empathy contrast models, which suggests correlation between client-therapist agreement and therapist empathy. As a probing step, we swap the client and therapist utterances to reverse the premise-hypothesis formulation and observe that it ($nli^A_{T\to C}$) leads to a substantial drop to -0.04 MCC, further illustrating the aforementioned correlation.

### 5.3 Analysis

To shed light on the impact of the OOD design choices we made in Section 4, we add a **control** group of OOD models that are trained without empathy contrast for comparison, as shown by the blue bars in the "OOD$_{(1)}$ w/o Contrast", "OOD$_{(2)}$ w/o Contrast" subplots. More specifically, We build 3 pairs: (*r/OMC* vs. *r/H*), (*ED* vs. *r/H*), and (*ED* vs. *r/OMC*), as we consider them **(empathy vs. empathy)** pairs from which an OOD model is not

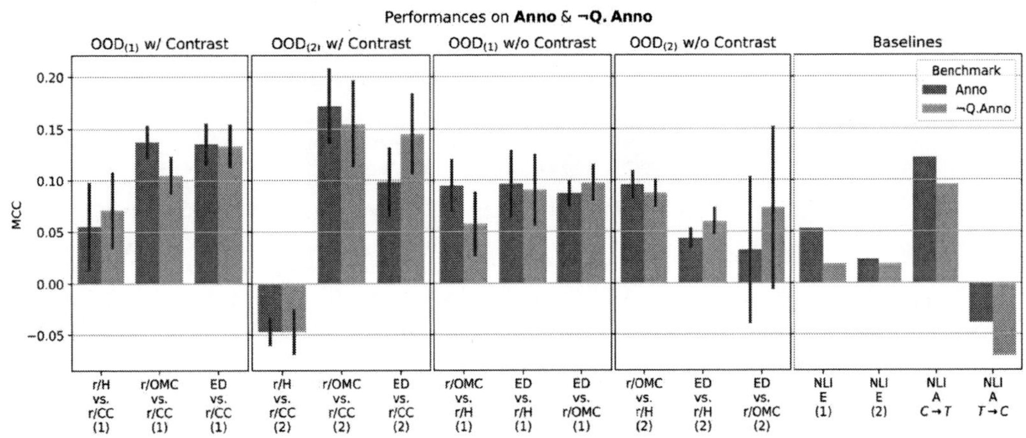


Figure 2: Results of all models on ANNO and ¬Q.ANNO, measured with Matthews correlation coefficient (Matthews, 1975). The names of the baseline models (shown in the rightmost subplot) are re-written in the figure for better visibility, e.g. "NLI\nE\n(1)" instead of $nli^{E}_{(1)}$). The first 4 subplots on the left show the performances of OOD-trained models. The first two show the performances of the 1- (e.g. *r/H* vs. *r/CC* (1)) and 2-utterance OOD models (e.g. *r/H* vs. *r/CC* (2)) trained on data **with** empathy contrast (e.g. *r/H* vs. *r/CC*, which is empathy vs. non-empathy), while the third and fourth show the performances of the 1- and 2-utterance OOD models trained on data **without** empathy contrast (e.g. *ED* vs. *r/H*, which is empathy vs. empathy). As explained in Section 5.1, for each OOD pair (e.g. *r/H* vs. *r/CC*), we randomly sample from the **class-unbalanced** OOD data 5 times to obtain 5 groups of **class-balanced** {train, dev, set} data, in order to address class imbalance and data selection bias. For each OOD pair, therefore, we train 5 models independently with the training data from their respective groups. Thus, the value of each rectangular bar indicates the mean of the scores of the 5 models from the 5 data groups of the corresponding OOD pair, and the error bar shows +/- one standard deviation from the mean.

able to learn **empathy vs. non-empathy** contrast. Additionally, we inspect the performances (orange bars) of all the models on ¬Q.ANNO to understand model behaviour in a less noisy context (i.e. question utterances removed).

Interestingly, the control group models score around 0.11 MCC and are not far behind empathy contrast models such as *r/OMC* vs. *r/CC* and *ED* vs. *r/CC* in the 1-utterance scenario, albeit with similarly large variation in their results. When it comes to 2-utterance, however, the lead of the empathy contrast models (except *r/H* vs. *r/CC*) becomes more obvious, with *r/OMC* vs. *r/CC* scoring over 0.15 MCC in contrast to *ED* vs. *r/OMC* recording less than 0.05. This shows that the benefit of learning from OOD empathy contrast, though small, does exist, and is more pronounced when **a)** compared against learning from no-empathy-contrast OOD data and **b)** more conversation context is taken into account by the models.

Finally, for the OOD contrast models, we notice mixed effects of removing questions from the benchmark dataset. It enables performance gains for *r/H* vs. *r/CC* (1) and *ED* vs. *r/CC* (2) but performance drops for the other OOD empathy contrast models. This shows that despite the annotations indicating that question therapist utterances are predominantly non-empathetic, whether a therapist utterance is a question generally does not substantially impact the empathy prediction of an OOD contrast model. One possible explanation, among others, is that the models simply did not learn to associate question with non-empathy during the OOD contrast training and instead learned to base its classification on semantic cues unrelated to question/non-question. Echoing Section 3.3, we argue that analysing non-questions at sentence level would be less noisy and better predictions would thus be possible, which we leave for future work.

## 6 Clinical Application & Impact

The motivation for this work was to minimise the annotation effort needed for training an utterance-level classifier of therapeutic empathy/non-empathy, based on the assumption that **1)** pre-trained language models can be fine-tuned to distinguish between empathy and non-empathy in general conversations, and **2)** the fine-tuned model can be leveraged to directly predict therapeutic empathy/non-empathy.

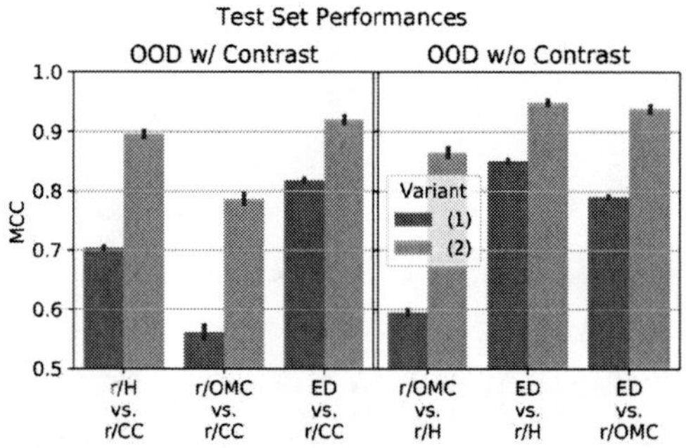


Figure 3: Test set performances (in MCC) of all OOD models. The first subplot on the left shows the test set performances of the 1- and 2-utterance OOD models trained on data with empathy contrast, and the second shows the test set performances of the 1- and 2-utterance models trained on data without empathy contrast. As explained in Figure 2, each OOD pair (e.g. *r/H* vs. *r/CC* (1) / (2)) corresponds to 5 groups of randomly sampled {train, dev, test} data and hence 5 trained models. Thus, the model trained on the training data of a group has a test set score associated with the test data of the group. Therefore, the value of each rectangular bar indicates the mean of the test set scores of the 5 models from the same OOD pair, and the error bar shows +/- one standard deviation from the mean.

Our results, for the most part, show that this simple OOD training approach did not sufficiently perform accurate classification, which limits its application in clinical settings. Compared to supervised learning of session-level empathy on sizeable corpora of well-annotated therapeutic conversations (Gibson et al., 2016), the task of utterance-level empathy classification with no in-domain training is more challenging and the models unsurprisingly fared worse. As discussed, the coarse, heuristic empathy labelling for the utterances in the training data and the domain gap between general conversation and therapeutic dialogue may have contributed considerably to the sub-optimal performance.

Nevertheless, we believe that this work is a meaningful step towards low-resource real-time assessment of empathy in counselling, and that the idea of utilising pre-trained language models for low-resource scenarios related to clinical psychology is still relevant. With smoothed domain gaps and more fine-grained annotation, future work can still use pre-trained language models to leverage parallels between empathy manifestations in general conversation and therapeutic dialogue. For instance, knowledge of empathy vs. non-empathy learned from well-annotated general conversations can serve as a bootstrapping step for empathy vs. non-empathy training on a minimal amount of well-annotated therapeutic conversations, since there can be a small to modest amount of therapeutic dialogue data available for a specialised domain instead of no data at all, which can take advantage of OOD empathy knowledge as a starting point for in-domain fine-tuning and thus maximise the benefit of OOD empathy training.

## 7 Conclusion

We find that our models trained to learn from empathy vs. non-empathy contrast in general conversation (i.e. out-of-domain w.r.t. counselling) are generally not reliable predictors of empathy/non-empathy in motivational interviewing. Upon probing, we observe that OOD empathy contrast learning is still marginally better than OOD learning without empathy contrast, particularly when more conversation context is available.

In future work, we plan to investigate more fine-grained empathy annotation and prediction, such as at sentence level, where we expect less noise and more accurate predictions. In addition, we will explore few-shot methods for the empathy classification task with out-of-domain empathy contrast training as a bootstrapping step.

## Ethics & Privacy

Empathy often involves deeply personal circumstances (e.g. distress & struggle) and computational studies on it therefore warrant ethical consideration. The greatest ethical risk of this work has been privacy implications, as the conversational data we used could contain large amounts of sensitive identifiable information. To mitigate this risk, we experimented with only de-identified data where mentions of information like name, date, and location are replaced with placeholders. As a counterbalance, this study has considerable benefit as the first investigation of using knowledge of general-conversation empathy to support low-resource computational analysis of MI empathy, and the findings can inspire future efforts in making research on therapeutic empathy more accessible.

## Acknowledgements

This work has been funded by the EC in the H2020 Marie Skłodowska-Curie PhilHumans project, contract no. 812882. The authors would also like to thank Dr. Mark Aloia for his guidance and support.

## References


Jie Cao, Michael Tanana, Zac E. Imel, Eric Poitras, David C. Atkins, and Vivek Srikumar. 2019. Observing dialogue in therapy: Categorizing and forecasting behavioral codes. In *Proceedings of the 57th Conference of the Association for Computational Linguistics, ACL 2019, Florence, Italy, July 28- August 2, 2019, Volume 1: Long Papers*, pages 5599–5611. Association for Computational Linguistics.

Zhuohao Chen, Karan Singla, James Gibson, Dogan Can, Zac E. Imel, David C. Atkins, Panayiotis G. Georgiou, and Shrikanth S. Narayanan. 2019. Improving the prediction of therapist behaviors in addiction counseling by exploiting class confusions. In *IEEE International Conference on Acoustics, Speech and Signal Processing, ICASSP 2019, Brighton, United Kingdom, May 12-17, 2019*, pages 6605–6609. IEEE.

Jacob Cohen. 1968. Weighted kappa: nominal scale agreement provision for scaled disagreement or partial credit. *Psychological bulletin*, 70(4):213.

Jacob Devlin, Ming-Wei Chang, Kenton Lee, and Kristina Toutanova. 2019. BERT: pre-training of deep bidirectional transformers for language understanding. In *Proceedings of the 2019 Conference of the North American Chapter of the Association for Computational Linguistics: Human Language Technologies, NAACL-HLT 2019, Minneapolis, MN, USA, June 2-7, 2019, Volume 1 (Long and Short Papers)*, pages 4171–4186. Association for Computational Linguistics.

Robert Elliott, Arthur C Bohart, Jeanne C Watson, and Leslie S Greenberg. 2011. Empathy. *Psychotherapy*, 48(1):43.

Joseph L Fleiss. 1971. Measuring nominal scale agreement among many raters. *Psychological bulletin*, 76(5):378.

Nikolaos Flemotomos, Victor R. Martinez, Zhuohao Chen, Karan Singla, Victor Ardulov, Raghuveer Peri, Derek D. Caperton, James Gibson, Michael J. Tanana, Panayiotis G. Georgiou, Jake Van Epps, Sarah P. Lord, Tad Hirsch, Zac E. Imel, David C. Atkins, and Shrikanth Narayanan. 2021. "am I A good therapist?" automated evaluation of psychotherapy skills using speech and language technologies. *CoRR*, abs/2102.11265.

Jacques Gaume, Gerhard Gmel, Mohamed Faouzi, and Jean-Bernard Daeppen. 2009. Counselor skill influences outcomes of brief motivational interventions. *Journal of substance abuse treatment*, 37(2):151–159.

James Gibson, David C. Atkins, Torrey Creed, Zac E. Imel, Panayiotis G. Georgiou, and Shrikanth S. Narayanan. 2018. Multi-label multi-task deep learning for behavioral coding. *CoRR*, abs/1810.12349.

James Gibson, Dogan Can, Bo Xiao, Zac E. Imel, David C. Atkins, Panayiotis G. Georgiou, and Shrikanth S. Narayanan. 2016. A deep learning approach to modeling empathy in addiction counseling. In *Interspeech 2016, 17th Annual Conference of the International Speech Communication Association, San Francisco, CA, USA, September 8-12, 2016*, pages 1447–1451. ISCA.

James Gibson, Nikolaos Malandrakis, Francisco Romero, David C. Atkins, and Shrikanth S. Narayanan. 2015. Predicting therapist empathy in motivational interviews using language features inspired by psycholinguistic norms. In *INTERSPEECH 2015, 16th Annual Conference of the International Speech Communication Association, Dresden, Germany, September 6-10, 2015*, pages 1947–1951. ISCA.

Bhanu Prakash Reddy Guda, Aparna Garimella, and Niyati Chhaya. 2021. Empathbert: A bert-based framework for demographic-aware empathy prediction. *CoRR*, abs/2102.00272.

Sepp Hochreiter and Jürgen Schmidhuber. 1997. Long short-term memory. *Neural computation*, 9(8):1735–1780.

Mahshid Hosseini and Cornelia Caragea. 2021. It takes two to empathize: One to seek and one to provide. *Proceedings of the AAAI Conference on Artificial Intelligence*. To appear.

Mike Lewis, Yinhan Liu, Naman Goyal, Marjan Ghazvininejad, Abdelrahman Mohamed, Omer Levy, Veselin Stoyanov, and Luke Zettlemoyer. 2020. BART: denoising sequence-to-sequence pre-training for natural language generation, translation, and comprehension. In *Proceedings of the 58th Annual Meeting of the Association for Computational Linguistics, ACL 2020, Online, July 5-10, 2020*, pages 7871–7880. Association for Computational Linguistics.

Zhaojiang Lin, Andrea Madotto, Jamin Shin, Peng Xu, and Pascale Fung. 2019. Moel: Mixture of empathetic listeners. In *Proceedings of the 2019 Conference on Empirical Methods in Natural Language Processing and the 9th International Joint Conference on Natural Language Processing, EMNLP-IJCNLP 2019, Hong Kong, China, November 3-7, 2019*, pages 121–132. Association for Computational Linguistics.

Zhaojiang Lin, Peng Xu, Genta Indra Winata, Farhad Bin Siddique, Zihan Liu, Jamin Shin, and Pascale Fung. 2020. Caire: An end-to-end empathetic chatbot. In *The Thirty-Fourth AAAI Conference on Artificial Intelligence, AAAI 2020, The Thirty-Second Innovative Applications of Artificial Intelligence Conference, IAAI 2020, The Tenth AAAI Symposium on Educational Advances in Artificial Intelligence, EAAI 2020, New York, NY, USA, February 7-12, 2020*, pages 13622–13623. AAAI Press.

Nurul Lubis, Sakriani Sakti, Koichiro Yoshino, and Satoshi Nakamura. 2018. Eliciting positive emotion through affect-sensitive dialogue response generation: A neural network approach. In *Proceedings of the Thirty-Second AAAI Conference on Artificial Intelligence, (AAAI-18), the 30th innovative Applications of Artificial Intelligence (IAAI-18), and the 8th AAAI Symposium on Educational Advances in Artificial Intelligence (EAAI-18), New Orleans, Louisiana, USA, February 2-7, 2018*, pages 5293–5300. AAAI Press.

Brian W Matthews. 1975. Comparison of the predicted and observed secondary structure of t4 phage lysozyme. *Biochimica et Biophysica Acta (BBA)-Protein Structure*, 405(2):442–451.

Jim McCambridge, Maria Day, Bonnita A Thomas, and John Strang. 2011. Fidelity to motivational interviewing and subsequent cannabis cessation among adolescents. *Addictive behaviors*, 36(7):749–754.

William R Miller, Theresa B Moyers, Denise Ernst, and Paul Amrhein. 2003. Manual for the motivational interviewing skill code (misc). *Unpublished manuscript. Albuquerque: Center on Alcoholism, Substance Abuse and Addictions, University of New Mexico*.

William R Miller and Stephen Rollnick. 2012. *Motivational interviewing: Helping people change*. Guilford press.

Theresa B Moyers, Lauren N Rowell, Jennifer K Manuel, Denise Ernst, and Jon M Houck. 2016. The motivational interviewing treatment integrity code (miti 4): rationale, preliminary reliability and validity. *Journal of substance abuse treatment*, 65:36–42.

Verónica Pérez-Rosas, Rada Mihalcea, Kenneth Resnicow, Satinder Singh, and Lawrence An. 2017. Understanding and predicting empathic behavior in counseling therapy. In *Proceedings of the 55th Annual Meeting of the Association for Computational Linguistics, ACL 2017, Vancouver, Canada, July 30 - August 4, Volume 1: Long Papers*, pages 1426–1435. Association for Computational Linguistics.

Verónica Pérez-Rosas, Xinyi Wu, Kenneth Resnicow, and Rada Mihalcea. 2019. What makes a good counselor? learning to distinguish between high-quality and low-quality counseling conversations. In *Proceedings of the 57th Annual Meeting of the Association for Computational Linguistics*, pages 926–935, Florence, Italy. Association for Computational Linguistics.

Hannah Rashkin, Eric Michael Smith, Margaret Li, and Y-Lan Boureau. 2019. Towards empathetic open-domain conversation models: A new benchmark and dataset. In *Proceedings of the 57th Conference of the Association for Computational Linguistics, ACL 2019, Florence, Italy, July 28- August 2, 2019, Volume 1: Long Papers*, pages 5370–5381. Association for Computational Linguistics.

Stephen Rollnick, William R Miller, and Christopher Butler. 2008. *Motivational interviewing in health care: helping patients change behavior*. Guilford Press.

Ashish Sharma, Adam S. Miner, David C. Atkins, and Tim Althoff. 2020. A computational approach to understanding empathy expressed in text-based mental health support. In *Proceedings of the 2020 Conference on Empirical Methods in Natural Language Processing, EMNLP 2020, Online, November 16-20, 2020*, pages 5263–5276. Association for Computational Linguistics.

Jamin Shin, Peng Xu, Andrea Madotto, and Pascale Fung. 2020. Generating empathetic responses by looking ahead the user's sentiment. In *2020 IEEE International Conference on Acoustics, Speech and Signal Processing, ICASSP 2020, Barcelona, Spain, May 4-8, 2020*, pages 7989–7993. IEEE.

Karan Singla, Zhuohao Chen, David C. Atkins, and Shrikanth Narayanan. 2020. Towards end-2-end learning for predicting behavior codes from spoken utterances in psychotherapy conversations. In *Proceedings of the 58th Annual Meeting of the Association for Computational Linguistics, ACL 2020, Online, July 5-10, 2020*, pages 3797–3803. Association for Computational Linguistics.

Michael Tanana, Kevin A Hallgren, Zac E Imel, David C Atkins, and Vivek Srikumar. 2016. A comparison of natural language processing methods for automated coding of motivational interviewing. *Journal of substance abuse treatment*, 65:43–50.

Anuradha Welivita and Pearl Pu. 2020. A taxonomy of empathetic response intents in human social conversations. In *Proceedings of the 28th International Conference on Computational Linguistics, COLING 2020, Barcelona, Spain (Online), December 8-13, 2020*, pages 4886–4899. International Committee on Computational Linguistics.

Adina Williams, Nikita Nangia, and Samuel Bowman. 2018. A broad-coverage challenge corpus for sentence understanding through inference. In *Proceedings of the 2018 Conference of the North American Chapter of the Association for Computational Linguistics: Human Language Technologies, Volume 1 (Long Papers)*, pages 1112–1122. Association for Computational Linguistics.

Thomas Wolf, Lysandre Debut, Victor Sanh, Julien Chaumond, Clement Delangue, Anthony Moi, Pierric Cistac, Tim Rault, Rémi Louf, Morgan Funtowicz, et al. 2019. Huggingface's transformers: State-of-the-art natural language processing. *arXiv preprint arXiv:1910.03771*.

Zixiu Wu, Rim Helaoui, Vivek Kumar, Diego Reforgiato Recupero, and Daniele Riboni. 2020. Towards detecting need for empathetic response in motivational interviewing. In *Companion Publication of the 2020 International Conference on Multimodal*

*Interaction, ICMI Companion 2020, Virtual Event, The Netherlands, October, 2020*, pages 497–502. ACM.

Bo Xiao, Daniel Bone, Maarten Van Segbroeck, Zac E. Imel, David C. Atkins, Panayiotis G. Georgiou, and Shrikanth S. Narayanan. 2014. Modeling therapist empathy through prosody in drug addiction counseling. In *INTERSPEECH 2014, 15th Annual Conference of the International Speech Communication Association, Singapore, September 14-18, 2014*, pages 213–217. ISCA.

Bo Xiao, Dogan Can, Panayiotis G. Georgiou, David C. Atkins, and Shrikanth S. Narayanan. 2012. Analyzing the language of therapist empathy in motivational interview based psychotherapy. In *Asia-Pacific Signal and Information Processing Association Annual Summit and Conference, APSIPA 2012, Hollywood, CA, USA, December 3-6, 2012*, pages 1–4. IEEE.

Bo Xiao, Dogan Can, James Gibson, Zac E. Imel, David C. Atkins, Panayiotis G. Georgiou, and Shrikanth S. Narayanan. 2016a. Behavioral coding of therapist language in addiction counseling using recurrent neural networks. In *Interspeech 2016, 17th Annual Conference of the International Speech Communication Association, San Francisco, CA, USA, September 8-12, 2016*, pages 908–912. ISCA.

Bo Xiao, Che-Wei Huang, Zac E. Imel, David C. Atkins, Panayiotis G. Georgiou, and Shrikanth S. Narayanan. 2016b. A technology prototype system for rating therapist empathy from audio recordings in addiction counseling. *PeerJ Comput. Sci.*, 2:e59.

Bo Xiao, Zac E. Imel, David C. Atkins, Panayiotis G. Georgiou, and Shrikanth S. Narayanan. 2015. Analyzing speech rate entrainment and its relation to therapist empathy in drug addiction counseling. In *INTERSPEECH 2015, 16th Annual Conference of the International Speech Communication Association, Dresden, Germany, September 6-10, 2015*, pages 2489–2493. ISCA.

Wenpeng Yin, Jamaal Hay, and Dan Roth. 2019. Benchmarking zero-shot text classification: Datasets, evaluation and entailment approach. In *Proceedings of the 2019 Conference on Empirical Methods in Natural Language Processing and the 9th International Joint Conference on Natural Language Processing, EMNLP-IJCNLP 2019, Hong Kong, China, November 3-7, 2019*, pages 3912–3921. Association for Computational Linguistics.

Peixiang Zhong, Chen Zhang, Hao Wang, Yong Liu, and Chunyan Miao. 2020. Towards persona-based empathetic conversational models. In *Proceedings of the 2020 Conference on Empirical Methods in Natural Language Processing, EMNLP 2020, Online, November 16-20, 2020*, pages 6556–6566. Association for Computational Linguistics.

Ke Zhou, Luca Maria Aiello, Sanja Scepanovic, Daniele Quercia, and Sara Konrath. 2021. The language of situational empathy. *Proc. of the ACM on Human-Computer Interaction*, 1:1–19.

Naitian Zhou and David Jurgens. 2020. Condolence and empathy in online communities. In *Proceedings of the 2020 Conference on Empirical Methods in Natural Language Processing, EMNLP 2020, Online, November 16-20, 2020*, pages 609–626. Association for Computational Linguistics.

Xianda Zhou and William Yang Wang. 2018. Mojitalk: Generating emotional responses at scale. In *Proceedings of the 56th Annual Meeting of the Association for Computational Linguistics, ACL 2018, Melbourne, Australia, July 15-20, 2018, Volume 1: Long Papers*, pages 1128–1137. Association for Computational Linguistics.

## A Data

We list the complete statistics of the general conversation datasets in Table 4 and those of ROLEPLAYMI in Table 5.

## B Implementation Details

All our pre-trained language models are implemented by the HuggingFace framework[9] (Wolf et al., 2019). All our models are implemented in PyTorch[10], while their evaluation is implemented with scikit-learn[11]. For $cls_{(1)}$, the input format to BERT is { `[CLS]` $u_m^L$ `[SEP]` } during training and { `[CLS]` $u_i^T$ `[SEP]` } during testing. Similarly, for $cls_{(2)}$, the input becomes { `[CLS]` $u_m^S$ `[SEP]` $u_m^L$ `[SEP]` } during training and { `[CLS]` $u_i^C$ `[SEP]` $u_i^T$ `[SEP]` } during testing.

During OOD training, we use a learning rate of 1*e*-5 and a batch size of 32, and evaluate every 500 steps on the development set. We choose the Matthews correlation coefficient (Matthews, 1975) (MCC) as the metric for validation. We stop the training if the performance has not improved in the most recent 10 validations, and select the best checkpoint w.r.t. the development set.

We formulate the input to $nli_{(1)}^E$ as { `[CLS]` $u_i^T$ `[SEP]` $h_{(1)}$ `[SEP]` }, and likewise { `[CLS]` $p_i$ `[SEP]` $h_{(2)}$ `[SEP]` } for $nli_{(2)}^E$ , { `[CLS]` $u_i^C$ `[SEP]` $u_i^T$ `[SEP]` } for $nli_{C\to T}^A$, and { `[CLS]` $u_i^T$ `[SEP]` $u_i^C$ `[SEP]` } for $nli_{T\to C}^A$.

[9]https://github.com/huggingface/transformers

[10]https://pytorch.org/

[11]https://scikit-learn.org/stable/

| | *r/Happy* ‡ | | | *r/OffMyChest* ‡ | | | *r/CasualConversation* ¶ | | | *EmpatheticDialogues* ‡ | | |
|---|---|---|---|---|---|---|---|---|---|---|---|---|
| Split | train | valid | test | train | valid | test | train | valid | test | train | valid | test |
| #Conv | 113.9K | 13.9K | 16.0K | 94.0K | 12.1K | 11.7K | 530.2K | 67.5K | 66.9K | 17.8K | 2.8K | 2.5K |
| $\mu$(#S-u./Conv) | 1.0 | 1.0 | 1.0 | 1.0 | 1.0 | 1.0 | 1.1 | 1.1 | 1.1 | 2.2 | 2.3 | 2.2 |
| $\mu$(#L-u./Conv) | 1.0 | 1.0 | 1.0 | 1.0 | 1.0 | 1.0 | 1.1 | 1.1 | 1.1 | 2.1 | 2.1 | 2.1 |
| $\mu$(S-u.Len.) | 30.8 | 30.2 | 30.4 | 48.9 | 51.0 | 47.8 | 42.8 | 42.9 | 43.2 | 17.6 | 19.4 | 21.2 |
| $\mu$(L-u.Len.) | 13.3 | 13.5 | 13.3 | 15.7 | 15.7 | 15.6 | 16.9 | 16.8 | 16.8 | 13.7 | 14.3 | 14.5 |

Table 4: Statistics of *PEC* (*r/Happy*, *r/OffMyChest*, and *r/CasualConversation*) & *EmpatheticDialogues*. For *PEC*, we utilise 2-interlocutor conversations only. #Conv: number of conversations in the data split. $\mu$(#S-u./Conv): average number of speaker turns per conversation. $\mu$(#L-u./Conv): average number of listener turns per conversation. $\mu$(S-u.Len.): average speaker utterance length (number of tokens), $\mu$(L-u.Len.): average listener utterance length (number of tokens). We consider *r/Happy*, *r/OffMyChest* and *EmpatheticDialogues* to consist of mostly empathetic (‡) listener utterances and *r/CasualConversation* to be comprised of predominantly non-empathetic (¶) ones. Note that the statistics of *PEC* are about the filtered dataset as described in Section 3.1.

| | **ROLEPLAYMI** | | **ANNO** | |
|---|---|---|---|---|
| MI Quality | High | Low | High | Low |
| #Conv | 152 | 101 | 7 | 14 |
| #T-u | 3928 | 1534 | 217 | 214 |
| $\mu$(#T-u/Conv) | 25.8 | 15.2 | 31.0 | 15.3 |
| $\mu$(#C-u/Conv) | 25.1 | 14.5 | 30.0 | 14.8 |
| $\mu$(T-uLen). | 33.5 | 31.1 | 33.2 | 32.9 |
| $\mu$(C-uLen.) | 28.5 | 20.6 | 24.4 | 21.6 |
| %(emp.T-u) | n/a | n/a | 38.7% | 2.3% |
| %($\neg$Q.T-u) | n/a | n/a | 71.9% | 73.8% |
| $p$(emp \| $\neg$Q, T-u) | n/a | n/a | **0.50** | **0.03** |
| $p$(emp \| Q, T-u) | n/a | n/a | 0.10 | 0.00 |

Table 5: Statistics of ROLEPLAYMI and ANNO. The abbreviation convention is similar to that in Table 4, while "T-u" is short for "Therapist Utterance(s)" and "C-u" for "Client Utterance(s)". #Conv: number of conversations in the subset. #T-u: number of therapist utterances in the subset. %(emp.T-u): percentage of empathetic therapist utterances. %($\neg$Q.T-u): percentage of non-question therapist utterances. $p$(emp | $\neg$Q, T-u): probability of a non-question therapist utterance being empathetic. $p$(emp | Q, T-u): probability of a question therapist utterance being empathetic.

# Towards Understanding the Role of Gender in Deploying Social Media-Based Mental Health Surveillance Models

**Eli Sherman**
Johns Hopkins University
esherman@jhu.edu

**Keith Harrigian**
Johns Hopkins University
kharrigian@jhu.edu

**Carlos Aguirre**
Johns Hopkins University
caguirr4@jhu.edu

**Mark Dredze**
Johns Hopkins University
mdredze@cs.jhu.edu

## Abstract

Spurred by advances in machine learning and natural language processing, developing social media-based mental health surveillance models has received substantial recent attention. For these models to be maximally useful, it is necessary to understand how they perform on various subgroups, especially those defined in terms of protected characteristics. In this paper we study the relationship between user demographics – focusing on gender – and depression. Considering a population of Reddit users with known genders and depression statuses, we analyze the degree to which depression predictions are subject to biases along gender lines using domain-informed classifiers. We then study our models' parameters to gain qualitative insight into the differences in posting behavior across genders.

## 1 Introduction

The United States Centers for Disease Control and Prevention estimates that 8% of American adults suffer from major depression at a given time (Brody et al., 2018). This represents a critical public health threat, as depression is associated with downstream physical health complications (Rush, 2007; Alboni et al., 2008) and an increased risk of suicide (Richards and O'Hara, 2014). Among the many efforts to address this crisis is a line of research at the intersection of language modeling, social media analysis, and mental health. Seminal papers by De Choudhury et al. (2013) and Coppersmith et al. (2014) demonstrated the general feasibility of predicting mental health status from social media data.

A major obstacle to the practical use of mental health surveillance models is differential performance for different subgroups of the population. This behavior can arise either because the training data is not sufficiently representative of the population, or because some groups are simply harder to predict given the same data. The former case is well-studied in the machine learning literature and can be addressed by careful data collection and training regimes. The latter case, however, is often more subtle and harder to address. *Not* identifying and addressing these differences in performance degrades the utility of the models. In particular, if the performance is worse for historically marginalized populations it can reinforce existing inequities such as under-diagnosis of depression (Elazar and Goldberg, 2018).

In this work we aim to assess the scope of the differential performance problem by studying the relationship between gender and predictions of depression. The most useful insight we could gain would be determining whether or not gender is a confounder for depression predictions; that is, whether gender both causally affects the way in which users post on Reddit and causally affects our predictions of the user's depression status. Unfortunately, testing whether this causal dynamic is true is very difficult with the purely observational data available to us. Towards testing this phenomena, we will instead test the slightly weaker hypotheses i) that depression predictions exhibit gender bias (i.e., there are differences in performance across genders) and ii) that these differences are due, at least in part, to differing uses of language between men and women in talking about their mental state. Together these hypotheses serve as a sort of associational version of the causal phenomenon we'd like to study. They can tell us whether depression predictions are correlated with gender and whether certain terms are likely to have different meanings based on the gender of the author.

We test hypothesis (i) quantitatively by fitting depression prediction models to a novel data set collected from Reddit with *ground truth* genders, derived from self-disclosures, and comparing the performances across genders. We test hypothesis (ii) qualitatively by looking at features strongly predictive of depression for each gender. We identify themes that are concordant across genders and consistent with the literature (De Choudhury et al., 2016) as well as themes that are discordant across genders and support our hypothesis that men and women use many terms differently to talk about (non-) depression. We follow these analyses with a discussion of open questions that follow from this work. In particular, we discuss the use of causal methodologies to assess our stronger hypothesis that gender confounds depression prediction. We highlight the types of methods that could be used and the data that is necessary to test the causal hypothesis. We conclude with a discussion of limitations and the ethical implications of this work.



*Proceedings of the Seventh Workshop on Computational Linguistics and Clinical Psychology*, pages 217–223
June 11, 2021. 

## 2 Related Work

Several existing papers have considered the role of demographics in mental health prediction. Elazar and Goldberg (2018) demonstrated that demographics are implicitly encoded in text data. Wood-Doughty et al. (2017) and Loveys et al. (2018) both studied differing language use across cultures. The former used a Twitter data set with *inferred* demographic labels, while the latter used a carefully-curated proprietary data set from 7 Cups of Tea. Amir et al. (2019) explored the role of cohort selection in assessing mental health disorder prevalence. Aguirre et al. (2021) is the closest to the present work. The authors characterized the biases present in depression prediction models by showing there are differences in performance for different demographic subgroups. This work studied biases that arise due to the specific data set used for training,focusing on the popular, publicly available data sets CLPsych (Coppersmith et al., 2015) and MULTITASK (Benton et al., 2017).

The present work differs from those cited in that we seek to quantify demographic bias in depression prediction using self-disclosures in a publicly available data set. This approach improves scalability and reproducibility compared to hand-labeled and proprietary data sets. Additionally, while self-disclosures are not perfect, they are not subject to the same degree of noise and error that is induced when using genders inferred by using a pre-trained model, trained on an auxiliary data source. Our estimates of the depression prediction performance across genders are therefore likely to be of a higher quality. Moreover, our analyses of features that are predictive of depression for each gender are also likely to be less noisy than they would be if we were also inferring genders from those same features.

## 3 Data Collection

To obtain a dataset with ground truth gender, we mined all posts and comments from the r/AskMen and r/AskWomen subreddits between January 1, 2019 and December 31, 2019 using the Pushshift API (Baumgartner et al., 2020). In total, we collected 251,487 original submissions and 4,481,354 comments.

For each post, we consider the flair – an optional tag users can apply to their posts to reveal information about themselves or the content of their post – to determine the ground truth gender of the post author. We considered the author of a post to be true-male if they used one of 'Male', 'male', 'Dude', or ♂ for their flair, and true-female if they used one of 'Female', 'female', ♀, or ♀♡. Of the mined posts, 1,002,079 had some sort of flair, while 660,684 had one of the male or female indicator flairs. This process yielded a data set of 15,140 unique male and 11,241 unique female users, as well as 59 users whose gender-related flair use was inconsistent (i.e. at least one post each with a male- and female-indicating flair). While people who identify as non-binary are known to have higher rates of depression (Budge et al., 2013; Wolohan et al., 2018) and thus could benefit from the studies like this one, we did not have a reliable method for identifying non-binary users beyond the list of inconsistent users and the sub-population in our cohort was too small to yield meaningful analysis. For the remainder of the paper we restrict attention to binary genders under the folk conception of gender (Larson, 2017).

For each of the 26,381 gender-binary users, we collected the user's entire Reddit posting and commenting history from January 1, 2019 to December 31, 2019, totaling 1,035,782 original submissions and 19,029,981 comments across 64,162 subreddits. Following the literature on social media-driven mental health surveillance (De Choudhury et al., 2013; Yates et al., 2017), we defined a user as true-depressed if they authored an original submission or comment in r/depression during the study period and true-control otherwise. The breakdown of gender and depression classes is 721 and 713 depressed males and females respectively, and 14,416 and 10,526 control males and females respectively. Replication data for this study can be found at `https://github.com/esherma/CLPsych2021_Gender_and_Depression` and is available under a data usage agreement.

## 4 Methods

We fit user-level models to predict depression status from our harvested Reddit data. To enable analysis of the impact of gender as a confounder, we fit separate models on two separate data sets: a random sample of the true-men users in our data set, and a random sample of the true-women users. To reduce noise induced by 'throwaway' or 'lurker' accounts, we excluded users who made fewer than 5 posts (submissions + comments) during the study period. This decision could reduce our results' generalizability since throwaway accounts may be owned by users with separate primary accounts and post with the throwaway differently (e.g. posting more personal information).

Because depression is a rare outcome in our data, our initial train and test sets had very few depressed individuals (109 train, 26 test). This proved too few to draw meaningful conclusions about the role of gender in depression prediction. We therefore report the performance of our models trained on data sets constructed by performing *balanced sampling* from the full data. The resulting class breakdowns are: 721 and 613 depressed males and females respectively, and 820 and 712 control males and females respectively.

We split each of these sampled data sets 80-20 into train and test sets, stratifying by user. We then constructed a Bag-of-Words (BoW) vocabulary from the submissions and comments for each user in the training sets. We included 1-, 2-, and 3-grams, as well as LIWC (Pennebaker et al., 2007) and TF-IDF (Jones, 1972) features. We imposed that features must be used by a minimum of 25 users to be included in the vocabulary. We also removed posts from the r/depression subreddit from each user's BoW vector and filtered out terms and subreddits commonly associated with self-disclosure of mental health disorders using the SMHD dataset (Cohan et al., 2018). To model depression, we used the scikit-learn implementation of regularized logistic regression (Pedregosa et al., 2011). At the end of training, we discarded all but the top 100,000 features using the pairwise mutual information criterion as an additional regularization step.

## 5 Results

### 5.1 Model Performance

The performance of each model on each test set is shown in Figure 1. The most striking result is that the performance of both models is considerably higher on the men-only test set than on the women-only test set (.770 vs. .702 and .758 vs. .707 respectively). This difference indicates that predicting depression among men is easier than among women. Looking at the distribution graphs, it appears that women are *over diagnosed* as depressed. Mechanically, this difference in predictions likely arises due to the existence of a few key features that indicate depression for one gender but not the other. We identify candidate features in the analysis below.

### 5.2 Feature Analysis

We extracted the regression coefficients from each of our models and generated a scatter plot in Figure 2 of the 50,967 features the two models had in common. Towards identifying strongly predictive features, we scored each feature using the sum of the absolute value of the coefficient from each model for that feature. In the figure, we labeled the 50 highest-scoring features in each plot quadrant.

**Concordant Depression Features (top right)** Even though we filtered out self-disclosure tokens (e.g. 'depression' and 'depressed'), we see that many of the most predictive features are consistent with themes discussed in the mental health surveillance literature (De Choudhury et al., 2016): emotion ('feel', LIWC affect, LIWC negemo), physical symptoms of depression ('sleep'), and indicators of social isolation ('alone', 'porn', and personal pronouns 'me', 'my', and 'I'). One notable feature is the token 'jews'. This feature could indicate that many depressed Jewish people of both genders frequently discuss their religious identity on Reddit, possibly in the context of their peoples' historically marginalized status (McCullough and Larson, 1999). Also plausible is that the token is indicative of anti-semitic tendencies which are correlated with depression (e.g. blaming one's personal struggles on a scapegoat minority group). This phenomenon has been documented in the largely-male 'incel' community (Hoffman et al., 2020) but we could not find a clear connection between anti-semitism and depression among women in the psychology or sociology literature.

**Concordant Control Features (lower left)** These feature themes are also consistent with findings in the literature. Features indicative of social interactions are quite common ('church', 'wedding', 'couple') as well as features that suggest positive affect regarding life activities ('fun', 'cool', LIWC leisure).

**Discordant Features (top left, lower right)** These features are of primary interest for identifying potential gender-based confounding. Here we find features that are predictive of depression in women but control in men or vice-versa. We observe that there are several terms that likely have different meanings for men and women users. Many of these pertain to social interactions.

For instance, 'gay', 'gay men' and 'my husband' are all strongly predictive of control for men. This suggests that men who are comfortable discussing non-straight sexualities online are also in a relatively healthy mental state. In contrast, these terms (along with 'my wife') indicate increased mental health struggles for (possibly gay) women. We suspect 'my husband' is neutral for women because there are roughly equal numbers of users praising and condemning their husbands.

Beyond sexuality, we see that some familial terms have differing predictive interpretations across genders. 'my mum' is predictive of depression for men and control for women, while the reverse is true for 'my son'. This suggests a substantial difference in parent-child relationships depending on the gender of each: each gender appears to have an affinity for family members of the same gender.

We also highlight a few features with broader societal interpretations. 'trump' is strongly predictive of depression among women but neutral for men. This is consistent with the well-known 'gender gap' phenomenon and could also indicate that mental health is in part a function of political climate. The LIWC category 'money' is slightly depression predictive for women and control-predictive for men. Similar to the above, this could be an artifact of the wage gap: money topics may be more stressful for women because they tend to earn less money for the same amount or more work.

## 6 Discussion

In this paper we showed that depression predictions do indeed exhibit gender bias. This was evidenced by a substantially better performance when predicting depression among males than when predicting among females. We also identified terms that are used differently between men and women, providing insight into the manifestations of depression beyond modeling dynamics.

### 6.1 Open Questions and Future Work

As hinted in the introduction, the key open question is **does gender confound depression predictions?** In

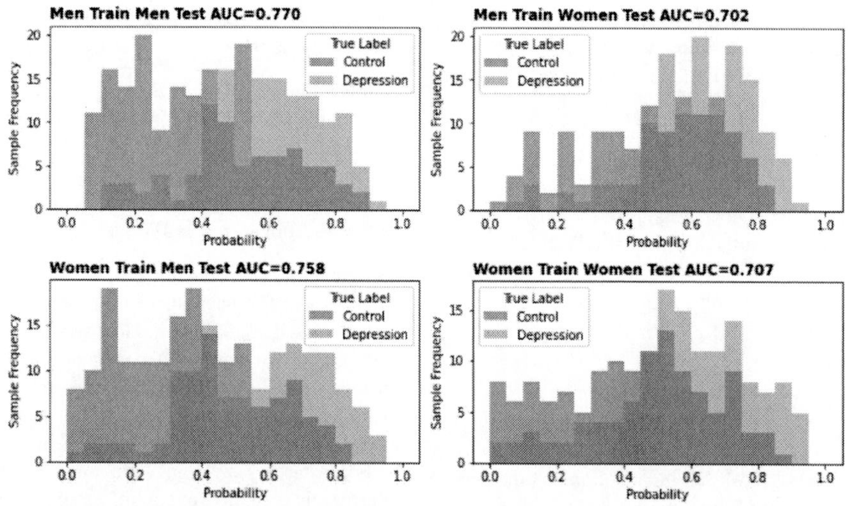


Figure 1: Performance of each model, trained to predict depression on either male users or female users only, when evaluated on each test set

other words, does gender *both* affect depression predictions *and* the features we use to predict it? There are numerous plausible explanations for why both of these causal relationships may hold or not hold, but without a rigorous causal analysis, it is not possible to rule any one explanation out in favor of another.

To properly evaluate whether a associational relationship is in fact causal, the causal framework requires 'intervening' on an independent variable while holding other variables in the system constant to see whether there are changes in the dependent variable. Here, that means intervening on gender, which is infeasible to carry out directly.

There may however, be some viable proxy approaches for simulating the intervention on gender. One such approach would entail fitting a model to predict the ground truth gender and then using a clustering algorithm to find male and female centroids based on the most predictive features in the gender prediction model. The analyst could then simulate an intervention on gender for the purposes of analyzing changes to depression prediction by replacing the user's feature vector in the depression inference model with each gender centroid vector. This approach will not permit a true causal interpretation but it could provide insights into the relationship between gender and depression prediction beyond those gained from the simple models studied in this work. Unfortunately this approach cannot be applied to analyzing the relationship between gender and the text features since it entails changing those text features.

Outside of the explicit question of confounding, we can ask **how do we correct for the performance differentials across demographic groups when predicting depression?**. As hinted earlier, an obvious approach with support in the literature (Amir et al., 2019) is to simply collect 'better' data. This is an unsatisfying answer, however, since good data is often hard to come by or expensive to collect. Instead, we can again turn to causal inference ideas to try to address data quality issues. We can potentially use methods from the causal fairness literature to impose constraints on depression models to ensure negligible differences in prediction performance. For instance, following (Nabi et al., 2019), we could impose a constraint that requires that the total effect of gender on depression predictions is zero, or, plainly, that there is no difference in model performance when we do or don't condition on gender.

## 6.2 Limitations

Aside from the limitations described above, i) all users in our cohort posted in r/AskMen or r/AskWomen (which we used to derive ground truth) and ii) we rebalanced our data sets due to insufficient numbers of depressed users in the 'representative' population. These decisions could reduce the generalizability of our results. One way to address this would be to collect data on more users by expanding the study period and by consulting other subreddits with gender self-disclosure such as r/relationships (Wang and Jurgens, 2018).

Additionally, while our use of self-disclosed genders increases scalability, this could induce bias in two ways. Users could be dishonest in their disclosure and, even if they aren't, users who choose to self-disclose could be fundamentally different from the general population. It's likely that the only solution is to collect data external to Reddit about Reddit users' genders as a more reliable supplement to our data.

Finally, our depression labels were not obtained via self-disclosures. Rather, they were defined based on

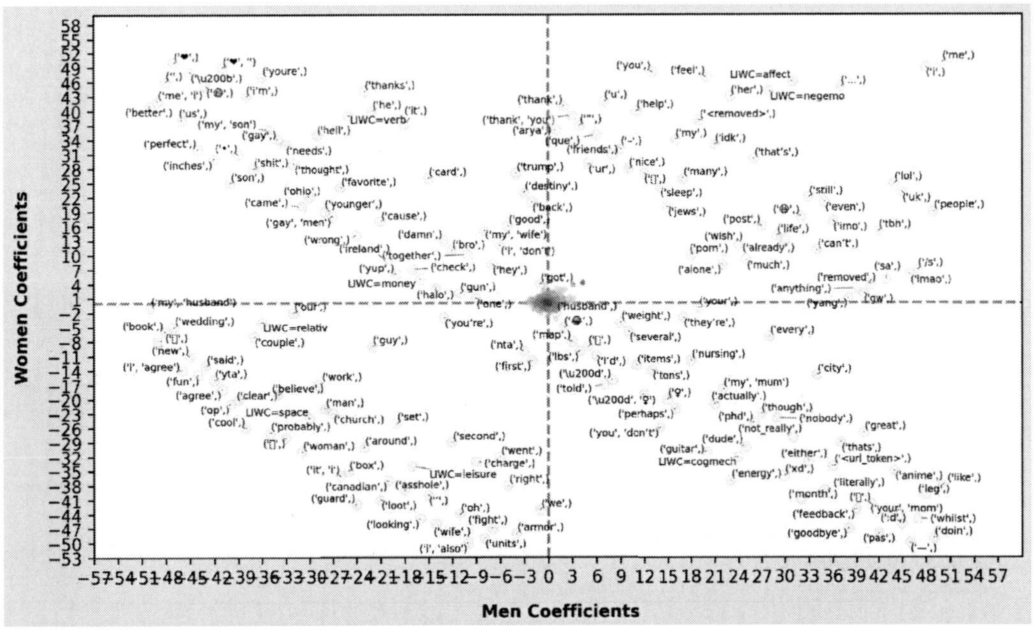


Figure 2: Features in common between the male- and female- trained models with the 50 highest scoring features in each quadrant labeled

whether the user posted in the r/depression subreddit. While this approach is consistent with data collection approaches from the literature (De Choudhury et al., 2013), it is likely to induce some noise. For instance, a user could post in the subreddit to seek support for a friend or relative, rather than for themself and would therefore be incorrectly labeled as depressed. One way to address this would be to take a more nuanced approach to labeling. For instance, we could use regular expressions matched on the text of r/depression posts to develop a more exclusive labeling policy that filters out users who are not seeking personal support.

### 6.3 Ethics

As in any applied setting it is necessary to weigh the potential advantages and harms of carrying out our research agenda. This work has the potential to cause harm in a couple key ways.

First, as previously mentioned, we restrict attention to users satisfying a narrow and dated 'folk' definition of gender in line with much of the existing research in the space of computational psychology. This is done at the cost of excluding non-binary individuals, who potentially stand to benefit the most from this work due to the increased prevalence of depression in gender non-conforming populations. Furthermore, excluding any marginalized population from a study of this type has the potential to reinforce existing biases. For instance, if our model had demonstrated improved prediction performance for the binary genders, that could lead to an incorrect assumption that the model will perform well on the general population, which includes non-binary genders. This could lead to *worse* performance for the unstudied groups.

Second, while we infer depression status from Reddit users with the goal of alleviating harms, these approaches could be harnessed with malice to identify and target already vulnerable individuals whose screen names and posting behavior are public.

On the other hand, there is great potential in this study and the work that will follow it. Identifying obstacles to model deployment for a restricted population will likely aid in correcting those obstacles for the entire population. This would substantially improve the performance and, more importantly, the clinical utility of mental health surveillance models. Given the potential benefits of this study we feel it is better to proceed, with care and transparency, rather than sit idle for lack of perfect answers to address the issues the work poses.

## Acknowledgements

The first author was sponsored by a Google PhD Fellowship.

## References


Carlos Aguirre, Keith Harrigian, and Mark Dredze. 2021. Gender and racial fairness in depression research using social media. In *Proceedings of the 16th Conference of the European Chapter of the Association for Computational Linguistics (EACL)*.

Paolo Alboni, Elisa Favaron, Nelly Paparella, Massimo Sciammarella, and Mario Pedaci. 2008. Is there an association between depression and cardiovascular mortality or sudden death? *Journal of Cardiovascular Medicine*, 9(4):356–362.

Silvio Amir, Mark Dredze, and John W Ayers. 2019. Mental health surveillance over social media with digital cohorts. In *Proceedings of the Sixth Workshop on Computational Linguistics and Clinical Psychology*, pages 114–120.

Jason Baumgartner, Savvas Zannettou, Brian Keegan, Megan Squire, and Jeremy Blackburn. 2020. The pushshift reddit dataset. In *Proceedings of the International AAAI Conference on Web and Social Media*, volume 14, pages 830–839.

Adrian Benton, Margaret Mitchell, and Dirk Hovy. 2017. Multi-task learning for mental health using social media text. *arXiv preprint arXiv:1712.03538*.

Debra J Brody, Laura A Pratt, and Jeffery P Hughes. 2018. Prevalence of depression among adults aged 20 and over: United states, 2013-2016.

Stephanie L Budge, Jill L Adelson, and Kimberly AS Howard. 2013. Anxiety and depression in transgender individuals: the roles of transition status, loss, social support, and coping. *Journal of consulting and clinical psychology*, 81(3):545.

Arman Cohan, Bart Desmet, Andrew Yates, Luca Soldaini, Sean MacAvaney, and Nazli Goharian. 2018. SMHD: a large-scale resource for exploring online language usage for multiple mental health conditions. In *Proceedings of the 27th International Conference on Computational Linguistics*, pages 1485–1497, Santa Fe, New Mexico, USA. Association for Computational Linguistics.

Glen Coppersmith, Mark Dredze, and Craig Harman. 2014. Quantifying mental health signals in twitter. In *Proceedings of the workshop on computational linguistics and clinical psychology: From linguistic signal to clinical reality*, pages 51–60.

Glen Coppersmith, Mark Dredze, Craig Harman, Kristy Hollingshead, and Margaret Mitchell. 2015. Clpsych 2015 shared task: Depression and ptsd on twitter. In *Proceedings of the 2nd Workshop on Computational Linguistics and Clinical Psychology: From Linguistic Signal to Clinical Reality*, pages 31–39.

Munmun De Choudhury, Michael Gamon, Scott Counts, and Eric Horvitz. 2013. Predicting depression via social media. In *Proceedings of the International AAAI Conference on Web and Social Media*, volume 7.

Munmun De Choudhury, Emre Kiciman, Mark Dredze, Glen Coppersmith, and Mrinal Kumar. 2016. Discovering shifts to suicidal ideation from mental health content in social media. In *Proceedings of the 2016 CHI conference on human factors in computing systems*, pages 2098–2110.

Yanai Elazar and Yoav Goldberg. 2018. Adversarial removal of demographic attributes from text data. *arXiv preprint arXiv:1808.06640*.

Bruce Hoffman, Jacob Ware, and Ezra Shapiro. 2020. Assessing the threat of incel violence. *Studies in Conflict & Terrorism*, 43(7):565–587.

Karen Sparck Jones. 1972. A statistical interpretation of term specificity and its application in retrieval. *Journal of documentation*.

Brian N Larson. 2017. Gender as a variable in natural-language processing: Ethical considerations.

Kate Loveys, Jonathan Torrez, Alex Fine, Glen Moriarty, and Glen Coppersmith. 2018. Cross-cultural differences in language markers of depression online. In *Proceedings of the Fifth Workshop on Computational Linguistics and Clinical Psychology: From Keyboard to Clinic*, pages 78–87.

Michael E McCullough and David B Larson. 1999. Religion and depression: A review of the literature. *Twin Research and Human Genetics*, 2(2):126–136.

Razieh Nabi, Daniel Malinsky, and Ilya Shpitser. 2019. Learning optimal fair policies. In *International Conference on Machine Learning*, pages 4674–4682. PMLR.

Fabian Pedregosa, Gaël Varoquaux, Alexandre Gramfort, Vincent Michel, Bertrand Thirion, Olivier Grisel, Mathieu Blondel, Peter Prettenhofer, Ron Weiss, Vincent Dubourg, et al. 2011. Scikit-learn: Machine learning in python. *the Journal of machine Learning research*, 12:2825–2830.

James W Pennebaker, Roger J Booth, and Martha E Francis. 2007. Liwc2007: Linguistic inquiry and word count. *Austin, Texas: liwc. net*.

C Steven Richards and Michael W O'Hara. 2014. *The Oxford handbook of depression and comorbidity*. Oxford University Press.

A John Rush. 2007. The varied clinical presentations of major depression disorder. *The Journal of clinical psychiatry*.

Zijian Wang and David Jurgens. 2018. It's going to be okay: Measuring access to support in online communities. In *Proceedings of the 2018 Conference on Empirical Methods in Natural Language Processing*, pages 33–45, Brussels, Belgium. Association for Computational Linguistics.

JT Wolohan, Misato Hiraga, Atreyee Mukherjee, Zeeshan Ali Sayyed, and Matthew Millard. 2018. Detecting linguistic traces of depression in topic-restricted text: Attending to self-stigmatized depression with nlp. In *Proceedings of the First International Workshop on Language Cognition and Computational Models*, pages 11–21.

Zach Wood-Doughty, Michael Smith, David Broniatowski, and Mark Dredze. 2017. How does twitter user behavior vary across demographic groups? In *Proceedings of the Second Workshop on NLP and Computational Social Science*, pages 83–89.

Andrew Yates, Arman Cohan, and Nazli Goharian. 2017. Depression and self-harm risk assessment in online forums. In *Proceedings of the 2017 Conference on Empirical Methods in Natural Language Processing*, pages 2968–2978, Copenhagen, Denmark. Association for Computational Linguistics.

# Understanding Patterns of Anorexia Manifestations in Social Media Data with Deep Learning

Ana Sabina Uban, Berta Chulvi and Paolo Rosso

Pattern Recognition and Human Language Technology (PRHLT),
Universitat Politècnica de València, València, Spain
ana.uban+acad@gmail.com, berta.chulvi@upv.es,
prosso@dsic.upv.es

## Abstract

Eating disorders are a growing problem especially among young people, yet they have been under-studied in computational research compared to other mental health disorders such as depression. Computational methods have a great potential to aid with the automatic detection of mental health problems, but state-of-the-art machine learning methods based on neural networks are notoriously difficult to interpret, which is a crucial problem for applications in the mental health domain. We propose leveraging the power of deep learning models for automatically detecting signs of anorexia based on social media data, while at the same time focusing on interpreting their behavior. We train a hierarchical attention network to detect people with anorexia, and use its internal encodings to discover different clusters of anorexia symptoms. We interpret the identified patterns from multiple perspectives, including emotion expression, psycho-linguistic features and personality traits, and we offer novel hypotheses to interpret our findings from a psycho-social perspective. Some interesting findings are patterns of word usage in some users with anorexia which show that they feel less as being part of a group compared to control cases, as well as that they have abandoned explanatory activity as a result of a greater feeling of helplessness and fear.

## 1 Introduction and Previous Work

Anorexia nervosa (AN) is a type of eating disorder that leads to multiple psychiatric and somatic complications and constitutes a major public health concern. It involves a restriction of energy intake in relation to needs, leading to significantly low body weight in relation to age, sex, developmental course and physical health. It includes among its typical symptomatology an intense fear of gaining weight or becoming fat and a distortion of one's body image (APA, 2014).

The incidence of AN, like that of other eating disorders (ED), has increased in recent decades. In a systematic literature review for the 2000-2018 period (Galmiche et al., 2019), the reported weighted means of lifetime ED (proportion of EDs at any point in life) were 8.4% (3.3–18.6%) for women and 2.2% (0.8–6.5%) for men. The authors also report that the weighted means of point ED prevalence increased over the study period from 3.5% for the 2000–2006 period to 7.8% for the 2013–2018 period. This highlights a real challenge for public health and healthcare providers.

In an attempt to understand the psychosocial origins of anorexia nervosa, some studies have investigated how body image is shaped in people suffering from this mental disorder (Giordani, 2006, 2009; Giacomozzi and da Silva Bousfield, 2011). From early research in social psychology we already know that body image, in an existential context, is the revelation of an identity that the subject constructs in the frame of concrete social relations (Goffman, 1963). From a sociological perspective, some research proposes to understand bodies attending the interaction with social forces (Turner, 2008). From anthropology, new uses of bodies (tattoos, piercings, etc) support Le Breton's idea about the study of body in modernity as an unfinished material, as "a place of self-presentation" (Le Breton, 2011). This body of research could be applied to the study of anorexia nervosa without forgetting the enormous symbolism of the act of eating. It is well established that eating with others (Dunbar, 2017) and eating the same food as the others is a major symbol of social integration (Harris, 1971; Young et al., 1971).

Mental health disorders in general, as a very significant public health matter (World Health Organization, 2012), have received attention in previous research in computational studies as well. The majority of research has focused on the study of depression (De Choudhury et al., 2013; Eich-



*Proceedings of the Seventh Workshop on Computational Linguistics and Clinical Psychology*, pages 224–236
June 11, 2021. ©2021 Association for Computational Linguistics

staedt et al., 2018; Abd Yusof et al., 2017; Yazdavar et al., 2017), but other mental illnesses have also been studied, including generalized anxiety disorder (Shen and Rudzicz, 2017), schizophrenia (Mitchell et al., 2015), post-traumatic stress disorder (Coppersmith et al., 2014, 2015), risks of suicide (O'dea et al., 2015), and self-harm (Losada et al., 2019; Yang et al., 2016).

For anorexia, there are very few studies approaching the problem from a computational perspective. To our knowledge, the only publicly available social media dataset dedicated to anorexia is the eRisk dataset (Losada et al., 2019). The winners of eRisk's shared task on anorexia detection (Mohammadi et al., 2019) used a hierarchical attention network and obtain a state-of-the-art F1 score of 0.71. In (Cohan et al., 2018) the authors introduce a dataset annotated for multiple mental disorders including anorexia. Another study (Amini and Kosseim) on the explainability of anorexia detection models analyzes attention weights of a neural network to show that attention at the user level correlates with the importance of individual texts for classification performance.

Explainability of machine learning models, especially in the field of mental health, is a very important issue. In practice, models based on neural networks are vastly successful for most NLP applications, even though they have been only briefly explored in existing computational studies on mental disorders. Nevertheless, neural networks are notoriously difficult to interpret. While there is increasing interest in the field of explainability of machine learning models including in NLP (Gilpin et al., 2018), there are fewer such studies for mental health disorder detection.

In the name of transparency, it is essential for any automatic system that can assist with mental health disorder detection to make its decision-making process understandable. Especially in the medical domain, using black-box systems can be dangerous for patients (Zucco et al., 2018; Holzinger et al., 2017). Moreover, recently, the need of explanatory systems is required by regulations like the General Data Protection Regulation (GDPR) adopted by the European Union.

While many quantitative studies in the computational analysis of mental health use features such as lexicons (Trotzek et al., 2017; De Choudhury et al., 2014) to study the manifestations of mental disorders in user-generated data, these models are very limited computationally in comparison to deep learning models. The behavior of powerful classifiers modelling complex patterns in the data has the potential to help uncover manifestations of the disease that are potentially difficult to observe with the naked eye, and thus be useful not only as tools for the detection of disorders, but also as analysis instruments for generating insights and potentially assisting clinicians in the diagnosis process.

In our study, we propose using deep learning as a tool to aid with a deeper investigation of anorexia manifestations in social media texts. We train a hierarchical attention network to classify people with anorexia against control cases based on their social media activity. This architecture has been shown to provide good results for anorexia detection, and additionally includes in the model a series of features that encode different levels of the language (style, emotions, topics etc). To our knowledge this has not been done in previous work, and allows us the advantage of a more interpretable model. We interpret the predictions of the network as well as its hidden layers as a way to identify different patterns of anorexia symptoms in social media users, which we analyze in view of the different features, and offer hypotheses on the different patterns observed from a psychological perspective. Thus, we aim to answer the following research questions:

**RQ1.** Is it possible to leverage complex deep learning models and their encoding power in order to identify different patterns of anorexia symptoms in social media texts?

**RQ2.** Can we characterize the differences between different groups of people with anorexia based on psycho-linguistic and emotion features, and measures of personality traits?

**RQ3.** Could we identify some features to explore the hypothesis that anorexia nervosa is a way to express some degree of conflict with one's own group?

## 2 Classification Experiments

In order to explore the proposed hypotheses, we start by building a deep learning model in order to automatically classify texts belonging to users with anorexia and control cases.

### 2.1 Dataset

**eRisk Reddit dataset on anorexia.** The eRisk CLEF lab[1] is focused on the early prediction of

[1] https://early.irlab.org/

mental disorder risk from social media data, focused on disorders such as depression (Losada et al., 2018), anorexia and self-harm tendencies (Losada et al., 2019, 2020). Data is collected from Reddit posts and comments selected from specific relevant sub-reddits. Users suffering from a mental disorder are annotated by automatically detecting self-stated diagnoses. Control cases are selected from participants in the same sub-reddits (having similar interests), thus making sure the gap between healthy and diagnosed users is not trivially detectable. A long history of posts are collected for the users included in the dataset, up to years prior to the diagnosis. The dataset on anorexia, released as part of eRisk 2019 (Losada et al., 2019), contains 823,754 posts collected from 1,287 users, of which 10.4% are anorexic users.

### 2.2 Model and Features

We choose a hierarchical attention network (HAN) as our model: a deep neural network with a hierarchical structure, including multiple features encoded with LSTM layers and two levels of attention. HANs have previously shown to be successful for the detection of anorexia in social media. (Amini and Kosseim; Mohammadi et al., 2019).

The HAN is made up of two components: a *post-level encoder*, which produces a representation of a post, and a *user-level encoder*, which generates a representation of a user's post history. The post-level encoder and the user-level encoder are modelled as LSTMs. The word sequences encoded using embeddings initialized with GloVe pre-trained embeddings (Pennington et al., 2014) and passed to the LSTM are then concatenated with the other features to form the hierarchical post encoding. The obtained representation is passed to the user-encoder LSTM, which is connected to the output layer. We use the train/test split provided by the shared task organizers, done at the user level, making sure users occurring in one subset don't occur in the other. Since individual posts are too short to be accurately classified, we construct our datapoints through concatenating groups of 50 posts, sorted chronologically. We use a weighted loss function to compensate for the class imbalance. The architecture of the model is shown in Figure 1.

We represent social media texts using features that capture different levels of the language (semantic, stylistic, emotions etc.) and train the model to predict anorexia risk for each user.

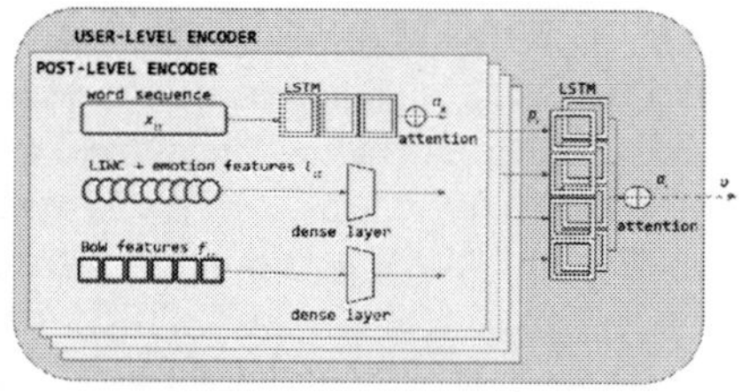


Figure 1: Model architecture.

**Content features**. We include a general representation of text content by transforming each text into word sequences, represented as embeddings.

**Style features**. The usage pattern of function words is known to be reflective of an author's style, at an unconscious level (Mosteller and Wallace, 1963). As stylistic features, we extract from each text a numerical vector representing function words frequencies as bag-of-words, which are passed through an additional dense layer of 20 units. We complement function word distribution features with other syntactical features extracted from the LIWC lexicon, as described below.

**LIWC features.** The LIWC lexicon (Pennebaker et al., 2001) has been widely used in computational linguistics as well as some clinical studies for analyzing how suffering from mental disorders manifests in an author's writings. LIWC is a lexicon mapping words of the English vocabulary to 64 lexico-syntactic features of different kinds, with high quality associations curated by human experts, capturing different levels of language: including style (through syntactic categories), emotions (through affect categories) and topics (such as money, health or religion).

**Emotions and sentiment.** We dedicate a few features to representing emotional content in our texts, since the emotional state of a user is known to be highly correlated with her mental health. Aside from the sentiment and emotion categories in the LIWC lexicon, we include a second lexicon: the NRC emotion lexicon (Mohammad and Turney, 2013), which is dedicated exclusively to emotion representation, with categories corresponding to a wider and a more fine-grained selection of emotions, containing the 8 Plutchik's emotions (Plutchik, 1984), as well as *positive/negative* sentiment categories: *anger, anticipation, disgust, fear, joy, sadness, surprise, trust*. We represent LIWC and NRC features by computing for each category the proportion of words in the input text which are associated with that category.

| Model | P | R | F1 | AUC |
|---|---|---|---|---|
| HAN | .60 | .63 | .60 | **.96** |
| RoBERTa | .64 | .69 | .70 | .83 |
| AlBERT | .78 | .54 | .65 | .77 |
| LogReg | .55 | .45 | .49 | .90 |

Table 1: Precision, recall, F1 (positive class) and AUC scores anorexia classification.

These are concatenated with the other features to form the post-level encodings, which are then stacked and passed to the final user-level LSTM which is connected to the output layer.

### 2.3 Classification Results

The results obtained with our neural network for the detection of anorexic users are shown in Table 1. As performance metrics we compute the F1-score of the positive class and the area under the ROC curve (AUC), which is more robust in the case of data imbalance. We compare the results of our model with baselines such as a logistic regression model with bag-of-word features, and transformer-based models including RoBERTa (Liu et al., 2019) and AlBERT (Lan et al., 2019) with word sequences as features.

Our HAN model achieves the best results in terms of AUC. In the following sections, we explore this model in more depth in order to explain its behavior and leverage it to discover insights on linguistic patterns associated with anorexia.

## 3 Explaining Predictions

In this section we present different analyses meant to uncover insights into how the model arrives at its predictions, first looking at the attention weights and abstract internal representations of the data in the layers of the neural network, and secondly providing several feature-focused analyses, using emotions and LIWC categories, as well as personality markers.

### 3.1 Attention Analysis

Attention is a mechanism frequently used in recurrent neural networks in order to weigh the parts of the input sequence differently according to their importance for prediction. Attention weights are learned by the network, and thus can be used as a means to interpreting its decisions. In our models we include recurrence at the user level, along with an attention layer, which can thus be used to infer the weight placed on each part in a user's post history by the neural network.

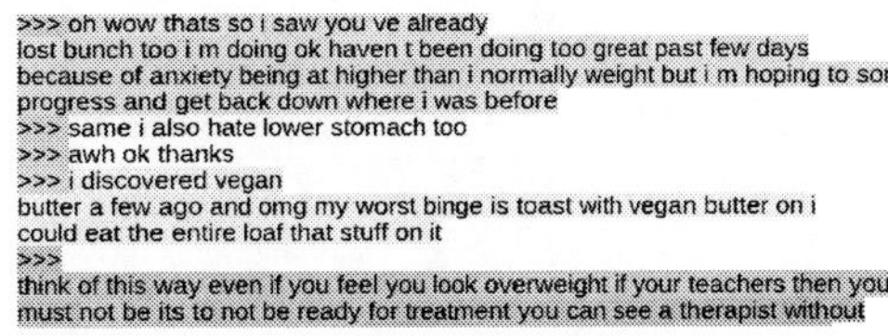


Figure 2: Attention weights example.

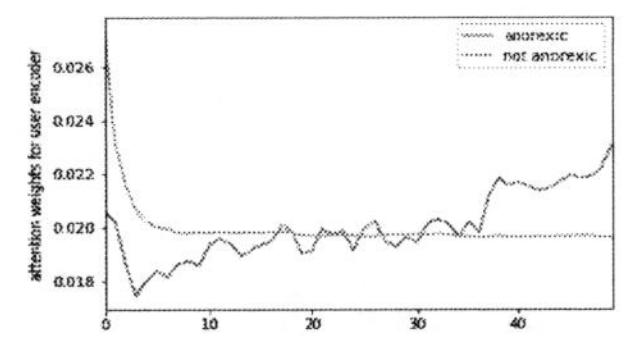


Figure 3: Attention weights over time

In Figure 2, we show an example post (with noise added in view of anonymizing the author) with words and sentences highlighted according to the attention weights provided by the neural network , showing in green the importance of words in each post, and in yellow the importance of each post. In Figure 3, we plot the attention weights for the user-level attention layer for each of the classifiers trained on the three datasets. For this experiment, we train the neural network on one datapoint for each user, so as to ensure attention weights consistently correspond to the same part of the post history for each training example. The plot shows a general increasing importance for users suffering from anorexia: posts in the end of a user's history are more heavily weighted. This is an interesting finding, since intuition, supported by findings such as those presented in the previous sections related to emotion evolution, would suggest a user's activity on social media becomes increasingly indicative of their mental state as time goes by.

Recent studies (Jain and Wallace, 2019; Serrano and Smith, 2019; Wiegreffe and Pinter, 2019) have questioned whether attention mechanisms necessarily help with the interpretability of neural network predictions. We further explore additional techniques in order to interpret the representations learned by the model.

### 3.2 User Embeddings

We continue explaining the model's behavior by analyzing the internal representations of the neural network. We can regard the final layer of the trained neural network as the most compressed representation of the input examples, which is, in terms of our trained model, the optimal representa-

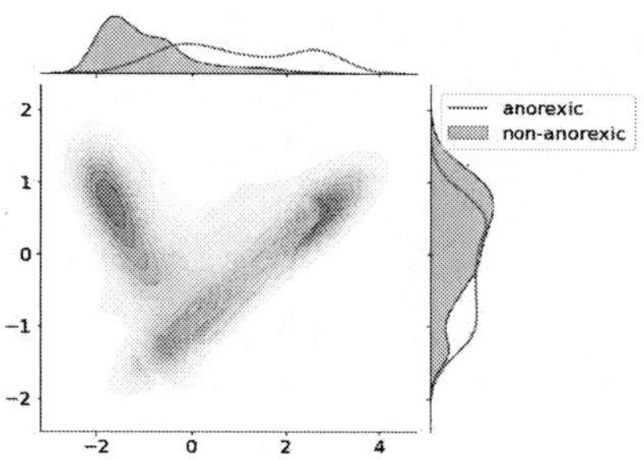


Figure 4: User embeddings in 2D.

(a) Cluster of users with anorexia **ANO1**.

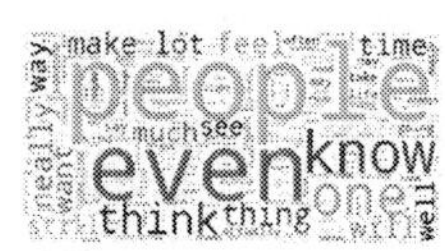


(b) Cluster of users with anorexia **ANO2**.

(c) Cluster of control cases.

Figure 5: Word clouds for the different clusters of anorexic and healthy users.

tion for distinguishing between control cases and those suffering from a disorder. Thus, the final layer (the output of the 32-dimensional user-level LSTM) can be interpreted as a 32-dimensional embedding for the input points, corresponding to the users to be classified.

We analyze the output of the *user embedding* layer by reducing it to 2 dimensions using principal component analysis (PCA) and visualizing it in 2D space with a kernel density estimate (KDE) plot to show the distribution of scores across the 2 dimensions (Figure 4). We make sure to train the PCA model on a balanced set of positive and negative users, then we extract 2D representations for all users in the test set. By looking at these representations, we can gain insight into the separability of the classes, from the perspective of the trained model, and better understand where it encounters difficulties in separating between the datapoints belonging to different classes.

We notice that, even in two dimensions, the groups of people with anorexia and control cases appear as separate clusters in the user embedding space, suggesting that the encodings generated by the model while training for its objective are powerful enough to separate the two groups.

We notice an interesting bimodal aspect in the distribution of positive users, which appear clustered into two distinct groups, while control cases' representations seem to be more compact. One of the clusters of people with anorexia is more clearly distinct from control cases in this space (ANO1), while the other cluster shows a higher overlap with control cases, presumably leading to false negatives in the model's predictions (ANO2). We suggest that the two observed clusters might represent two different groups of people with anorexia and of patterns of symptoms of anorexia, which the model is able to capture during the training process.

The problem of false negative predictions in particular is important to study, since false negatives can lead to missing cases of people with high risk of anorexia, and, left undetected, the disorder might further develop. In the following sections we take a deeper dive into the three identified clusters and attempt to interpret from different perspectives the differences among the clusters of people with anorexia and control cases.

## 4 Patterns of Anorexia

### 4.1 Clustering Method

We begin the analysis of the different patterns of anorexia manifestations by attempting to automatically identify the clusters of people with anorexia in the user embedding space. We perform k-means clustering on the user embeddings for users with anorexia in the test set, and cluster the user embeddings into 2 groups. We obtain 31 users in one cluster which we denote as **ANO1**, and 43 in the second, denoted as **ANO2**.

As a first analysis, we take a look at the vocabulary of the three groups. Figure 5 illustrates word clouds corresponding to the vocabulary used by the 3 groups. As we can see, in both groups where users with anorexia are present, the word *people* has a greater presence suggesting it would be interesting to go deeper into analyzing the dynamics related to social relations among users with anorexia. In order to move in this direction, we will use different strategies that try to describe in more depth these three clusters and to identify the most common narratives in these clusters.

### 4.2 Emotions and Psycho-linguistic Features across Clusters

We analyze the three groups from the perspective of usage of emotion words and psycho-linguistic

|  | ANO1 | ANO2 | Control |
|---|---|---|---|
| **ingest*** ** | 1.15 | 0.82 | 0.38 |
| **bio*** ** | 3.47 | 2.82 | 1.49 |
| **health*** ** | 1.01 | 0.81 | 0.39 |
| **body** ** | 0.88 | 0.77 | 0.51 |
| **anx*** ** | 0.47 | 0.37 | 1.19 |
| **negemo** ** | 2.78 | 2.53 | 1.66 |
| **disgust** ** | 1.25 | 1.31 | 0.90 |
| **fear*** | 1.81 | 1.66 | 1.71 |

Table 2: Features about typical symptoms of anorexia and negative emotional states, percentage of average usage per cluster.
***Statistically significant difference across the three clusters
**Statistically significant difference between people suffering from anorexia and control users.
*Statistically significant difference between **ANO1** and others

categories in the LIWC lexicon. Similarly to the way we encode these features as inputs to the deep learning model, we compute the average values of prevalence (as percentages of overall word usage) for words in each category, separately for the texts in each cluster. We then identify features where there are statistically significant differences among the groups (using a *t*-test).

We identify separately features which show significant differences in usage across all three clusters, or just between users with anorexia and control cases. In general, we observe a pattern of **ANO2** having intermediate values between users with anorexia and control cases, for most of the significant features (36 out of 75 categories show statistically significant differences among all three clusters, and for 30 of them the values for **ANO2** are situated between **ANO1** and control cases).

To obtain a deeper interpretation of the observed differences we select features which are most relevant to anorexia symptoms (see Table 2) and the features which refer to negative emotions. Interestingly, these features also show a distinct pattern from an error analysis perspective: if we select those categories which have lower values in misclassified examples in a statistically significant way, we obtain these four LIWC categories: *anx*, *health*, *bio* and *ingest*.

### 4.3 Personality Analysis

As a separate analysis, we try to analyze the different clusters from the perspective of personality types using the Five Factor Model. The Five Factor Model is a process of attributing certain psychological characteristics to an individual according to the so-called 'Big Five' taxonomy that has been developed into a laborious research paradigm initiated by the social psychologist Gordon Allport (Allport, 1937). Allport (1937) formulated *The lexical hypothesis* proposing that most of the socially relevant and salient personality characteristics have become encoded in the natural language (John et al., 1999). After decades of research, the field is approaching consensus on a general taxonomy of personality traits, the "Big Five" personality dimensions. The five factors are openness to experience (1) conscientiousness (2) agreeableness (3) extraversion (4) and neuroticism (5), as emotional stability. Exploiting this theoretical framework to extract information about users' personality from their posts means identifying such semantic associations and mapping the text around the five factors according to the words referring to them. An effective approach to do this consists in the one proposed by Neuman and Cohen (2014): the evidences of a particular personality trait are summarised into a score, which is calculated as the semantic similarity between the context-free word embedding representations respectively of the text written by the author and of the set of the benchmark adjectives (i.e., the terms empirically observed to be able to encode each of the five personality aspects according to the 'Big Five' framework). In more detail, for each trait, Neuman and Cohen (2014) define a positive and a negative sub-dimension, which correspond respectively to the possession of a sufficient degree of a given factor or, vice versa, the evidence of the exact opposite characteristic. Neuman and Cohen (2014) associate a small series of benchmark adjectives to all the 19 sub-dimensions. In the Appendix we list in full the adjectives that make up the vectors associated with each personality trait.

The set of the benchmark adjectives for the personality traits proposed by Neuman and Cohen has been successfully employed in other tasks such as profiling fake news spreaders (Giachanou et al., 2020). We use a similar approach, and measure personality scores by computing the overlap of the words in the defined vectors for each trait with the words used in each text in our dataset, normalized by text length. Following this approach we found significant differences (p-value <.005) between users with anorexia and control cases in three factors: *Agreeableness (+) (-)*, *Extraversion (-)*, *Neuroticism (+)*.

As we can see in Table 3, in *Extraversion (-)* the difference is statistically significant when we

| | ANO1 | ANO2 | Control |
|---|---|---|---|
| **EXT+** | 0.0037 | 0.0076 | 0.0038 |
| **EXT-**** | 0.23 | 0.82 | 0.41 |
| **AGR+**** | 0.79 | 0.82 | 0.41 |
| **AGR-**** | 0.84 | 1.07 | 0.68 |
| **NEUR+***** | 0.47 | 0.28 | 0.11 |
| **NEUR-** | 0.0081 | 0.18 | 0.10 |
| **CON+** | 0.28 | 0.24 | 0.22 |
| **CON-** | 0.10 | 0.12 | 0.14 |
| **OPN+** | 0.25 | 0.42 | 0.29 |
| **OPN-** | 0.12 | 0.20 | 0.13 |

Table 3: Personality-related words, usage per-mille across the clusters.
***Statistically significant difference across the three clusters
**Statistically significant difference between people suffering from anorexia and control users.
*Statistically significant difference between **ANO1** and others

| | ANO1 | ANO2 | Control |
|---|---|---|---|
| **work**** | 1.22 | 1.47 | 2.31 |
| **money**** | 0.41 | 0.50 | 0.86 |
| **leisure**** | 1.12 | 1.15 | 1.88 |
| **pronoun***** | 17.41 | 16.20 | 11.54 |
| **I***** | 6.95 | 5.52 | 3.49 |
| **we*** | 0.33 | 0.46 | 0.51 |
| **friend**** | 0.22 | 0.25 | 0.15 |
| **family**** | 0.27 | 0.28 | 0.22 |
| **humans**** | 0.82 | 0.92 | 0.72 |

Table 4: Features about everyday activities and social relations, percentage of average usage per cluster.
***Statistically significant difference across the three clusters
**Statistically significant difference between people suffering from anorexia and control users.
*Statistically significant difference between **ANO1** and others

compare users suffering from anorexia with control cases, but not between the two clusters of people with anorexia. It seems that more introverted personality traits characterize users who suffer from anorexia. The same applies to the factor *Agreeableness*, in positive and in negative sense, the difference in agreeableness is statistically significant among people with anorexia and control cases. Users suffering from anorexia show more characteristic traits of a pleasant personality and unplesant personality than those who do not suffer from this disorder. The explanation for this difference, in positive and in negative traits, may be related to a greater manifestation of emotions and feelings among people with anorexia.

With the factor measuring *Neuroticism (+)* we find statistically significant differences among the three clusters suggesting that users in **ANO1** are at a more severe stage of this mental disorder than **ANO2** because they have higher scores in this factor. Neuroticism speaks about emotional instability that leads to frequent experiences of negative emotions and which is said to result from a low ability to handle stress or strong external stimuli.

## 5 Identifiying Different Narratives and Cognitive Styles

In **RQ3** we raised the possibility of exploring if some features from LIWC and emotion lexicons allow us to identify among users with anorexia a higher degree of conflict with their own group and some degree of social isolation.

As we can see in Table 4, users with anorexia talk less about *work*, *money* and *leisure* (LIWC categories) than control cases. The absence of words from these three categories tells us about a certain degree of social isolation. These results connect with a pattern that we find in the use of *personal pronouns* among the different clusters.

As we can observe, users with anorexia employ significantly fewer words under the category *we* (*we*, *us*, *our*), a clear linguistic marker of a sense of belonging. Users in **ANO1** use it significantly less than users in **ANO2** and than control cases, suggesting that a higher degree of conflict with one's own group or a higher degree of social isolation may be linked to the more severe manifestations of anorexia. The opposite pattern is found in the use of the first person pronouns that LIWC collects under the category *I* (*I*, *me*, *mine*): users suffering from anorexia use it significantly more than control cases and cluster **ANO1** uses it significantly more than cluster **ANO2**.

Three other features expressing social processes, *family*, *friends* and *humans* are more present in the narratives of users with anorexia than among control cases. The greater presence of these linguistic categories may indicate a greater centrality of social relations in the identity of these subjects suffering from anorexia. We would need to design new strategies to go deeper into this interpretation, but this greater centrality of social relations categories in people with anorexia could be derived, precisely, from a greater degree of conflict with the social environment.

We also observed in Table 5 some differences among clusters in relation to some LIWC categories related to cognitive and perceptual processes. These differences may indicate the existence of different cognitive styles between users who suffer from anorexia and those who do not.

*Cognitive style* is a concept used in cognitive

psychology to describe the way individuals think, perceive, and remember information (Grigorenko and Sternberg, 1995). Research in psychology suggests that some cognitive styles are more prevalent in some patients suffering from depression and anorexia (Lo et al., 2008; Kaye et al., 1995).

As we can see in Table 5, people with anorexia use in their narratives more words that LIWC classifies as *cognitive processes* (*cogmech*) than control cases and the difference is also statistically significant between clusters **ANO1** and **ANO2**. A greater presence of these traits among users with anorexia is indicative of a special effort to reason about reality, which is also a characteristic of conflictual states. We could also consider that this effort to understand involves the subject at a personal level because we see a higher presence of words belonging to the *feel* category among users with anorexia. *Feel* is, among the perceptual process in LIWC, the one that involves the subject at a deeper level.

Within *cognitive processes* we find it interesting to analyze certain features such as *certainty*, *tentative* and *causation* where there are significant differences among clusters. *Certainty* expresses a rigid or absolute style of thinking and *tentative* expresses a more flexible or less absolute style of thinking. We find significant differences in these two categories between users with anorexia and control cases. *Certainty* is more used in narratives from users suffering anorexia and could indicate a major degree of cognitive conflict. We see the opposite pattern with *tentative*, that just expresses a flexible style. It could be expressing the two sides of the same phenomenon.

With *causation* we only found statistically significant differences between **ANO1** and the other two clusters (similarly to what we found for the expression of fear, as seen in Table 2). Reasoning about causes indicates an effort to understand the world that shows a healthier position of the subjects, and one possible interpretation is that users in **ANO1** have abandoned this explanatory activity as a result of a greater feeling of helplessness and feeling more fear.

## 6 Connection to Clinical Research

Trying to identify different groups of patients who claim to suffer from anorexia nervosa or have compatible symptomatology may be a way to develop a better understanding of this complex pathology (Clinton et al., 2004). Research such as that of Viborg et al. (2018) shows a relationship between six clusters of young adolescents suffering from anorexia and higher levels of psychological difficulties and lower levels of body esteem. However, these clusters rely exclusively on symptomatology linked to eating behavior. We think that our results show that a deep learning model applied to social media texts allows us to identify clusters of patients considering more variables than those related to the eating disorder itself.

From these initial results we can provide some insights to clinical discussion. More and more clinicians (Gutiérrez and Carrera, 2021) ask themselves about the intractability of anorexia nervosa, including the disconcerting aspect of the recovery of a significant number of patients not receiving formal treatment (Keski-Rahkonen, 2014). As Gutiérrez and Carrera (2021) state in a recent review of this issue, Bruch's proposal regarding the charactization of typical anorexia in terms of *Body Image Disturbances* (BID) (Bruch et al., 1974) and the relevance of this construct in different editions of the *Diagnostic and Statistical Manual of Mental Disorders* (APA, 2014), may have over-directed clinical practice, rendering other aspects of psychopathology invisible and increasing the numbers of patients diagnosed as atypical cases.

Analyzing the language of people suffering from anorexia, we found traces of problems that could guide new approaches indicating the existence of a conflict between the patient and his or her own group: lower use of the first person plural and a greater presence of features expressing social processes. Some research focusing on the discourse analysis of people suffering from anorexia points to the existence of a link between acting on one's own body as a mechanism to take control over their lives (Malson and Ussher, 1996).

Our results may indicate that a possible origin of this need of control could be the social conflict with one's own group and the inability to communicate this conflictual situation. In this sense, Botta and Dumlao (2002) have shown the relationship between parent-child conflict and family communication styles and the development of eating disorders. In addition, Davies et al. (2012) have demonstrated in their experimental research the relevance of a verbal expression of emotions in patients with anorexia nervosa. More research is needed, for instance it may be interesting to revisit one of Bruch's ideas in her seminal work which has not been as

|  | ANO1 | ANO2 | Control |
|---|---|---|---|
| **cogmech*** | 16.43 | 15.88 | 14.58 |
| **feel** | 0.86 | 0.86 | 0.43 |
| **certain** | 1.55 | 1.69 | 1.04 |
| **tentative** | 3.09 | 3.01 | 3.13 |
| **causation* | 1.71 | 1.85 | 1.87 |

Table 5: Features about cognitive styles (cognitive processes and perceptual processes), percentage of average usage per cluster.
***Statistically significant difference across the three clusters
**Statistically significant difference between people suffering from anorexia and control users.
*Statistically significant difference between **ANO1** and others

successful as her concept of BID: the description of anorexia as "a communicative disorder" which is experienced as a means of taking control over one's body as a pseudosolution to intra- and interpersonal difficulties (Bruch, 1978). Following this idea, we think that deep learning models applied to social media data can open an interesting avenue to explore the language of people suffering from anorexia and provide elements for further clinical discussion.

## 7 Ethical Considerations

Powerful machine learning models that can be trained to detect or predict the development of mental health disorders can be very valuable, but any deployment of a tool for mental disorder detection should take into account potential ethical concerns. If such tools are used by third parties (such as employers seeking to filter candidates based on their mental health profile), this could compromise the privacy of the subjects. We suggest that the development of an ethical standard is necessary, and that launching such tools could be accompanied by an ethical statement to constrain its use.

Moreover, it is ethically necessary, and recently even required by regulations in some countries (such as countries in the European Union) that artificial intelligence models used in the mental health domain have interpretable behavior. We hope that we were able to take a step forward in this direction, by providing an in-depth explanation of the representations generated by our neural network, and thus facilitating trust in the predictive model.

## 8 Conclusions and Future Work

In this study, we have approached the problem of detecting people suffering from anorexia in social media thorugh training a deep learning model, and taken it a step further by explaining the behavior of the model. Based on this, we identified different clusters of users suffering from anorexia with regards to the manifested symptoms, as encoded by the trained model (**RQ1**). We presented several analyses for interpreting the decisions of the model trained to profile social media users at risk of developing anorexia, going beyond more common techniques such as attention weight analysis, and including hidden layer analysis and error analysis at different levels of the language for better understanding how mental disorders manifest in social media data (**RQ2**).

We have shown that we can interpret the detected clusters from the point of view of the social behavior of people with anorexia (**RQ3**), and provided in-depth interpretations of these patterns as a socio-psychological phenomenon. To our knowledge, our approach and findings are novel in the domain of the computational study of anorexia, and encourage us to go deeper into the analysis of these patterns, such as looking into the use of pronouns in relation with emotions, which could help identify more clearly the existence of a social conflict.

From a technical perspective, a more sophisticated analysis of the features included here (LIWC categories, emotions and personality vectors) could be achieved by using more semantically rich representations than bag-of-words. One such approach would be using word embeddings to identify sub-emotions (Aragón et al., 2019) starting from Plutchik's 8 emotions.

Moreover, including a temporal dimension as a variable could also reveal additional insights such as cases of patients with symptoms moving from one cluster to another over time. The possibility to categorize people with anorexia into different groups according to their symptoms might help with identifying those people at a higher risk of more serious developments of their disorder and also could give some inputs for for a more in-depth discussion of symptomatology in clinical forums.

Finally, it would also be interesting to explore the connection between anorexia and other mental disorders or manifestations such as suicide attempts.

## Acknowledgements

The authors thank the EU-FEDER Comunitat Valenciana 2014-2020 grant IDIFEDER/2018/025. The work of Paolo Rosso was in the framework of the research project PROMETEO/2019/121 (DeepPattern) by the Generalitat Valenciana.

## References


Noor Fazilla Abd Yusof, Chenghua Lin, and Frank Guerin. 2017. Analysing the causes of depressed mood from depression vulnerable individuals. In *Proceedings of the International Workshop on Digital Disease Detection using Social Media 2017 (DDDSM-2017)*, pages 9–17.

Gordon Willard Allport. 1937. Personality: A psychological interpretation.

Hessam Amini and Leila Kosseim. Towards explainability in using deep learning for the detection of anorexia in social media. *Natural Language Processing and Information Systems*, 12089:225.

APA. 2014. *Diagnostic and Statistical Manual of Mental Disorders (DSM–5)*. American Psychiatric Association.

Mario Ezra Aragón, Adrian Pastor López-Monroy, Luis Carlos González-Gurrola, and Manuel Montes. 2019. Detecting depression in social media using fine-grained emotions. In *Proceedings of the 2019 Conference of the North American Chapter of the Association for Computational Linguistics: Human Language Technologies, Volume 1 (Long and Short Papers)*, pages 1481–1486.

Renee Botta and Rebecca Dumlao. 2002. How do conflict and communication patterns between fathers and daughters contributeto or offset eating disorders? *Health communication*, 14:199–219.

Hilde Bruch. 1978. *The golden cage: The enigma of anorexia nervosa*. Harvard University Press.

Hilde Bruch et al. 1974. *Eating disorders. Obesity, anorexia nervosa, and the person within*. Routledge & Kegan Paul.

David Clinton, Eric Button, Claes Norring, and Robert Palmer. 2004. Cluster analysis of key diagnostic variables from two independent samples of eating-disorder patients: Evidence for a consistent pattern. *Psychological Medicine*, 34(6):1035.

Arman Cohan, Bart Desmet, Andrew Yates, Luca Soldaini, Sean MacAvaney, and Nazli Goharian. 2018. Smhd: a large-scale resource for exploring online language usage for multiple mental health conditions. In *Proceedings of the 27th International Conference on Computational Linguistics*, pages 1485–1497.

Glen Coppersmith, Mark Dredze, and Craig Harman. 2014. Quantifying mental health signals in twitter. In *Proceedings of the workshop on computational linguistics and clinical psychology: From linguistic signal to clinical reality*, pages 51–60.

Glen Coppersmith, Mark Dredze, Craig Harman, Kristy Hollingshead, and Margaret Mitchell. 2015. Clpsych 2015 shared task: Depression and ptsd on twitter. In *Proceedings of the 2nd Workshop on Computational Linguistics and Clinical Psychology: From Linguistic Signal to Clinical Reality*, pages 31–39.

Helen Davies, Nicola Swan, Ulrike Schmidt, and Kate Tchanturia. 2012. An experimental investigation of verbal expression of emotion in anorexia and bulimia nervosa. *European Eating Disorders Review*, 20(6):476–483.

Munmun De Choudhury, Scott Counts, Eric J Horvitz, and Aaron Hoff. 2014. Characterizing and predicting postpartum depression from shared facebook data. In *Proceedings of the 17th ACM conference on Computer supported cooperative work & social computing*, pages 626–638.

Munmun De Choudhury, Michael Gamon, Scott Counts, and Eric Horvitz. 2013. Predicting depression via social media. In *Seventh international AAAI conference on weblogs and social media*.

R. I. M. Dunbar. 2017. Breaking bread: the functions of social eating. *Adaptive Human Behavior and Physiology*, 3:198–211.

Johannes C Eichstaedt, Robert J Smith, Raina M Merchant, Lyle H Ungar, Patrick Crutchley, Daniel Preoţiuc-Pietro, David A Asch, and H Andrew Schwartz. 2018. Facebook language predicts depression in medical records. *Proceedings of the National Academy of Sciences*, 115(44):11203–11208.

Marie Galmiche, Pierre Déchelotte, Gregory Lambert, and Marie Tavolacci. 2019. Prevalence of eating disorders over the 2000-2018 period: a systematic literature review. *The American journal of clinical nutrition*, 109:1402–1413.

Anastasia Giachanou, Esteban A Ríssola, Bilal Ghanem, Fabio Crestani, and Paolo Rosso. 2020. The role of personality and linguistic patterns in discriminating between fake news spreaders and fact checkers. In *International Conference on Applications of Natural Language to Information Systems*, pages 181–192. Springer.

Andréia Isabel Giacomozzi and Andréa Bárbara da Silva Bousfield. 2011. Representação social do corpo de participantes de comunidades pró-anorexia do orkut. *Psicologia, Saúde e Doenças*, 12(2):255–266.

Leilani H Gilpin, David Bau, Ben Z Yuan, Ayesha Bajwa, Michael Specter, and Lalana Kagal. 2018. Explaining explanations: An overview of interpretability of machine learning. In *2018 IEEE 5th International Conference on data science and advanced analytics (DSAA)*, pages 80–89. IEEE.

Rubia C.F. Giordani. 2006. A auto-imagem corporal na anorexia nervosa: uma abordagem sociológica. *Psicologia & Sociedade*, 18:81–88.

Rubia C.F. Giordani. 2009. O corpo sentido e os sentidos do corpo anoréxico. *Revista de Nutrição*, 22:809–821.

Ervin. Goffman. 1963. *Stigma: Notes on the management of spoiled identity*. Englewood Cliffs, N.J: Prentice-Hall.

Elena L Grigorenko and Robert J Sternberg. 1995. Thinking styles. In *International handbook of personality and intelligence*, pages 205–229. Springer.

Emilio Gutiérrez and Olaia Carrera. 2021. Severe and enduring anorexia nervosa: Enduring wrong assumptions? *Frontiers in Psychiatry*, 11:1–19.

Marvin Harris. 1971. *Culture, man, and nature: An introduction to general anthropology*. Crowell.

Andreas Holzinger, Chris Biemann, Constantinos S Pattichis, and Douglas B Kell. 2017. What do we need to build explainable ai systems for the medical domain? *arXiv preprint arXiv:1712.09923*.

Sarthak Jain and Byron C Wallace. 2019. Attention is not explanation. In *Proceedings of the 2019 Conference of the North American Chapter of the Association for Computational Linguistics: Human Language Technologies, Volume 1 (Long and Short Papers)*, pages 3543–3556.

Oliver P John, Sanjay Srivastava, et al. 1999. *The Big-Five trait taxonomy: History, measurement, and theoretical perspectives*, volume 2. University of California Berkeley.

Walter H. Kaye, Andrea M. Bastiani, and Howard Moss. 1995. Cognitive style of patients with anorexia nervosa and bulimia nervosa. *International Journal of Eating Disorders*, 18:287–290.

Anna et al. Keski-Rahkonen. 2014. Factors associated with recovery from anorexia nervosa: a population-based study. *The International journal of eating disorders*, 47(2):117–23.

Zhenzhong Lan, Mingda Chen, Sebastian Goodman, Kevin Gimpel, Piyush Sharma, and Radu Soricut. 2019. Albert: A lite bert for self-supervised learning of language representations. *arXiv preprint arXiv:1909.11942*.

David Le Breton. 2011. *Anthropologie du corps et modernité*. Presses universitaires de France.

Yinhan Liu, Myle Ott, Naman Goyal, Jingfei Du, Mandar Joshi, Danqi Chen, Omer Levy, Mike Lewis, Luke Zettlemoyer, and Veselin Stoyanov. 2019. Roberta: A robustly optimized bert pretraining approach. *arXiv preprint arXiv:1907.11692*.

Cola Lo, Samuel M.Y.Ho, and Steven D.Hollonb. 2008. The effects of rumination and negative cognitive styles on depression: A mediation analysis. *Behaviour Research and Therapy*, 46(4):487–495.

David E Losada, Fabio Crestani, and Javier Parapar. 2018. Overview of erisk: early risk prediction on the internet. In *International Conference of the Cross-Language Evaluation Forum for European Languages*, pages 343–361. Springer.

David E Losada, Fabio Crestani, and Javier Parapar. 2019. Overview of erisk 2019 early risk prediction on the internet. In *International Conference of the Cross-Language Evaluation Forum for European Languages*, pages 340–357. Springer.

David E. Losada, Fabio Crestani, and Javier Parapar. 2020. Overview of erisk 2020: Early risk prediction on the internet. In *L. Cappellato, C. Eickhoff, N. Ferro, and A. Névéol (eds.) CLEF 2020 Labs and Workshops, Notebook Papers. CEUR Workshop Proceedings.CEUR-WS.org*, volume 2696.

Helen Malson and Jane M Ussher. 1996. Body polytexts: Discourses of the anorexic body. *Journal of community & applied social psychology*, 6(4):267–280.

Margaret Mitchell, Kristy Hollingshead, and Glen Coppersmith. 2015. Quantifying the language of schizophrenia in social media. In *Proceedings of the 2nd workshop on Computational linguistics and clinical psychology: From linguistic signal to clinical reality*, pages 11–20.

Saif M Mohammad and Peter D Turney. 2013. Nrc emotion lexicon. *National Research Council, Canada*, 2.

Elham Mohammadi, Hessam Amini, and Leila Kosseim. 2019. Quick and (maybe not so) easy detection of anorexia in social media posts. In *L. Cappellato, N. Ferro, D. Losada and H. Müller (eds.) CLEF 2019 Labs and Workshops, Notebook Papers. CEUR Workshop Proceedings.CEUR-WS.org*, volume 2380.

Frederick Mosteller and David L Wallace. 1963. Inference in an authorship problem: A comparative study of discrimination methods applied to the authorship of the disputed federalist papers. *Journal of the American Statistical Association*, 58(302):275–309.

Yair Neuman and Yochai Cohen. 2014. A vectorial semantics approach to personality assessment. *Scientific reports*, 4(1):1–6.

Bridianne O'dea, Stephen Wan, Philip J Batterham, Alison L Calear, Cecile Paris, and Helen Christensen. 2015. Detecting suicidality on twitter. *Internet Interventions*, 2(2):183–188.

James W Pennebaker, Martha E Francis, and Roger J Booth. 2001. Linguistic inquiry and word count: Liwc 2001. *Mahway: Lawrence Erlbaum Associates*, 71(2001):2001.

Jeffrey Pennington, Richard Socher, and Christopher D Manning. 2014. Glove: Global vectors for word representation. In *Proceedings of the 2014 conference on empirical methods in natural language processing (EMNLP)*, pages 1532–1543.

Robert Plutchik. 1984. Emotions: A general psychoevolutionary theory. *Approaches to emotion*, 1984:197–219.

Sofia Serrano and Noah A Smith. 2019. Is attention interpretable? In *Proceedings of the 57th Annual Meeting of the Association for Computational Linguistics*, pages 2931–2951.

Judy Hanwen Shen and Frank Rudzicz. 2017. Detecting anxiety through reddit. In *Proceedings of the Fourth Workshop on Computational Linguistics and Clinical Psychology—From Linguistic Signal to Clinical Reality*, pages 58–65.

Marcel Trotzek, Sven Koitka, and Christoph M Friedrich. 2017. Linguistic metadata augmented classifiers at the clef 2017 task for early detection of depression. In *L. Cappellato, N. Ferro, L. Goeuriot and T. Mandl (eds.) CLEF 2017 Labs and Workshops, Notebook Papers. CEUR Workshop Proceedings.CEUR-WS.org*, volume 1866.

Bryan Turner. 2008. *The body & society: Explorations in social theory*. SAGE Publications Ltd.

Njördur Viborg, Margit Wångby-Lundh, Lars-Gunnar Lundh, Ulf Wallin, and Per Johnsson. 2018. Disordered eating in a swedish community sample of adolescent girls: Subgroups, stability, and associations with body esteem, deliberate self-harm and other difficulties. *Journal of Eating Disorders*, 6:3–11.

Sarah Wiegreffe and Yuval Pinter. 2019. Attention is not not explanation. In *Proceedings of the 2019 Conference on Empirical Methods in Natural Language Processing and the 9th International Joint Conference on Natural Language Processing (EMNLP-IJCNLP)*, pages 11–20.

WHO World Health Organization. 2012. Depression: A global crisis. world mental health day, october 10 2012. *World Federation for Mental Health, Occoquan, Va, USA*.

Zichao Yang, Diyi Yang, Chris Dyer, Xiaodong He, Alex Smola, and Eduard Hovy. 2016. Hierarchical attention networks for document classification. In *Proceedings of the 2016 conference of the North American chapter of the association for computational linguistics: human language technologies*, pages 1480–1489.

Amir Hossein Yazdavar, Hussein S Al-Olimat, Monireh Ebrahimi, Goonmeet Bajaj, Tanvi Banerjee, Krishnaprasad Thirunarayan, Jyotishman Pathak, and Amit Sheth. 2017. Semi-supervised approach to monitoring clinical depressive symptoms in social media. In *Proceedings of the 2017 IEEE/ACM International Conference on Advances in Social Networks Analysis and Mining 2017*, pages 1191–1198.

Michael W Young et al. 1971. Fighting with food. leadership, values and social control in a massim society. *Fighting with food. Leadership, values and social control in a Massim society*.

Chiara Zucco, Huizhi Liang, Giuseppe Di Fatta, and Mario Cannataro. 2018. Explainable sentiment analysis with applications in medicine. In *2018 IEEE International Conference on Bioinformatics and Biomedicine (BIBM)*, pages 1740–1747. IEEE.

## Appendix

## A Hyperparameter configurations for the neural network

### A.1 Hierarchical attention network

- LSTM units (post encoder) = 128
- dense BoW units = 20
- dense lexicon units = 20
- LSTM units (user encoder) = 32
- dropout = 0.3
- $l_2 = 0.00001$
- optimizer = Adam
- learning rate = 0.0001
- early stopping patience = 20
- epochs = 20
- maximum sequence length = 256
- posts per chunk = 50

## B Adjective vectors for each personality dimension

**EXT+:** dominant assertive authoritarian forceful assured confident firm persistent
**EXT-:** nervous modest quiet forceless afraid shy calm indecisive
**AGR+:** tender gentle soft kind affectionate helpful sympathetic friendly
**AGR-:** cruel unfriendly negative mean brutal inconsiderate insensitive cold
**CON+:** organized orderly tidy neat efficient persistent systematic straight careful reliable
**CON-:** distracted unreliable incompetent wild inefficient disloyal chaotic confused messy disorganized
**NEU+:** worried stressed anxious nervous fearful touchy guilty insecure restless emotional
**NEU-:** balanced stable confident fearless calm easy-going relaxed secure comforted peaceful
**OPN+:** philosophical abstract imaginative curious reflective literary questioning individualistic unique open
**OPN-:** narrow-minded concrete ordinary incurious thoughtless ignorant uneducated common conventional restricted

# Author Index

**Association for Computational Linguistics**
209 N. Eighth Street
Stroudsburg, Pennsylvania 18360

CURRAN ASSOCIATES INC.
proceedings.com

ISBN 978-1-7138-3019-1